Sustainable Fish Farming

Sustainable Fish Farming

Editor:

Dan Piestun, Ph.D.

Delve Publishing
www.delvepublishing.com

Sustainable Fish Farming

Editor: Dan Piestun, Ph.D.

© 2017 Delve Publishing LLC

708 3rd Avenue, 6th Floor

New York, NY 10017

United States of America

www.delvepublishing.com

ISBN: 978-1-68095-857-7

Library of Congress Control Number: 2016951610

This book contains information obtained from highly regarded resources. Reprinted material sources are indicated. Copyright for individual articles remains with the authors as indicated. A Wide variety of references are listed. Reasonable efforts have been made to publish reliable data and views articulated in the chapters are those of the individual contributors, and not necessarily those of the editors or publishers. Editors or publishers are not responsible for the accuracy of the information in the published chapters or consequences of their use. The publisher believes no responsibility for any damage or grievance to the persons or property arising out of the use of any materials, instructions, methods or thoughts in the book. The editors and the publisher have attempted to trace the copyright holders of all material reproduced in this publication and apologize to copyright holders if permission has not been obtained. If any copyright holder has not been acknowledged, please write to us so we may rectify.

Notice: Registered trademark of products or corporate names are used only for explanation and identification without intent of infringement.

Delve Publishing publishes wide variety of books and eBooks. For more information about Magnum Publishing and its products, visit our website at **www.delvepublishing.com**

About the Editor

Dan Piestun, Ph.D.

Dan Piestun (PhD) is currently a startup entrepreneur in Israel working on the interface of Agriculture and Biomedical Sciences and was formerly president-CEO of the National Institute of Agricultural Research (INIA) in Uruguay. Dan is a widely published scientist who has received many honours during his career including being a two-time recipient of the Amit Golda Meir Prize from the Hebrew University of Jerusalem, his areas of expertise includes stem cell molecular biology, plant and animal genetics and bioinformatics. Dan's passion for applied science and technological solutions did not stop him from pursuing a deep connection to the farmer, his family and nature. Among some of his interest and practices counts enjoying working as a beekeeper and onboard fishing.

Contents

List of Contributors

Geir Lasse Taranger
Institute of Marine Research, PB 1870, N-5817 Bergen, Norway

Ørjan Karlsen
Institute of Marine Research, Austevoll Research Station, N-5392 Storebø, Norway

Raymond John Bannister
Institute of Marine Research, PB 1870, N-5817 Bergen, Norway

Kevin Alan Glover
Institute of Marine Research, PB 1870, N-5817 Bergen, Norway

Vivian Husa
Institute of Marine Research, PB 1870, N-5817 Bergen, Norway

Egil Karlsbakk
Institute of Marine Research, PB 1870, N-5817 Bergen, Norway

Bjørn Olav Kvamme
Institute of Marine Research, PB 1870, N-5817 Bergen, Norway

Karin Kroon Boxaspen
Institute of Marine Research, PB 1870, N-5817 Bergen, Norway

Pål Arne Bjørn
Institute of Marine Research, PB 6404, N-9294 Tromsø, Norway

Bengt Finstad
Norwegian Institute for Nature Research, PB 5685 Sluppen, N-7485 Trondheim, Norway

Abdullah Sami Madhun
Institute of Marine Research, PB 1870, N-5817 Bergen, Norway

H. Craig Morton
Institute of Marine Research, PB 1870, N-5817 Bergen, Norway

Terje Svåsand
Institute of Marine Research, PB 1870, N-5817 Bergen, Norway

Etienne Bezault
UMR110 Cirad-Ifremer INTREPID, Montpellier, France
INRA, UMR1313 Génétique animale et biologie intégrative, Jouy-en-Josas, France
Division of Aquatic Ecology, Institute of Ecology & Evolution, University of Bern, Bern, Switzerland
EAWAG, Swiss Federal Institute of Aquatic Science and Technology, Centre of Ecology, Evolution and Biogeochemistry, Department of Fish Ecology and Evolution, 6047 Kastanienbaum, Switzerland

Patricia Balaresque
Department of Genetics, University of Leicester, Leicester, UK
CNRS/ FRE2960 Laboratoire AMIS (Anthropologie Moléculaire et Imagerie de Synthèse), Toulouse, France

Aboubacar Toguyeni
Institut du Développement Rural, Université Polytechnique de Bobo-Dioulasso, Bobo-Dioulasso, Burkina Faso

Yves Fermon
18 rue J. Richepin, F-91120 Palaiseau, France

Hitoshi Araki
EAWAG, Swiss Federal Institute of Aquatic Science and Technology, Centre of Ecology, Evolution and Biogeochemistry, Department of Fish Ecology and Evolution, 6047 Kastanienbaum, Switzerland

Jean-François Baroiller
UMR110 Cirad-Ifremer INTREPID, Montpellier, France

Xavier Rognon
INRA, UMR1313 Génétique animale et biologie intégrative, Jouy-en-Josas, France
AgroParisTech, UMR1313 Génétique animale et biologie intégrative, Paris, France

Tim Ellis
Centre for Environment, Fisheries and Aquaculture Science (Cefas), Weymouth Laboratory, Barrack Road, The Nothe, Weymouth, Dorset DT4 8UB, UK

James F. Turnbull
Institute of Aquaculture, University of Stirling, Stirling, Stirlingshire FK9 4LA, UK

Toby G. Knowles
University of Bristol, Cabot Institute and School of Veterinary Science, Langford, Bristol BS40 5DU, UK

Jeff A. Lines
Silsoe Livestock Systems Ltd, Wrest Park, Silsoe, Bedford, Bedfordshire MK45 4HS, UK

Neil A. Auchterlonie
Centre for Environment, Fisheries and Aquaculture Science (Cefas), Weymouth Laboratory, Barrack Road, The Nothe, Weymouth, Dorset DT4 8UB, UK

Pallab K. Sarker
Environmental Studies Program, Dartmouth College, Hanover, NH 03755, United States of America

Anne R. Kapuscinski
Environmental Studies Program, Dartmouth College, Hanover, NH 03755, United States of America

Alison J. Lanois
Environmental Studies Program, Dartmouth College, Hanover, NH 03755, United States of America

Erin D. Livesey
Environmental Studies Program, Dartmouth College, Hanover, NH 03755, United States of America

Katie P. Bernhard
Environmental Studies Program, Dartmouth College, Hanover, NH 03755, United States of America

Mariah L. Coley
Environmental Studies Program, Dartmouth College, Hanover, NH 03755, United States of America

Marcel Martinez-Porchas
Departamento de Tecnología de Alimentos de Origen Animal, Centro de Investigación en Alimentación y Desarrollo, Km. 0.7 Carretera a La Victoria, Hermosillo, SON, Mexico

Luis R. Martinez-Cordova
Departamento de Investigaciones Científicas y Tecnológicas de la Universidad de Sonora, Boulevard Luis Donaldo Colosio s/n, 83000 Hermosillo, SON, Mexico

Paulo Branco
CEF – Centro de Estudos Florestais, Instituto Superior de Agronomia, Universidade Te´cnica de Lisboa, Lisboa, Portugal

Jose´ M. Santos
CEF – Centro de Estudos Florestais, Instituto Superior de Agronomia, Universidade Te´cnica de Lisboa, Lisboa, Portugal

Christos Katopodis
Katopodis Ecohydraulics Ltd., Winnipeg, Manitoba, Canada

Anto´ nio Pinheiro
CEHIDRO – Centro de Estudos de Hidrossistemas, Instituto Superior Te´cnico, Universidade Te´cnica de Lisboa, Lisboa, Portugal

Maria T. Ferreira
CEF – Centro de Estudos Florestais, Instituto Superior de Agronomia, Universidade Te´cnica de Lisboa, Lisboa, Portugal

Ngo Phu Thoa
Research Institute for Aquaculture No.1, Dinh Bang, Tu Son, Bac Ninh, Vietnam
Faculty of Science, Health, Education and Engineering, University of the Sunshine Coast, Maroochydore QLD 4558, Australia

Nguyen Huu Ninh
Research Institute for Aquaculture No.1, Dinh Bang, Tu Son, Bac Ninh, Vietnam

Wayne Knibb
Faculty of Science, Health, Education and Engineering, University of the Sunshine Coast, Maroochydore QLD 4558, Australia

Nguyen Hong Nguyen
Faculty of Science, Health, Education and Engineering, University of the Sunshine Coast, Maroochydore QLD 4558, Australia

Ondr˘ej Slavı´k
Department of Zoology and Fisheries, Faculty of Agrobiology, Food and Natural Resources, Czech University of Life Sciences Prague, Prague, Czech Republic

Pavel Horky
Department of Zoology and Fisheries, Faculty of Agrobiology, Food and Natural Resources, Czech University of Life Sciences Prague, Prague, Czech Republic

Libor Za´vorka
Department of Zoology and Fisheries, Faculty of Agrobiology, Food and Natural Resources, Czech University of Life Sciences Prague, Prague, Czech Republic
Department of Zoology, University of Gothenburg, Go¨teborg, Sweden

Sarah E. Lester
Marine Science Institute, University of California Santa Barbara, Santa Barbara, California, United States of America
Bren School of Environmental Science and Management, University of California Santa Barbara, Santa Barbara, California, United States of America

Christopher Costello
Bren School of Environmental Science and Management, University of California Santa Barbara, Santa Barbara, California, United States of America

Andrew Rassweiler
Marine Science Institute, University of California Santa Barbara, Santa Barbara, California, United States of America

Steven D. Gaines
Bren School of Environmental Science and Management, University of California Santa Barbara, Santa Barbara, California, United States of America

Robert Deacon
Department of Economics, University of California Santa Barbara, Santa Barbara, California, United States of America

Michael R. Lowe
Department of Coastal Sciences, University of Southern Mississippi, Ocean Springs, Mississippi, United States of America

Wei Wu
Department of Coastal Sciences, University of Southern Mississippi, Ocean Springs, Mississippi, United States of America

Mark S. Peterson
Department of Coastal Sciences, University of Southern Mississippi, Ocean Springs, Mississippi, United States of America

Nancy J. Brown-Peterson
Department of Coastal Sciences, University of Southern Mississippi, Ocean Springs, Mississippi, United States of America

William T. Slack
United States Army Engineer Research and Development Center, Waterways Experiment Station, Vicksburg, Mississippi, United States of America

Pamela J. Schofield
United States Geological Survey, Southeast Ecological Science Center, Gainesville, Florida, United States of America

Moussa Gue`ye[1]
[1]IRD, route des hydrocarbures, Dakar, Se´ne´gal,
[2]Max Planck Institute for Molecular Genetics, Berlin, Germany,
[3]IRD, LEMAR - Institut Universitaire Europe´en de la Mer, Plouzane´, France,
[4]Institut Fondamental d'Afrique Noire, Universite´ Cheikh Anta Diop, Dakar, Senegal,

Mbaye Tine
IRD, route des hydrocarbures, Dakar, Se´ne´gal
Max Planck Institute for Molecular Genetics, Berlin, Germany

Justin Kantoussan
IRD, route des hydrocarbures, Dakar, Se´ne´gal
IRD, LEMAR - Institut Universitaire Europe´en de la Mer, Plouzane´, France

Papa Ndiaye
Institut Fondamental d'Afrique Noire, Universite´ Cheikh Anta Diop, Dakar, Senegal

Omar Thiom Thiaw
Laboratoire de Biologie cellulaire et mole´culaire, Reproduction et Ge´ne´tique, Universite´ Cheikh Anta Diop de Dakar, Faculte´ des Sciences et Techniques, Dakar-Fann, Se´ne´gal

JeanJacques Albaret
IRD, route des hydrocarbures, Dakar, Se´ne´gal

Chao Ran
Department of Biological Sciences, Auburn University, Auburn, Alabama, United States of America

Abel Carrias
Department of Biological Sciences, Auburn University, Auburn, Alabama, United States of America

Malachi A. Williams
Department of Biological Sciences, Auburn University, Auburn, Alabama, United States of America

Nancy Capps
Department of Biological Sciences, Auburn University, Auburn, Alabama, United States of America

Bui C. T. Dan
Novus International Inc., Novus Aqua Research Center, Ho Chi Minh City, Vietnam

Joseph C. Newton
Department of Pathobiology, Auburn University, Auburn, Alabama, United States of America

Joseph W. Kloepper
Department of Entomology & Plant Pathology, Auburn University, Auburn, Alabama, United States of America

Ei L. Ooi
Novus International Inc., Novus Aqua Research Center, Ho Chi Minh City, Vietnam

Craig L. Browdy
Novus International Inc., Charleston, South Carolina, United States of America

Jeffery S. Terhune
Department of Fisheries and Allied Aquacultures, Auburn University, Auburn, Alabama, United States of America

Mark R. Liles
Department of Biological Sciences, Auburn University, Auburn, Alabama, United States of America

Amelia Paredes-Trujillo
Laboratorio de Parasitología, Centro de Investigación y de Estudios Avanzados del Instituto Politécnico Nacional, Unidad Mérida Km. 6 Carretera Antigua a Progreso, Cordemex, Mérida, Yucatán 97310, México

Iván Velázquez-Abunader
Laboratorio de Parasitología, Centro de Investigación y de Estudios Avanzados del Instituto Politécnico Nacional, Unidad Mérida Km. 6 Carretera Antigua a Progreso, Cordemex, Mérida, Yucatán 97310, México

Edgar Torres-Irineo
Laboratorio de Parasitología, Centro de Investigación y de Estudios Avanzados del Instituto Politécnico Nacional, Unidad Mérida Km. 6 Carretera Antigua a Progreso, Cordemex, Mérida, Yucatán 97310, México

David Romero
Departamento de Geografía, Universidad Nacional Autónoma de México, Circuito Interior, Ciudad Universitaria 04510, México, D.F, México

Víctor Manuel Vidal-Martínez
Laboratorio de Parasitología, Centro de Investigación y de Estudios Avanzados del Instituto Politécnico Nacional, Unidad Mérida Km. 6 Carretera Antigua a Progreso, Cordemex, Mérida, Yucatán 97310, México

Patrick B Hamilton
Biosciences, College of Life and Environmental Sciences, University of Exeter, Exeter EX4 4QD, UK

Elizabeth Nicol
Institute for the Environment, Brunel University, Uxbridge, Middlesex UB8 3PH, UK

Eliane SR De-Bastos
Biosciences, College of Life and Environmental Sciences, University of Exeter, Exeter EX4 4QD, UK

Richard J Williams
Centre for Ecology and Hydrology, Oxfordshire OX10 8BB, UK

John P Sumpter
Institute for the Environment, Brunel University, Uxbridge, Middlesex UB8 3PH, UK

Susan Jobling
Institute for the Environment, Brunel University, Uxbridge, Middlesex UB8 3PH, UK

Jamie R Stevens
Biosciences, College of Life and Environmental Sciences, University of Exeter, Exeter EX4 4QD, UK

Charles R Tyler
Biosciences, College of Life and Environmental Sciences, University of Exeter, Exeter EX4 4QD, UK

L.C.Kreutz
Laboratório de Virologia e Imunologia, Programa de Pós-Graduação em Bioexperimentação, Universidade de Passo Fundo, Passo Fundo, RS, Brasil

T.R. Pavan
Laboratório de Virologia e Imunologia, Programa de Pós-Graduação em Bioexperimentação, Universidade de Passo Fundo, Passo Fundo, RS, Brasil

A.G. Alves
Laboratório de Virologia e Imunologia, Programa de Pós-Graduação em Bioexperimentação, Universidade de Passo Fundo, Passo Fundo, RS, Brasil

A.G. Correia
Laboratório de Virologia e Imunologia, Programa de Pós-Graduação em Bioexperimentação, Universidade de Passo Fundo, Passo Fundo, RS, Brasil

B. Barriquel
Laboratório de Virologia e Imunologia, Programa de Pós-Graduação em Bioexperimentação, Universidade de Passo Fundo, Passo Fundo, RS, Brasil

E.D. dos Santos
Laboratório de Virologia e Imunologia, Programa de Pós-Graduação em Bioexperimentação, Universidade de Passo Fundo, Passo Fundo, RS, Brasil

L.J.G. Barcellos
Laboratório de Virologia e Imunologia, Programa de Pós-Graduação em Bioexperimentação, Universidade de Passo Fundo, Passo Fundo, RS, Brasil

Preface

Our goal is to provide a broad treatment of topics directly related with Sustainable Fish Farming with a simple resource addressing from natural resources, wild fish populations, through fish-food human needs. Furthermore, this book focuses attention on sustainability through the lens of threats to fish health and management, up to water ecosystem dynamics as affected by human activity and rising opportunities for sustainable fish farming with finite-resource awareness.

Audience

The overall theme of the book is aimed at oceanography and veterinary professionals most likely involved in fish farming and aquaculture and aware of the fragility of our relationship with the freshwater and marine ecosystems. As technological advances in sustainable and fish-raising farms evolve, there is an increasing need for permanent education for professionals working in the field and advanced university and technical students.

The following readers may benefit from this resource:

Students in professional oceanography training programs

Students in veterinary education programs

Organization

"Sustainable Fish Farming" is organized into five sections:

Section 1 - Finite-Resource Awareness And Strategies To Meet Increasing Demand For Fish-Food. This section presents a relevant assessment of aquaculture "domesticated" species on local wild marine populations, environmental factors roles on diversification and adaptation of wide regions. Besides, it analyzes the challenge of increasing fish-food demand in a context of aquaculture expansion, new more responsible farming methods, and fish genetics choices. Risk assessment of the environmental impact of Norwegian Atlantic salmon farming: The authors present the potential hazards of Atlantic Salmon to influence the surrounding environment and wild populations, in the context of a recent assessment of the Institute of Marine Research of Norway. Further, the study explores into the risks of genetic introgression, lice infection, fish viral disease outbreaks, into the wild populations, as well as local organic loading of the ecosystems. Spatial and temporal variation in population genetic structure of wild Nile tilapia (Oreochromis niloticus) across Africa: Analyzes how the factors (climate, habitat, ecology) configuring the genetic diversity of populations can play a pivotal role in the course of diversification, speciation and adaptation. Trends during development of Scottish salmon farming: An example of sustainable intensification?: discusses strategies to meet the challenge of augmenting demand for food, highlighting strategies like aquaculture expansion, adopting farming methods, selecting genotypes better adapted to farm conditions, a shift in human diets to species from less vulnerable trophic levels and maximizing farming systems relative to resource inputs and outputs with a holistic view.

Section 2 - Sustainable Fish Nutrition with Fish-Ingredient-Free Sources. The relevance of the consequence of feed production and utilization not only on the target fish species but in the whole ecosystem, is exemplified in the alternatives explored in the following research work. Towards Sustainable Aquafeeds: Complete Substitution of Fish Oil with Marine Microalga Schizochytrium sp. Improves Growth and Fatty Acid Deposition in Juvenile Nile Tilapia (Oreochromis niloticus): To improve reformulation of aquaculture feeds, this research group evaluates the marine microalga, Schizochytrium sp as a substitute for fish oil in tilapia diets.

Section 3 - Ecosystem Economy, Responsible Fish Farming and Vulnerabilities of Natural Resources. The following section occupies with the question whether fish farming could be a truly sustainable activity. World Aquaculture: Environmental Impacts and Troubleshooting Alternatives: Presents a review of the world aquaculture and its environmental impacts and responsible management opportunities. Pool-Type Fishways: Two Different Morpho-Ecological Cyprinid Species Facing Plunging and Streaming Flows._: Studied solutions for local populations of fish to bypass dams, thus restoring connectivity of the fresh water stream. The explored solutions are species specific, and vary according to swimming ability, size, physiologic state and are poised to rescue fish resources from extinction. Does selection in a challenging environment produce Nile tilapia genotypes that can thrive in a range of production systems?: In this work the authors explore how selection for high growth rate in a challenging environment of moderate salinity renders tilapia genotypes that can perform well in freshwater systems. In determining this, they examined the effect of genotype by environment interaction for traits like sexual maturity, deformity, survival and food conversion rate. Energy Costs of Catfish Space Use as Determined by Biotelemetry: This study intended to define how fish utilized energy in the home range and core area and observed the energetic expenses of space use in a freshwater predator catfish, Silurus glanis, using physiological sensors. Encourage Sustainability by Giving Credit for Marine Protected Areas in Seafood Certification: Marine protected areas (MPAs), one of the most important implements for protecting the ocean's ecosystems, can be effectively promoted from the sustainable fisheries and consumer markets through credits for compliance. This paper discusses how better to implement incentives for consumer driven certification of sustainability. Survival, Growth and Reproduction of Non-Native Nile Tilapia II: Fundamental Niche Projections and Invasion Potential in the Northern Gulf of Mexico.: The authors created a predictive model of Nile tilapia survival, growth, and reproduction in the Mississippi basin. Besides, they studied the niche of this species by integrating spatial – environmental data and the potential to invade other freshwater streams on the northern Gulf of Mexico. Furthermore, look at how climate variability may disturb tilapia distribution.

Section 4 - Fish Farming Intensification, Threats To Fish Health And Alternative Fish Health Management. In here we compiled articles stressing the importance of strengthening the fish populations' defenses by new nonchemical approaches and exemplify with a case of a parasitic threat to extensively grown fish species. The research studies are Identification of Bacillus Strains for Biological Control of Catfish Pathogens: While indiscriminate administration of antibiotics in fish feeds has become an extended practice to fight bacterial diseases, this study fostered the utilization of probiotics (Bacillus strains) in catfish to control enteric septicemia. Geographical distribution of protozoan and metazoan parasites of farmed Nile tilapia Oreochromis niloticus (L.) (Perciformes: Cichlidae) in Yucatán, México: The authors described the geographical dispersal of protozoan and helminth parasites of farmed Nile tilapia in Yucatan and analyzed the possible statistical associations between the prevalence and average abundance of these parasites with management and environmental clues.

Section 5 - Ecotoxicology On Fish Habitats And Ecosystems. Populations of a cyprinid fish are self-sustaining despite widespread feminization of males: This study evaluates whether sections of river highly polluted with estrogenic effluents have compromised breeding populations of roach (Rutilus rutilus L.). To accomplish this, the authors undertook an analysis of population genetic structures of roach in the region using DNA microsatellite analysis. Increased immunoglobulin production in silver catfish (Rhamdia quelen) exposed to agrichemicals.: These researchers studied the kinetics of immunoglobulin secretion and whether the joint effect of water temperatures and boosting vaccination, and the presence of atrazine and glyphosate-based herbicides would have any impact on the production of antibodies to pathogenic bacteria.

It is our wish that readers not only learn or deepen on principles of Sustainable Fish Farming but moreover renew and actualize their awareness in broad concepts and new findings on fish science and water ecosystems.

Editor
Dan Piestun, Ph.D.

INTRODUCTION

Fish farming means 'raising fish commercially in tanks, ponds or other enclosures for the purpose of producing food'. Commercial fish farming has already established as a profitable business venture throughout the world. Fish is a great source of food and protein. The demand and price of fish and fish related products are also increasing rapidly, in accordance with rapid population growth. This is the main reason of increasing this business around the world. Almost all countries around the world are somehow suitable for fish farming business.

But the countries with coastal areas are very suitable for this business. There are many areas, where fish farming business is the only way of livings for the people. However, here we are describing more about the advantages of fish farming and the steps for starting this business commercially.

Why Fish Farming?

There are many advantages of starting commercial fish farming business. Here we are describing the main advantages of fish farming business.

- According to the demand, commercial fish farming business allows for large supplies of fish. Catching fish from the wild can't always fulfill the consumer's demand. In such cases commercial fish farming can meet up this demand.
- Fish can be raised in tanks until they are ready for marketing, and they don't require the extensive capture of wild fish. Thus commercial fish farming helps to preserve natural ecosystems.
- Compared to the wild brethren, some farm raised fish species are more nutritious. Fish are usually fed a wide variety of protein and nutrient enriched foods or pellets on commercial fish farms. So farm fish become healthier than the wild fish.
- Various types of fish species are available throughout the world. So you can choose your desired species for your fish farming business.
- Fish are very popular as food around the world. So there is an established fish market, and you don't have to worry about marketing your products.
- You can start fish farming business in both large and small scale production purpose.
- If you don't have enough capital for starting this business, then you can apply for bank loans. Many banks will allow loans for starting this business commercially.
- Fish farming business is a great source of employment. More than 1 billion people around the world depend on fish as their primary protein source. And most of this people are directly or indirectly involved with fish products or fish farming business. As a result, fish farming creates a great income and employment source for the people. Global fish exportation business is now earning more money every year, than any other food commodity.
- Even you can meet up your daily family nutritional demands through small scale fish farming in tanks or ponds.

How to Start Fish Farming Business?

For maintaining a profitable fish farming business, you have to go through some step by step process. The steps for starting a fish farming business includes selecting suitable farm land or area, fish farm type (cage, tank or pond), cage or pond construction, selecting fish species, feeding, care & management, harvesting and marketing. We are shortly describing all the steps below. For running a successful fish farming business, follow every steps very carefully.

Select A Suitable Farm Land/Area

Selecting a suitable land or area is very important for starting commercial fish farming business. All the areas are not suitable for profitable business. And some areas have plenty of natural resources, which are very effective for fish farming business. Especially coastal areas and the areas near big rivers or stream are very suitable for establishing fish farming business. Consider the followings while selecting land or areas for your business.

- Select relatively level land and avoid steeply sloped lands.
- Consider your future business plan, while selecting the land. It will be better if you can select a large piece of land, where you can perform all types of necessary farm activities.
- Avoid flooding and polluted areas, because flooding area can harm your business seriously.
- Don't select fish farming land near the crop fields. Farmers usually use a lots of fertilizers and pesticides in their field for better production. These chemicals get mixed with water and the polluted water can affect your fish farm.
- It will be better,if your selected land become slightly lower than the main water source. It also help to reduce the cost of filling your farm land with water. Natural gravity will fill the land without any cost.
- Ensure good transportation system is available in your selected area. Good transportation system will be very effective for marketing your products and purchasing necessary commodities from the market.

Type of Your Farm

There are numerous specific type of fish farms in both intensive and extensive fish farming system. You can start fish farming business by choosing any type. You can choose cage system, tank system or pond system. In cage system, make a suitable cage and place it in lakes, ponds, bayous or oceans and start feeding the fish until they reach marketing age. In case of raising fish in tanks, make a or a few tanks and stock fish there. Here we are describing more about fish farming in pond system.

Pond Design & Construction

Construct a suitable pond, after selecting your farm area. Before constructing, make a good design and make the pond according to your desired design. While designing the pond, ensure the availability of all types of necessary facilities for maintaining a profitable fish farming business. Although the design of a pond depends on the fish species you intend to raise and your location. You can consult with your nearest fisheries institute to learn more about specific pond design for specific fish species. Always try to maintain a good environment in the pond. Good environment helps to live and grow the fish well, and it directly involved with better production and maximum profits. See pond management.

Suitable Fish Species

Selecting suitable fish species is very important for maximizing profits form fish farming business. Select those breeds, which have a huge demand and high price in your local market. The most important fish species used in fish farming throughout the world are carp, salmon, tilapia and catfish. All of these fish species have many varieties and suitable for farming in all types of agro-climatic conditions. Select the fish species for farming, depending on your local facilities, demand and price.

Feeding

Feeding is the most important part of commercial fish farming business. Always try to provide your fish high quality and nutritious food. High quality food not only ensures maximum production but also help to keep the fish healthy. So provide your fish supplementary fish feed along with natural food. There are various types of prepared commercial fish feed available in the market for specific fish species. You can purchase those commercial feed from the market or prepare it by your own. Learn more about preparing supplementary fish feed, if you want to prepare it by your own. Don't forget to add all types of necessary nutrient elements, such as vitamins, minerals, salt etc. Feed your fish several times a day, depending on the fish species.

Care & Management

Always try to provide your fish fresh and nutritious foods. It will be better, if you can change water from the pond occasionally. If not possible, then you can use some chemicals according to the suggestion of an expert. Monitor the health of your fish on a regular basis. Do all your necessary farm tasks timely. Keep the pond environment clean and suitable for proper growth. Test the water and soil quality of your pond on a regular basis. Always stock some necessary drugs on your farm. Prevent all types of predators, including frogs, snakes etc.

Harvesting

After a certain period, the fish become suitable for harvesting. Although this time depends on the fish species. Start harvesting, when a major numbers of fish reach marketing age. You can use net for harvesting fish or by removing water from the pond. Try to harvest during morning or afternoon, when temperature is low. After harvesting, send the fish to the market as soon as possible.

Marketing

Marketing is the easiest step of fish farming business. There are numerous markets available where you can sell your products. And all types of fish have a huge demand in the market. After harvesting, you can easily sell the fish at any of your nearest local market. Even there are many companies available who export fish to the foreign countries. So don't worry about marketing the products, just focus on the other steps.

MAKING FISH FARMING MORE SUSTAINABLE

The Food and Agriculture Organization of the U.N. (FAO) projects that by 2030, aquaculture, one of the fastest growing methods of producing food in the world, will be responsible for almost two-thirds of the fish we eat.

The most common type of aquaculture is farming in net pens or cages anchored to the sea floor in the ocean near the coast. There are also closed systems of tanks or ponds that float on water or operate on land. The FAO estimates that over 600 aquatic species are produced globally in a variety of aquaculture systems using freshwater, brackish water or salt water.

In the U.S., over 91 percent of the seafood we eat is imported. China is the largest exporter of fish globally, the third largest importer of fish and the biggest aquaculture producer. As early as 2500 B.C., the Chinese practiced fish farming, putting carp in rice paddies where they ate insects and weeds, fertilized the rice, and then were eaten. Today, 88 percent of the world's aquaculture comes from Asia.

While fish are known for their omega-3 fatty acid benefits, they do not actually produce omega-3s themselves. Microscopic algae that live in fresh or salt water or sediments produce them. Herbivorous fish and forage fish like sardines, anchovies and herring obtain omega-3s by eating the microalgae. Larger carnivorous fish such as salmon or sea bass then eat the forage fish. Because salmon and other popular carnivorous fish need omega-3s to grow, 30 to 50 percent of the fish feed traditionally used for these species consists of fishmeal (ground fish) and fish oil. Over 50 percent of the world's fish oil is used in feed for farmed salmon.

This is one reason that aquaculture has a reputation as being unsustainable. In 1997, it took almost 3 tons of forage fish to produce one ton of salmon. A third of the global fish harvest still goes toward making fish meal and fish oil. As a result, forage fish are being overfished, and some populations have crashed, which has implications for the entire food web since larger fish depend on them for food.

Most fish farming methods are harmful to the ecosystem in other ways as well.

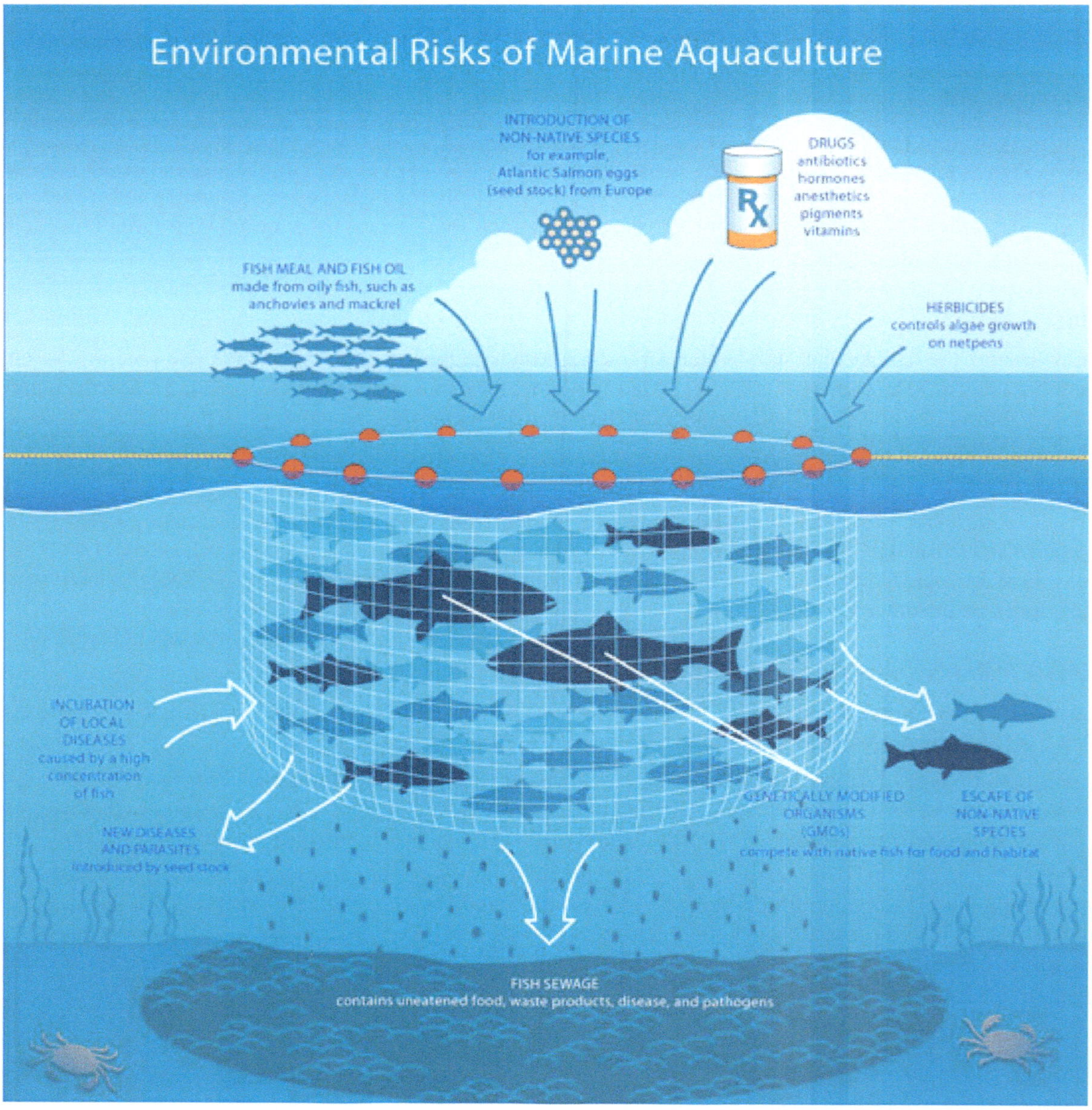

Fish waste and left over food spill out from nets into the ocean, causing nutrient pollution. This may lead to oxygen depletion in the water, which can stress or kill aquatic creatures. In addition, antibiotics or pesticides used on farmed fish can affect other marine life or human health. These nutrients and chemicals also sink to the ocean floor, where they may impact the biodiversity there.

Fish crowded together in nets or pens are more susceptible to stress, which can foster disease and parasites that may then spread to wild species.

Farmed fish sometimes escape into the ocean, breeding with wild species and affecting the population's overall genetic diversity.

Land-based closed systems minimize the amount of waste and nutrients expelled to the environment, eliminate fish escapes and limit the spread of disease; but pumping the water through them requires a great deal of energy, and the wastewater must still be disposed of properly.

Aquaculture has also resulted in land conversion for feed and the destruction of ecosystems. To grow the soybeans used in herbivorous and other fish feed, vast areas of South America, almost 4 million hectares of forest, are razed yearly and converted to agricultural land. Growing the soybeans also depends on water availability.

Shrimp farming, which is usually done in salty coastal waters, has been responsible for the destruction of 38 percent of the world's mangroves. Mangroves have critical ecological functions including providing food and habitat for many species, preventing erosion, sequestering carbon, and offering protection from storms. Moreover, many shrimp ponds accumulate shrimp waste, antibiotics and pesticides, and without mangroves to filter them, eventually become unusable.

But Pete Malinowski, former aquaculture director of the New York Harbor School and director of the Billion Oyster Project, which aims to restore one billion live oysters to New York Harbor, thinks aquaculture's bad reputation is undeserved.

"Most aquaculture operations are far more sustainable than animal food production," he said. "Aquaculture gets a bad rap because we see it as an alternative to natural environments like coasts that haven't been developed, as opposed to comparing it with agriculture. Actually aquaculture is a much more sustainable way to get protein than agriculture."

"A lot of damage is being caused by aquaculture, but it's getting much better," said Malinowski. "People are getting educated, and it's becoming unacceptable to farm fish in a way that's harmful to the environment."

How Aquaculture is Becoming More Sustainable

One strategy involves moving aquaculture out into the open ocean where the water is pristine and currents are strong and steady enough to continually flush the farms of fish waste and pests such as sea lice. The open ocean also provides farmed fish with more consistent salinity and temperature. This means they are less stressed and less vulnerable to disease, which promotes better growth and minimizes the need for antibiotics or vaccines.

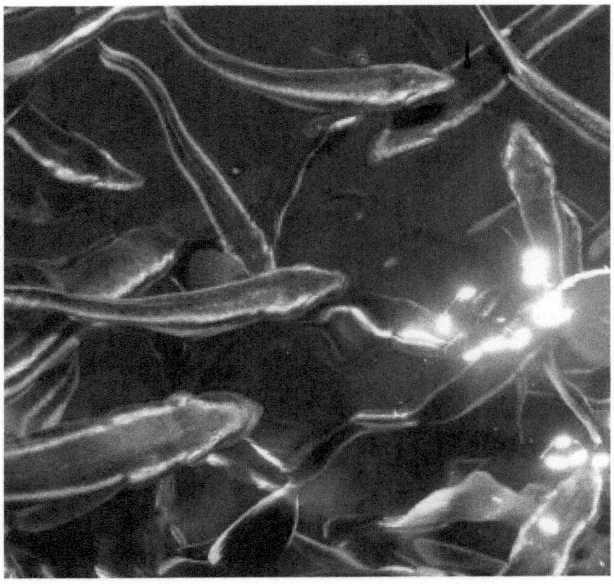

SECTION I - FINITE-RESOURCE AWARENESS AND STRATEGIES TO MEET INCREASING DEMAND FOR FISH-FOOD.

CHAPTER 1

RISK ASSESSMENT OF THE ENVIRONMENTAL IMPACT OF SALMON FARMING

Geir Lasse Taranger, Ørjan Karlsen, Raymond John Bannister, Kevin Alan Glover, Vivian Husa, Egil Karlsbakk, Bjørn Olav Kvamme, Karin Kroon Boxaspen, Pål Arne Bjørn, Bengt Finstad, Abdullah Sami Madhun, H. Craig Morton and Terje Svåsand

Keywords: Environmental impact, Eutrophication, Genetic interaction, Organic load, Pathogens, risk assessment, Salmon lice

BACKGROUND

The Atlantic salmon (*Salmo salar* L.) farming industry was first started in Norway in the early 1970s and has now grown to become one of the country's largest export industries by economic value. In addition to Atlantic salmon, which is by far the most significant species farmed in Norway, there are also commercial farming of rainbow trout [*Oncorhynchus mykiss* (Walbaum)] and other marine species such as Atlantic cod (*Gadus morhua* L.) and halibut [*Hippoglossus hippoglossus*(L.)]. In 2012, the production of Atlantic salmon and rainbow trout in Norway was 1 232 095 and 74 583 tons, respectively, and a total of 1006 marine farms was licensed. These farms are distributed along much of Norway's coastline.

Aquaculture of salmonids in Norway, and other countries where these species are farmed in significant numbers, is primarily based around the production of eggs and juveniles in freshwater facilities on land, combined with grow out of fish in open marine cages. During the last decades, technical standards for the production of aquaculture infrastructure has improved dramatically. However, the primary methods for cultivation of finfish have remained similar, with the size of sea cages (up to 160 m in circumference) and the number of stocked smolt (up to 200 000 individuals per cage) increasing.

The rapid expansion of the aquaculture industry, both in Norway and other regions where this form of open-cage production has increased, has not occurred without environmental challenges. However, although a significant body of evidence suggests various environmental impacts of aquaculture, the rapid expansion of this industry means that management guidelines and targets to address potential negative effects have generally not developed in association with the rapid expansion of the industry. Therefore, there is a need for more coordinated efforts to identify hazards related to open sea cage farming and evaluate environmental risks.

Risk Analysis

Several approaches have been suggested and discussed for risk analysis of marine ecosystems and marine aquaculture

Citation: Geir Lasse Taranger, Ørjan Karlsen, Raymond John Bannister, Kevin Alan Glover, Vivian Husa, Egil Karlsbakk, Bjørn Olav Kvamme, Karin Kroon Boxaspen, Pål Arne Bjørn, Bengt Finstad, Abdullah Sami Madhun, H. Craig Morton and Terje Svåsand, Risk assessment of the environmental impact of salmon farming, doi: 10.1093/icesjms/fsu132

activities (Anon., 2006, 2010;Nash, 2007; GESAMP, 2008; Samuel-Fitwi *et al.*, 2012), and similar approaches of risk analysis and assessment have been adapted to animal welfare including welfare of farmed fish (e.g. EFSA, 2012). According to GESAMP (2008), a risk analysis should first involve hazard identification, then risk assessment of these hazards including the assessment of release, exposure, and consequences, followed by risk estimation/evaluation. The latter preferably related to politically defined thresholds of acceptability or level of protection. Subsequently, this can be followed up by appropriate risk management and appropriate risk communication.

A full risk analysis is based on the ability to quantify both the probability of a certain event and its consequences, but in biological systems it is normally very difficult to quantify these factors precisely. Hence, risk analyses in biological systems are often conducted using broad qualitative categories, by scoring the probability and consequences from low to high (e.g. GESAMP, 2008). This can in turn be based on some semi-quantitative assessment or on expert opinion as suggested by Anon. (2006).

In 2009, the Norwegian government established a set of environmental goals for sustainability in the "Strategy for an Environmentally Sustainable Norwegian Aquaculture Industry" (Anon., 2009b; Table 1). In response to this, the Institute of Marine Research, Norway, initiated a risk assessment of Norwegian salmon farming in 2010, and yearly since (Taranger *et al.*, 2011a, b, 2012a, 2013, 2014). These risk assessments were based on identified hazards and specific endpoints or proxies related to environmental impacts of salmon farming (Table 2). The endpoints/proxies were in turn derived from the governmental goals for environmental sustainability mentioned above. Moreover, evaluation thresholds for some of these endpoints/proxies (acceptance levels of impact or level of protection) were proposed (Taranger *et al.*, 2012b), and subsequently used in the risk assessments in 2013 and 2014. Here, we describe the way in which these assessments have been conducted, the methodological limitations and challenges, as well as future needs to data and analytical tools.

Table 1. The five primary goals for the future development of the Norwegian aquaculture industry as established by the Norwegian government in 2009.

Goals	
Goal 1: Disease	Disease in fish farming will not have a regulating effect on stocks of wild fish, and as many farmed fish as possible will grow to slaughter age with minimal use of medicines.
Goal 2: Genetic interaction	Aquaculture will not contribute to permanent changes in the genetic characteristics of wild fish populations.
Goal 3: Pollution and discharges	All fish farming locations in use will maintain an acceptable environmental state and will not have higher emissions of nutrient salts and organic materials than the receiving waters can tolerate.
Goal 4: Zoning	The aquaculture industry will have a location structure and zoning which reduces impact on the environment and the risk of infection.
Goal 5: Feed and feed resources	The aquaculture industry's needs for raw materials for feed will be met without overexploitation of wild marine resources.

Table 2. Identified hazards, process of concern, and endpoint of concern for goals 1–3 for the future development of the Norwegian aquaculture industry as established by the Norwegian government in 2009.

Hazard	Process of concern	Endpoint of concern
Genetic interaction (Goal 2)	Farmed escaped salmon successfully interbreed with wild salmon populations	Changes observed in the genetic characteristics of wild salmon populations
Salmon lice (Goal 1)	Salmon lice from fish farming affects wild fish	Salmon lice from fish farming significantly increase the mortality of wild salmonids
Viral diseases (Goal 1)	Disease transmission from fish farming affects wild fish	Viral transmission from fish farming significantly increase the mortality of wild salmonids
Discharges of organic material: (i) local effects (ii) regional effects (Goal 3)	Emissions of organic materials to the surrounding environment	(i) Unacceptable change in sediment chemistry and faunal communities in the production zone (ii) Significant change in bottom communities beyond the production zone—regional impact
Discharges of nutrients: (i) local effects (ii) regional effects (Goal 3)	Emissions of nutrients to the surrounding environment	(i) Nutrients from fish farms results in local eutrophication (ii) Nutrients from fish farms results in regional eutrophication

Hazard Identification

The first step in a risk assessment is to identify the most important hazards. A range of criteria for hazard identification was proposed by GESAMP (2008). This includes an analysis on how potential hazards relates to undesirable changes in the environment/ecosystem. To this end, potential hazards are evaluated for their possible severity, extent and duration of the change, either based on past experiences, analogue situations, or models. Some of the environmental challenges (i.e. hazards) identified include ecosystem and benthic community effects of organic loading and nutrients (Buschmann *et al.*, 2006;

Kutti *et al.*, 2008;Bannister *et al.*, 2014), transfer of parasites to native populations (Krkošek*et al.*, 2005, 2013a, b; Jackson *et al.*, 2013; Skilbrei *et al.*, 2013; Torrissen*et al.*, 2013; Serra-Llinares *et al.*, 2014), disease interactions (Glover *et al.*, 2013b; Madhun *et al.*, 2014a), and ecological (Jonsson and Jonsson, 2006) and genetic interactions with wild populations (Crozier, 1993; Clifford *et al.*, 1998b; Skaala *et al.*, 2006; Glover *et al.*, 2012, 2013a).

Based on the accumulating evidence of the severity, geographical extent and duration and/or reversibility of the various impacts related to open sea cage salmon farming in Norwegian coastal waters, we have based the current risk assessment on the following hazards: (i) genetic introgression of escaped farmed salmon into wild populations, (ii) impact of salmon lice (*Lepeophtheirus salmonis*) on wild salmonid populations, (iii) potential disease transfer from farmed salmon to wild salmonid populations, and (iv) local and regional impacts of organic load and nutrients from marine salmon farms.

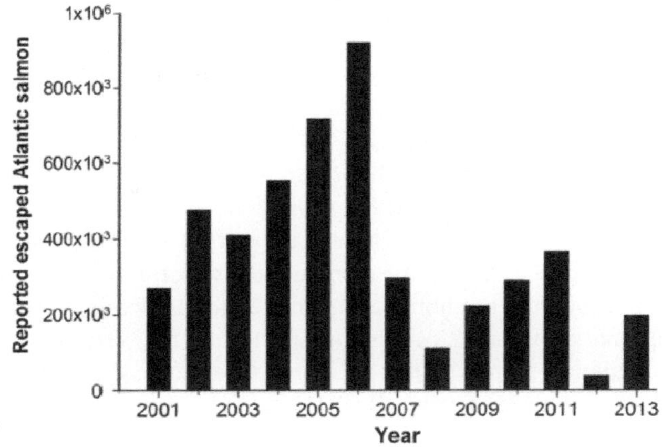

Figure 1. The number of farmed salmon escapes reported to the Norwegian Directorate of Fisheries by fish farmers for the period 2001–2013. Data were taken from the Norwegian Directorate ofhttp://www.fiskeridir.no/.

IMPACT OF FARMED ESCAPEES ON THE GENETIC INTEGRITY OF WILD NORWEGIAN POPULATIONS

Risk Assessment

In the following chapter, we have considered the following elements of risk assessment; release, exposure, and consequences, in the following manner. Release assessment is defined as the number of farmed salmon escaping into the natural environment, both as reported and unreported numbers of escapees. Exposure assessment is defined as the physical mixing of farmed escaped salmon on the spawning grounds of wild populations, and the subsequent level of genetic introgression resulting from successful spawning. Finally, we have defined consequence assessment as the consequence of genetic introgression for both the short fitness consequences and the long evolutionary consequences on the native populations.

One of the challenges to the continued development of a sustainable aquaculture industry is containment, and each year, thousands or hundreds of thousands of farmed salmon escape into the natural environment in Norway (Figure 1). Furthermore, the official statistics for numbers of escapees reported to the Norwegian Directorate of Fisheries underestimate the real number of escapees. This has been documented through extensive simulated release experiments and statistical modelling (Skilbrei *et al.*, 2015), and is clearly supported by the fact that the legal authorities in Norway have implemented DNA tracing methods to identify the farm of origin for escapees where they have not been reported (Glover*et al.*, 2008; Glover, 2010; Zhang *et al.*, 2013). While the majority of escapees disappear post-escape (Hansen, 2006; Skilbrei, 2010a, b, 2013), each year, significant numbers of farmed salmon are nevertheless observed in rivers inhabited by wild populations (Fiske *et al.*, 2006, Fiske, 2013). It is therefore considerable potential for genetic interaction between these escapees and native populations.

The Atlantic salmon displays considerable population genetic structure throughout its native range. This variation is partitioned in a hierarchical manner, with the largest genetic differences being observed between populations located in different continents or countries, and the smallest differences being observed among neighbouring populations within regions (Ståhl, 1987; Taggart *et al.*, 1995; Verspoor *et al.*, 2005). This structure reflects various processes, for example recolonization patterns, genetic isolation by distance (Glover *et al.*, 2012), and landscape features which modify population connectivity within regions (Dillane *et al.*, 2008). In addition to differences in allele frequencies of molecular genetic markers, Atlantic salmon populations display different life history characteristics. While much of this phenotypic variation is environmentally caused, some of these differences are influenced by underlying genetic variation, and it is generally accepted that these differences potentially reflect adaptations to their native rivers (Taylor, 1991; Garcia de Leaniz *et al.*, 2007; Fraser *et al.*, 2011).

Norwegian farmed Atlantic salmon dominates global production, originates from over 40 Norwegian rivers, and has been subject to approximately ten or more generations of domestication selection (Gjedrem, 2010). Breeding programmes have successfully selected for fish that outgrow their wild counterparts multiple times under farming conditions (Glover *et al.*, 2009; Solberg *et al.*, 2013a, b). In addition to traits that have been directly selected for in the breeding programmes, genetic changes in non-targeted traits have also been observed, for example in predator awareness (Einum and Fleming, 1997), stress tolerance (Solberg *et al.*, 2013a), and gene transcription (Roberge *et al.*, 2006). In addition, decreased genetic variation, as revealed by molecular genetic markers (Norris *et al.*, 1999; Skaala *et al.*, 2004), and lower estimates of heritability for growth (Solberg *et al.*, 2013a), has been observed in farmed populations. Reduced genetic variation in molecular genetic markers reflects founder effects and genetic drift, driven by limited farmed population sizes, while reduction in heritability for growth is likely to be a result of successful directional selection for this trait over multiple generations.

An early study in Ireland estimated introgression of farmed escaped salmon in a native population based upon escapement from a nearby farm (Clifford *et al.*, 1998b). However, despite the fact the genetic changes in native wild populations have been observed in molecular genetic markers as a result of farmed salmon introgressing in Canadian (Bourret *et al.*, 2011), Irish (Crozier, 1993, 2000; Clifford *et al.*, 1998a, b), and Norwegian rivers (Skaala *et al.*, 2006; Glover *et al.*, 2012), until a recent breakthrough in a study of 20 Norwegian rivers (Glover *et al.*, 2013a), the cumulative level of introgression of farmed salmon has not been calculated for any native Atlantic salmon population. This is due to the fact that estimation of cumulative introgression of farmed salmon is statistically challenging (Besnier *et al.*, 2011). In the Norwegian study of 20 rivers (Glover *et al.*, 2013a), a combination of Approximate Bayesian Computation, and genetic data for wild-historical, wild-contemporary and a diverse pool of farm samples that were genotyped for a set of collectively informative single-nucleotide polymorphic markers (Karlsson *et al.*, 2011) was used to estimate introgression for a period of 2–4 decades.

Overall, the study by Glover *et al.* (2013a) revealed less introgression of farmed Atlantic salmon in many Norwegian populations (Figure 2) than may be expected based upon the reported numbers of escapees in these populations, and estimations from introgression models (Hindar *et al.*, 2006). The authors concluded that spawning success of farmed escaped salmon has been generally low in many Norwegian rivers, a suggestion consistent with earlier estimates of spawning success in controlled experiments (Fleming *et al.*, 1996, 2000). Nevertheless, results from the study demonstrated high levels of admixture in some native populations, and together with an earlier study using microsatellites, reported decreased genetic differentiation among populations over time (Glover *et al.*, 2012, 2013a). The latter of which is consistent with suggestions that widespread introgression of farmed salmon will lead to erosion of population genetic structure among native populations (Mork, 1991).

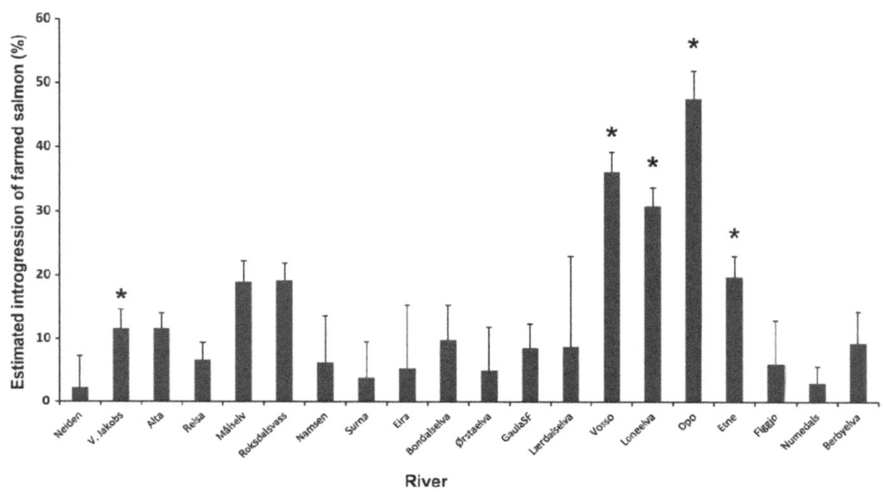

Figure 2. Estimated cumulative introgression of farmed Atlantic salmon in 20 Norwegian Atlantic salmon populations in the period 1970–2008 based upon Approximate Bayesian computation using genetic data. Figure is produced using estimations of admixture from Table 3 in Glover *et al.*(2013a). The computed median level of introgression is 9.1%.

Estimating the genetic consequences of farmed salmon introgression on life history traits, population fitness, and long-term evolutionary capacity of wild populations is more challenging than estimating introgression. This is in part due to the fact that wild populations display large natural variation in, for example, marine survival, and at the same time are influenced by a wide range of anthropogenic factors (Parrish *et al.*, 1998), which may potentially mask biological changes caused by introgression of farmed salmon. Nevertheless, comparative studies in Ireland and Norway have demonstrated additive genetic variation for fitness in the wild, with offspring of farmed salmon displaying lower survival than fish of native origin (McGinnity *et al.*, 1997, 2003; Fleming *et al.*, 2000; Skaala *et al.*, 2012). Similar studies conducted on other salmonid species in response to releases of hatchery fish have also arrived at similar conclusions (Araki *et al.*, 2008; Araki and Schmid, 2010).

In summary, the presence of farmed escaped salmon on the spawning grounds of native populations, and the potential for genetic interactions between escapees and wild conspecifics, is of concern. This is because farmed escapees may be genetically different from the recipient wild population for several reasons. (i) Farmed salmon usually do not originate from the same wild population into which they migrate post-escape and will therefore display population genetic differences to the native population. (ii) Farmed salmon have been subject to directional selection and thus differ to all wild salmon for those traits. (iii) Through relaxation of natural selection and inadvertent adaption to the domestic environment, farmed salmon have undergone domestication selection and will also differ to wild salmon.

Risk Estimation

As part of a national strategy for an environmentally sustainable aquaculture industry (Anon., 2009b), the Norwegian government established the following management goal to prevent genetic interactions of farmed escapees with wild salmon populations: "Aquaculture will not contribute to permanent genetic changes in the genetic characteristics of wild fish stocks" (Table 1). This political target, which forms the basis for the risk estimation, is clearly open for scientific interpretation. However, it was interpreted in a conservative sense for the estimation of risk (Table 2). Thus, any observed genetic change in allele frequencies of molecular genetic markers caused by introgression of farmed salmon would be regarded as permanent genetic change, and therefore in violation of the management goal for sustainability established by the government. The rationale behind this interpretation was first and foremost because molecular genetic markers would be required to directly measure genetic changes in the wild populations. Furthermore, while natural selection will influence the genetic composition of any population, including those where farmed salmon have successfully introgressed, it is unlikely that natural selection will revert the population back to its exact genetic composition before introgression of farmed salmon. This is despite the possibility that natural selection may potentially restore fitness in the natural population.

The documentation of genetic change in a wild population is most directly achieved through the analysis of molecular genetic markers. However, while introgression of farmed Atlantic salmon has been estimated for 20 wild salmon populations in Norway in a 3–4 decade period from 1970 onwards (Glover *et al.*, 2013a), genetic data to estimate introgression of farmed salmon does not exist for the great majority of Norwegian populations. Furthermore, the analysis was being used to address risk of continued and future genetic changes in relation to today's aquaculture industry rather than changes that have already occurred through historical introgression. Therefore, the frequency of farmed escaped salmon observed in wild populations, which is correlated with genetic introgression of farmed escapees over time (Glover *et al.,* 2012, 2013a) was chosen as the indicator to estimate risk of further genetic changes in each wild population for the estimation of risk (Taranger *et al.*, 2012a).

From 2012 onwards, risk was estimated against the below categories for probability of further genetic changes in wild populations caused by introgression of farmed salmon:

- No or low risk of genetic change: <4% incidence of farmed salmon
- Moderate risk of genetic change: 4–10% incidence of farmed salmon
- High risk of genetic change: >10% incidence of farmed salmon

The threshold values were set according to knowledge of natural straying (reviewed by Stabell, 1984) and knowledge about the present correlation between frequency of farmed fish and corresponding genetic introgression (Glover *et al.*, 2012, 2013a). Stabell (1984) showed that most fish returned to their natal river, although in two of the experiences referred to, straying rates were as high as 10% and nearly 20%. We have chosen a threshold value for no or low risk for genetic change at a frequency of farmed fish in the river corresponding to the lower part of the natural straying estimates (4%), while 10% as high risk of genetic change from the upper part of the distribution. Especially, the threshold value for the upper limit is uncertain and might be modified according to new knowledge about the corresponding correlation between frequency of farmed fish and actual introgression (Glover *et al.*, 2013a; Taranger *et al.*, 2014).

The frequency of farmed salmon in each river surveyed was based upon autumn data where the frequency has been reported for a series of Norwegian rivers (Fiske, 2013). To estimate risk, the frequency of farmed salmon observed in autumn survey was recomputed into an "incidence of farmed salmon" per population using a formula for normalizing data from summer angling catches and autumn surveys (Diserud *et al.*, 2010). This was done because the percentage of farmed salmon in autumn samples is usually higher than in summer angling catches (Fiske *et al.*, 2006), which is in part because farmed salmon enter rivers later than wild salmon. Thus, without normalization of data, the frequency of farmed salmon in summer and autumn surveys are not directly comparable. While many rivers have both summer and autumn estimates, some only have one or the other estimate and therefore require transforming into what has been defined as the "incidence of farmed salmon". The formula for normalizing data from summer and autumn surveys to create the "incidence of farmed salmon" were obtained from Diserud *et al.* (2010) and are presented below:

$$\arcsin(\sqrt{(\text{incidence of farmed salmon}})=0.116+0.888\times\arcsin(\sqrt{(\text{summer frequency})})$$

$$\arcsin(\sqrt{(\text{incidence of farmed salmon}})=0.044+0.699\times\arcsin(\sqrt{(\text{autumn frequency})})$$

For the risk assessment, the mean incidence of farmed salmon was estimated in 34 rivers distributed along the Norwegian coast using autumn survey data collected in the period 2010–2012. Only rivers having autumn survey data from a minimum of 2 of the 3 years in this survey period were included in the risk assessment. Based on these data, the risk for genetic changes as a result of farmed salmon introgression was low, moderate, or high for 13 (38%), 11 (32%), and 10 (29%) surveyed rivers, respectively (Figure 3).

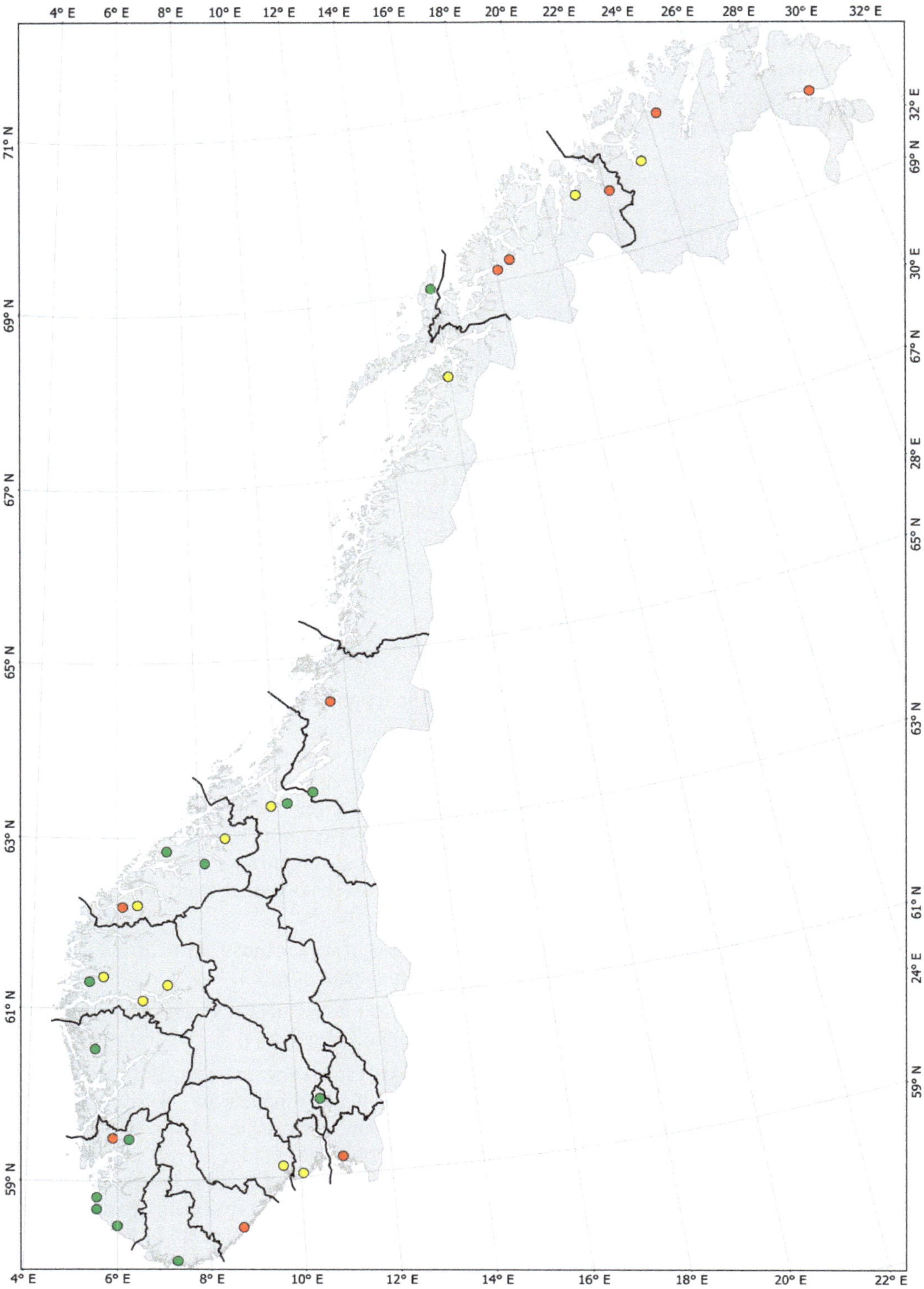

Figure 3. Assessment of risk for genetic changes in 34 wild Atlantic salmon populations in Norway in the period 2010–2012. Green point = No or low risk, yellow point = moderate risk, and red point = high risk.

Limitations

There are a number of challenges and limitations to the conducted risk assessment and estimation, its approach, and calibration against threshold values and potential impacts. One limitation is the fact that the observed frequency of farmed escapees in rivers has been used as the proxy for potential genetic changes in wild populations caused by introgression of the farmed salmon. This is a limitation because the correlation between the frequency of escapees observed in a river, and documented genetic introgression is only modest (R^2 = 0.47, p = 0.0007) (Glover *et al.*, 2013a). As a result, some rivers display higher and lower levels of genetic introgression from farmed than would be estimated by analysis of the frequency of escapees on the spawning grounds. The consequence of this is that the observed frequency of farmed escapees in each population will not accurately reflect the true risk of genetic changes for all populations, and only by using genetic methods directly will the risk be able to be quantified accurately.

The underlying causes of the lack of a strong relationship between the observed frequency of escapees and genetic introgression are important to identify to help improve the accuracy of the risk assessment in the future. From their genetic study of introgression in 20 Norwegian rivers, Glover *et al.* (2013a) identified both technical and biological elements that are likely to influence the strength of the relationship between the observed frequency of escapees and genetic introgression. Two of the primary components suggested were potential inaccuracies in the frequency of escapees reported for each river (e.g. limited, biased, or non-standardized sampling or reporting), and the fact that the density of the native population, especially on the spawning grounds, may also influence relative success of farmed escapees through spawning competition (Fleming *et al.*, 1996, 2000). In the future, models may be used which include covariables in addition to just the incidence of farmed salmon to predict genetic changes.

A further limitation of the present risk assessment is that it was only conducted for 34 rivers for the 2013 risk assessment. These rivers were chosen as they had autumn survey data published (Fiske, 2013) and thus readily available for assessment of risk for the period 2010–2012. In Norway, there are over 400 salmon rivers and for ~220 of them the status of the stocks are assessed (Anon., 2013b). Thus, the rivers investigated in the current risk assessment only represent a small proportion of those in Norway. Therefore, it is not possible to make clear regional inferences regarding introgression of farmed salmon, only for a small number of specific rivers. Data for the frequency of farmed salmon exist for a larger number of rivers than are currently included in this risk assessment. However, the quality of some of these data, the reporting and availability of the data are highly variable. It was for this reason that the risk assessment was only conducted for the 34 rivers.

Clearly, there is a significant need to increase efforts to expand and improve the monitoring of escaped salmon in a larger number of Norwegian rivers using data gathered and reported in a standardized manner. This will initially be able to improve estimates of the proportion of escaped salmon and will also provide a better foundation for the collection of representative samples for subsequent use in genetic analysis to validate introgression in rivers. An effort to coordinate data collection of escapees has been initiated within Norway in 2014, and in the future it is predicted that the risk assessment will be conducted in a much larger number of rivers.

In addition to technical and data availability challenges linked with the risk assessment and its implementation, there are gaps in current knowledge which limit the ability to identify threshold tolerance limits for introgression of escapees and the level of potentially detrimental effects on the wild populations. Current knowledge points toward a potential negative effect of introgression of farmed Atlantic salmon on the fitness and future evolutionary capacity of recipient wild populations. This is when taking into consideration data available from experimental comparisons of farmed and wild salmon especially in the natural environment (McGinnity *et al.*, 1997, 2003; Fleming *et al.*, 2000; Skaala *et al.*, 2012), background knowledge of salmon biology, life history and ecology, and extensive information from hatchery-fish supplementation for both Atlantic salmon as well as other salmonid species in both the Atlantic and Pacific (Araki *et al.*, 2008; Araki and Schmid, 2010). Nevertheless, significant gaps in understanding of the biological consequences of introgression of farmed salmon remain. These need to be quantified in the future to make a full assessment of risk of biological consequences following introgression of escapes. The major points are summarized in Table 3.

Table 3. Some identified gaps in current knowledge with respect to understanding the potential negative consequences of introgression of farmed salmon in native populations.

Question	Hypothesis to be tested
What is the fitness differential between the offspring of wild, hybrid, and farmed salmon, including multiple generation back-crossed individuals, in different rivers and environments?	To what degree can the results from the few comparative experiments of survival in the wild be generalized for all types of rivers and populations, and is it possible to generalize results?
How strong is natural selection, and will the offspring of farmed salmon readapt to the natural environment?	Lower survival of the offspring of farmed salmon and hybrids in the wild also implies that natural selection will purge poorly adapted individuals from the recipient population, but how strong is natural selection and what is the time-scale of this potential re-adaptation process?
What are the threshold limits of introgression?	What degree of genetic introgression will be tolerated in wild populations before biological, life history and ecological characteristics of the population, and population productivity are compromised?
What is the underlying genomic architecture of domestication and local adaptation in salmonids?	What genetic changes have occurred during domestication, is it possible to identify genetic markers linked with these changes, and fitness in the wild to provide more accurate measurements of functional genetic changes in native populations?

SALMON LICE IMPACT ON WILD SALMO-NIDS

Salmon lice (*Lepeophtheirus salmonis*) from salmon farms are recognized as an important hazard to wild anadromous salmonids in Norwegian coastal waters (Serra-Llinares *et al.*, 2014). Salmon lice on farmed salmon produce large amounts of planktonic larvae stages that spread via the water currents and can infect migrating Atlantic salmon post-smolts, as well as sea trout (*Salmo trutta*) and Arctic charr (*Salvelinus alpinus*) that stay in coastal waters (Jones and Beamish, 2011). Hydrodynamic models coupled with biological data show that salmon lice can be transported up to 200 km over a 10-d period, although most dispersed 20–30 km (Asplin*et al.*, 2011; Serra-Llinares *et al.*, 2014). The number of salmon lice allowed on farmed salmon is tightly controlled by Norwegian legislation (www.mattilsynet.no). However, the large number of farmed salmon, with ~300 million smolts put into sea cages every year along the Norwegian coast, results in worse case releases in the order of more than a billion salmon lice larvae daily from salmon farms in Norway (Taranger *et al.*, 2014).

New analyses reveal strong correlation between salmon farms and lice infections on wild salmonids in Norwegian coastal waters (Helland *et al.*, 2012; Serra-Llinares *et al.*, 2014). The Norwegian salmon lice monitoring programme on wild salmonids demonstrate annual lice epidemics, most likely connected to the density of salmon farms in the surrounding areas as well as the seasonal dynamics of salmon lice infections on farmed salmon (Jansen *et al.*, 2012; Serra-Llinares *et al.*, 2014; Taranger *et al.*, 2014). A series of experiments has shown that salmon lice may affect anadromous salmonids (reviewed in Finstad and Bjørn, 2011; Anon., 2012;Torrissen *et al.*, 2013).

To assess the risk of salmon lice infection on wild populations, we have considered the following elements of risk assessment: release, exposure, and consequences in the following manner. The release assessment is based on estimating the production and distribution of infectious salmon lice. The exposure assessment is based on estimating the lice infection on wild salmonid populations using different methods for direct measurements of salmon lice infections on salmon and sea trout. The consequence assessment is the effect salmon lice have on salmonid populations in terms of estimated likelihood of increased marine mortality and/or reduced reproduction based on the exposure assessment.

Risk Assessment

All salmon farming sites in Norway report the numbers of salmon lice on the fish weekly when the temperature is >4°C. This is reported together with fish biomass and number of individual salmon per cage (reported each month). Based on the number of sexually mature female salmon lice on the fish on each farming site, the number of infectious salmon lice larvae produced from the different salmon farms are calculated (Jansen *et al.*, 2012). However, at present we do not have enough information and validated models to accurately estimate the impact of the salmon lice infections on wild populations based on reported data from the fish farms (Taranger *et al.*, 2013, 2014).

The lice infection on wild salmonid populations is estimated using different methods as part of a national monitoring programme (Serra-Llinares *et al.*, 2014; Taranger *et al.*, 2014). These methods encompass catch of sea trout and Arctic charr in traps or nets (Bjørn *et al.*, 2011a), and salmon post-smolts caught in special surface trawls in fjord systems (Holm *et al.*, 2000; Bjørn *et al.*, 2007b; Holst *et al.*, 2007). In addition, groups of small sentinel cages containing on average 30 farmed salmon post-smolts are placed in the fjords to monitor the salmon lice infection rate. The fish are kept in the cages for 3 weeks before lice are counted on all the fish, and the procedure repeated three times during spring and summer (Bjørn *et al.*, 2011a, 2013). The sampling programme is focused on areas with high salmon farming activity (Figure 4), as well as some fjords that are protected against salmon farming. The assessment in the period 2010–2013 is based on data from 1 to 5 sites per fjord in 13–16 fjord systems annually, and with increasing numbers of fish sampled at each site in the later years (Table 4). More details about the national salmon lice monitoring programme are provided elsewhere (Helland *et al.*, 2012; Bjørn *et al.*, 2013; Serra-Llinares *et al.*, 2014).

Table 4. The number of sampling locations and number of wild-caught salmonids (mostly sea trout) investigated for salmon lice infestations in the Norwegian salmon lice monitoring programme during 2010–2013.

	Sites and number sampled in Period 1 (only fish < 150 g)				Sites and number sampled in Period 2 (all fish sizes)			
Year	2010	2011	2012	2013	2010	2011	2012	2013
Locations	26	31	29	23	26	32	29	23
n	218	422	944	1711	623	806	1144	2368

Period 1 covers smolt migration in spring and early summer, whereas Period 2 covers summer period to assess the accumulated effects on sea trout and Arctic charr. Both periods are adjusted different timing of smolt migration and seawater residence along the Norwegian coast, with an earlier sampling window in the southern part of Norway and later further north.

To conduct the risk assessment on the potential impact on salmonid smolts that migrate from the rivers in spring and early summer on the one hand, and the risk of sea trout and Arctic charr that stay in fjords and coastal waters during summer, the national monitoring programme covers two different periods. These periods are adjusted for different timing of smolt migration and seawater residence along the Norwegian coast (Anon., 2011), with an earlier sampling window in the southern part of Norway and later further north. This corresponds to two assumed "critical periods", the first during spring when the salmonid smolts leave the rivers and enters the estuaries and fjords (Period 1), and the second period to estimate the accumulated infection rate on sea trout and Arctic charr that remain in fjords and on the coast during summer (Period 2). See alsoAnon. (2011) for further information on median migration dates and migration speeds for Atlantic salmon post-smolts in Norwegian fjords.

A range of laboratory studies demonstrate the impact of salmon lice on salmon post-smolts (Grimnes and Jakobsen, 1996; Finstad *et al.*, 2000,2010; Heuch *et al.*, 2005; Wagner *et al.*, 2008). It has been shown that 0.04–0.15 lice per g fish weight can increase stress levels, reduce swimming ability and create disturbances in water and salt balance in Atlantic salmon (Nolan *et al.*, 1999; Wagner *et al.*, 2003, 2004; Tveiten *et al.*, 2010). In sea trout, ~50 mobile lice are likely to give direct mortality (Bjørn and Finstad, 1997), and only 13 mobile lice, or~0.35 lice per g fish weight might cause physiological stress in sea trout (weight range of 19–70 g; Wells *et al.*, 2006, 2007). Moreover, ~0.05–0.15 lice per g fish weight were found to negatively affect sexually maturing Arctic charr (Tveiten *et al.*, 2010).

According to a review by Wagner *et al.* (2008), infections of 0.75 lice per gram fish weight, or ~11 salmon lice per fish, can kill a recently emigrated wild smolt of ~15 g if all the salmon lice develop into preadult and adult stages. Studies of naturally infected wild salmon post-smolts indicate that only those with <10 lice survived the infection (Holst *et al.*, 2007). This is consistent with field studies on salmon lice infections in salmon post-smolts in the Norwegian Sea. Over a decade of surveys, no post-smolts was found with >10 salmon lice, and fish with up to 10 mobile lice were observed to be in poor condition with low blood count and poor growth (Holst *et al.*, 2007). New studies of naturally infected wild salmon post-smolts also show that sea lice are fatal at high infections (Berglund Andreassen, 2013). More work in this field is in progress.

Population-wise effects of salmon lice have been demonstrated on wild salmonids in Ireland and Norway. This was studied by protecting individually tagged Atlantic salmon smolts against salmon lice (using Substance EX or Slice) before they were released into the sea near their respective home rivers (Finstad and Jonsson, 2001; Hazon *et al.*, 2006;Skilbrei and Wennevik, 2006; Hvidsten *et al.*, 2007; Skilbrei *et al.*, 2008,2013; Jackson *et al.*, 2011, 2013; Anon., 2012; Gargan *et al.*, 2012;Krkošek *et al.*, 2013a, b; Vollset *et al.*, 2014). These studies suggest that salmon lice infections increase the marine mortality in areas with intensive salmon farming activity.

Risk Estimation

The risk assessment on salmon lice is based on Goal 1 in the policy document on sustainable aquaculture by the Norwegian government from 2009 (Anon., 2009b, Table 1) stating that "Disease in fish farming will not have regulating effect on populations of wild fish…". This target has in this risk estimation been interpreted as "Salmon lice from fish farming significantly increase the mortality of wild salmonids" (Table 2).

A salmon lice risk index, attempting to estimate the increased mortality due to salmon lice infections, was proposed by Taranger *et al.* (2012a) and are based on the assumption that small salmonid post-smolts (<150 g body weight) will suffer 100% lice-related marine mortality, or return prematurely to freshwater for sea trout, in the wild if they are infected with >0.3 lice g^{-1} fish weight. Furthermore, the lice-related marine mortality is estimated to 50%, if the infection is between 0.2 and 0.3 lice g^{-1} fish weight, 20% if the infection rate is between 0.1 and 0.2 lice g^{-1} fish weight, and finally 0% lice-related mortality if the salmon lice infection is <0.1 lice g^{-1} fish weight.

For larger salmonids (over 150 g), we assume that lice-related mortality or compromised reproduction will be 100% in the group if they have >0.15 lice g^{-1} fish weight, 75% for lice infections between 0.10 and 0.15 lice g^{-1}fish weight, 50% for lice infections between 0.05 and 0.10 lice g^{-1} fish weight, 20% for lice infections between 0.05 and 0.01 lice g^{-1} group, and finally 0% if the salmon lice infection is <0.01 lice g^{-1} fish weight.

For both indices, increased mortality risk or compromised seawater growth or reproduction at population level are calculated as the sum of the increased mortalities/compromised reproduction for the different "infection classes" in the sample, reflecting the distribution of the intensity of salmon lice infections of the different individuals sampled. This assumes that individuals caught with traps, gillnets or trawls are representative for the various salmonid populations in that fjord area. The risk was further scored according to the system proposed by Taranger *et al.* (2012a); as low (<10% estimated increase in mortality; green colour), moderate (yellow) for those with between 10 and 30% increase, and high (red) if the increase is calculated as >30%.

The current assessment is based on these scorings, and we use data from the national monitoring programme on salmon infections in wild salmonids (Bjørn *et al.*, 2010, 2011b, 2012, 2013). Separate result tables are presented for the risk for Atlantic salmon smolts (Period 1; Table 5) and for sea trout and the Arctic charr populations (Period 2; Table 6). The results are sorted by county from south to north and by fjord system.

Table 5. Estimated risk for lice-related mortality (%) of Atlantic salmon post-smolts based on lice levels on sea trout caught in traps or gillnets and with weight <150 g in Period 1 at the sites from south to north in Norway in 2010–2013.

County	Fjord	Site	2010	2011	2012	2013
Aust-Agder	Sandnesfjord	Sandnes	0	0	0	
Rogaland	Ryfylke	Hellvik		0	0	0
		Vikedal[a2012]		36	20	
		Nedstrand				3
		Forsand		0	0	
Hordaland	Hardanger	Granvin	0	0		
		Ålvik		54	51	0
		Rosendal	0	69	53	13
		Etne	0	0	16	1
Sogn og Fjordane	Sognefjorden	Balestrand	0	0	2	0
		Vik				0
		Brekke/Dingja	0	35	23	0
Møre og Romsdal	Romsdal	Eresfjord	0	0	0	
		Sandnesbukta				22
		Isfjord	0	0	0	0
		Bolsøy [a2010–2012]	2	10	22	15
		Vatnefjorden				0
		Frænfjorden				7
	Storfjord	Sylte	0	0	37	
		Sykkylven	0	0	0	
		Ørsta	0	5	9	
Sør-Trøndelag	Trondheimsfjorden	Skatval [a2010]	6	0	2	0
		Agdenes	0	90	94	0
		Hitra	0	5	0	0
Nord-Trøndelag	Namsen	Tøtdal	0	0	0	0
		Sitter [a2010–2011]	32	24	71	15
		Vikna			7	98
Nordland	Eidsfjord	Vik [a2010–2012]	0	32	50	
	Folda	Ballkjosen	0	25	13	
		Sagfjord	0	7	7	
	Vefsn	Fagervika	0	3		
		Leirfjord	0	0		
	Velfjord	Indre Velfjord [a]			4	
		Ytre Velfjord [a]			4	
Troms	Salangen	Løksa[a2012]		0	4	
Finnmark	Altafjord	Talvik[a2012]	0	0	1	12
		Skillefjord	5	0	3	26
	Bugøyfjord	Bugøyfjord				
	Porsanger	Handelsbukt	0	0		0
		Kåfjord/Repvåg [a2013]	0	0		0

The colour code refers to the assessment of potential population-reducing effect (red ¼ high, yellow ¼ moderate, and green ¼ low).

[a]Small number of fish caught, all fish were used in the assessment irrespective of body size.

Table 6. Estimated risk for salmon lice-related mortality or compromised reproduction (%) of sea trout based on observations in Period 2 at the various sites from south to north in Norway in 2010–2013.

County	Fjord	Site	2010	2011	2012	2013
Aust-Agder	Sandnesfjord	Sandnes	0	2	2	
Rogaland	Ryfylke	Hellvik		0	0	0
		Vikedal		15	0	
		Nedstrand				7
		Forsand		3	0	
Hordaland	Hardanger	Granvin	0	14		
		Ålvik		17	40	32
		Rosendal	55	67	87	38
		Etne	54	3	74	32
Sogn og Fjordane	Sognefjorden	Balestrand	1	0	3	0
		Vik				
		Brekke / Dingja	46	19	72	19
Møre og Romsdal	Romsdal	Eresfjord	0	21	8	
		Sandnesbukta				71
		Isfjord	7	15	25	26
		Bolsøy	14	13	32	73
		Vatnefjorden				98
		Frænfjorden				81
	Storfjord	Sylte	0	0	0	
		Sykkylven	3	10	34	
		Ørsta	25	16	35	
Sør-Trøndelag	Trondheimsfjorden	Skatval	0	13	2	11
		Agdenes	26	40	34	35
		Hitra	8	47	88	41
Nord-Trøndelag	Namsen	Tøtdal	0	9	14	55
		Sitter	65	40	59	62
		Vikna			57	60
Nordland	Eidsfjord	Vik	34	54	59	
	Folda	Ballkjosen	52	45	66	
		Sagfjord	4	52	32	
	Vefsn	Fagervika	19	4		
		Leirfjord	3	0		
	Velfjord	Indre Velfjord			34	
		Ytre Velfjord			28	
Troms	Salangen	Løksa		22	20	
Finnmark	Altafjord	Talvik	3	47	5	18
		Skillefjord	4	55	24	51
	Bugøyfjord	Bugøyfjord		10		
	Porsanger	Handelsbukt	0	0		5
		Kåfjord/Repvåg	0	0		27

Mortality estimates are based on all fish sampled in the period, with different thresholds for small (,150 g) and large (.150 g) fish. The colour code refers to the assessment of potential population-reducing effect (red ¼ high, yellow ¼ moderate, and green ¼ low).

The results indicate considerable variation in risk between years and sampling locations. Moreover, these data strongly indicate a much higher risk for sea trout (and also Arctic charr in the Northern regions) compared with Atlantic salmon post-smolts and reveal moderate-to-high risk of population-reducing effects on sea trout in most counties with high salmon farming activity. The risk of population-reducing effects for Atlantic salmon varies much more between years and sites, and was low at most sites in 2010 and 2013, but moderate and high at several sites 2011 and 2012.

Limitations

The assessment in the period 2010–2013 is based on data from 13 to 16 fjord systems annually. Despite large field effort (Bjørn *et al.*, 2011b,2012, 2013), the geographical coverage is insufficient in terms of the distribution of salmon farms and wild salmonid populations along the Norwegian coast. There are also problems considering how well the different sampling methods are representative for the different anadromous populations in that area. We have limited data on salmon lice infections in migrating Atlantic salmon smolts, so the risk assessment for salmon is mainly conducted by the use data on salmon lice infections on sea trout caught in traps and gillnets in Period 1 as proxy for the risk to Atlantic salmon post-smolts. Lice infections on trout may not be directly proportional to lice infections on migrating salmon smolt. It is likely that differences in, for example, migratory behaviour and marine ecology exposes salmon and sea trout smolts for different sea lice infection risk (Anon., 2011), even within the same fjord system (Bjørn *et al.*, 2007a,2011b, c, 2013; Serra-Llinares *et al.*, 2014). The link between individual lice infections and population effects is also very uncertain. There is therefore uncertainty of the current risk assessment both for Atlantic salmon, and for sea trout and Arctic charr. Moreover, the current data are presented without any estimates of uncertainty, which must be included in future analyses.

Future

The "Strategy for an environmentally sustainable aquaculture industry" (Anon., 2009b) states that no disease, including lice, should have a regulatory effect on wild fish. The monitoring of salmon lice infection of wild salmonids is an important verification of whether this goal is achieved, and whether the measures taken are appropriate and sufficient. An indicator system that allows detection of possible problems needed therefore to be established. Hitherto, this assessment has been based on data from the national monitoring of sea lice. An analysis of the historical data in this monitoring series (2004–2010) shows that both the extent and nature of the data have had some weaknesses that limit the ability to analyse and understand the observed variation in infections on wild fish (Helland *et al.*, 2012; Serra-Llinares *et al.*, 2014). Therefore, monitoring and risk assessment based only on lice counts on wild salmonids is not considered sufficient.

Consequently, a rather radical change in the monitoring, advisory and management system for lice has therefore been proposed (Taranger *et al.*, 2012b, 2013, 2014; Bjørn *et al.*, 2013; Serra-Llinares *et al.*, 2014). This system is based on (i) detection and forecasting of increased production of infectious salmon lice using models, (ii) verification of infection pressure through risk-based and adapted surveillance on wild salmonids and (c) extended risk assessment based on a considerably larger dataset and fine scale hydrodynamic lice dispersal modelling to assess the effect on wild populations, which then allows adjustment of measures taken by management and industry to reduce this effect to levels within the objective of the strategy. This is now possible due to better knowledge about the relationship between intensive salmon farming activity and infection pressure (Helland *et al.*, 2012; Jansen *et al.*, 2012; Serra-Llinares*et al.*, 2014; Taranger *et al.*, 2014), and better and more accessible farming and environmental data (Jansen *et al.*, 2012; Taranger *et al.*, 2014).

A preliminary analysis indicates that under realistic conditions of lice infections, water currents, temperature and salinity, and relative lice infections may be predicted using a coupled hydrodynamic-biological lice dispersion model. With further calibrations and validation, such a system can probably be developed at least for specific areas along the Norwegian coast. As validation, calibration, and implementation of such a risk-based monitoring system is done, more of the monitoring could be based on the model and less on catch of wild salmonids. The preliminary model results are encouraging in terms of validation and calibration the model predictions against observed infection of wild salmonids. However, considerable research and development remains, where the main challenges are:

- The system for detecting problem areas based on farming data and sea lice infections of notification of problem areas need to be further developed and operationalized, and systems for risk-based and adapted surveillance on wild salmonids must be developed.
- Coupled hydrodynamic-biological lice dispersion models must be validated and calibrated against observed infection levels on wild salmonids in the field.
- Knowledge about the ecological effects of a given infection pressure on stocks of wild salmonids (population-reducing effects) must be increased so that more precise predictions can be developed.

DISEASE TRANSFER FROM FARMED SALMON TO WILD FISH

Background

Infectious diseases represent a major problem in Norwegian fish farming, despite successful development and application of vaccines against a range of pathogens. In addition to lice (considered separately above), viral diseases currently represent the largest disease problems in Norwegian aquaculture (Johansen, 2013). In the period 2005–2012, the four most frequent viral diseases [infectious pancreatic necrosis (IPN), pancreas disease (PD), heart and skeletal muscle inflammation (HSMI), and cardiomyopathy syndrome (CMS)] had 400–500 outbreaks annually (Johansen, 2013). The main reason for the dominance of viral diseases is the lack of effective vaccines. Bacterial diseases, on the other hand, cause only ~20 outbreaks annually, reflecting that the currently used bacterial vaccines provide good protection (Austin and Austin, 2007). Among the parasites, parvicapsulosis due to the myxosporean *Parvicapsula pseudobranchicola* is a problem mainly in northern Norway, whereas heavy gill infections with the microsporidian *Paranucleospora theridion* and the amoeba *Paramoeba perurans* occur mostly in the southern parts of the country. The significance of infections with the former two parasites is unclear, while amoebic gill disease (AGD) has so far been detected only during fall in 3 years (2006, 2012–2013).

For most pathogens, clear evidence for transmission from farmed to wild fish is limited (Raynard *et al.*, 2007). Most of the diseases that currently cause problems in fish farms are likely enzootic, originating from wild fish. This implies that these infections occur or occurred in the past at some "background" level in wild stocks. Such considerations complicate an estimation of the impact of aquaculture, since the "normal" prevalence range of many important disease agents is unknown. However, in two cases exotic pathogens have been introduced in association with farming activities. These have clearly affected wild Atlantic salmon populations.

The ectoparasite *Gyrodactylus salaris* (Monogenea) was first detected in Norway in 1975 (Johnsen *et al.*, 1999). There have been several introductions of *G. salaris* to Norway (Hansen *et al.*, 2003) linked to the import of salmonids from Sweden. Later, the parasite has spread (or has been spread) to many rivers (Johnsen *et al.*, 1999). By 2005, *G. salaris* had been detected in 45 rivers and 39 freshwater farms (Mørk and Hellberg, 2005). Norwegian Atlantic salmon stocks are very susceptible to *G. salaris*, and gyrodactylosis in farmed salmon may lead to 100% mortality if not treated (Bakke, 1991; Bakke *et al.*, 1992; Bakke and MacKenzie, 1993). Mortality in rivers is high, with the density of Atlantic salmon parr being reduced by 50–99% (Johnsen *et al.*, 1999).

Aeromonas salmonicida, the causative agent of furunculosis, was introduced to Norway in 1964, when furunculosis was detected in a single farm that received rainbow trout from Denmark. The disease then spread to other farms and wild fish within a limited area, being detectable there until

1979. A second introduction occurred in 1985, in connection with an import of Atlantic salmon smolts from Scotland. The disease then spread rapidly to farms and wild fish, and in 1992 a total of 550 salmon farms and 74 river systems were affected (Johnsen and Jensen, 1994). This rapid spread of the disease was likely facilitated by frequent escapement events involving infected fish (Johnsen and Jensen, 1994). Mortality due to furunculosis was registered in many rivers among escaped salmon, wild salmon, and trout. Mortality in farmed fish was high, reaching 50%, but the disease was first controlled by antibiotics and subsequently effectively with oil-based vaccines (Sommerset *et al.*, 2005; Johansen, 2013).

These two examples show the devastating effects that introductions of exotic pathogens can have. Even when disregarding agents only known from non-salmonids, there is a large number of potential pathogens (i.e. hazards) infecting salmonids elsewhere that could have significant impact on both salmon farming and wild fish populations in Norway if introduced (Raynard *et al.*, 2007; Brun and Lillehaug, 2010). Import of live fish represents the major threat to both fish farming and wild stocks, since this may lead to the introduction of exotic pathogens. However, *G. salaris*infections have not been detected in Atlantic salmon hatcheries in recent years (Hytterød *et al.*, 2014 and references therein), and the parasite does not survive in seawater. Furunculosis outbreaks in farms are rare, since most farmed salmon is protected through vaccination. Regarding disease transfer from farmed salmon to wild salmonids, these diseases are currently considered to be under control.

The detection of disease in wild fish and estimating disease impact on wild populations is difficult. Clinically affected fish usually disappear quickly in nature (e.g. predated). Epizootics with mass mortality of fish are rare, but have occurred in Norway (Bakke and Harris, 1998; Sterud *et al.*, 2007) and elsewhere (Hyatt *et al.*, 1997; Gaughan *et al.*, 2000). Such episodes are usually caused either by an exotic pathogen introduced to naïve host populations (Bakke and Harris, 1998) or by exceptional environmental conditions such as high temperature (Sterud *et al.*, 2007). However, infection with native (enzootic) agents under normal environmental conditions can cause disease in individuals and affect an individual's survival or investment in reproduction. Hence, all pathogens may contribute to the regulation of wild populations at some level (May and Anderson, 1979; May, 1983), although the impact may vary and is often the result of a complex interaction between hosts, pathogens, environment, and predators (Dobson and Hudson, 1986; Combes, 2001).

Risk Assessment

There is relatively little data available on the infection status of Norwegian wild salmonid stocks with respect to the most important pathogens that affect farmed salmon (e.g. viral agents). The available data mainly concerns returning adult salmon and some local sea trout populations screened with molecular methods (Kileng *et al.*, 2011; Garseth *et al.*, 2012, 2013a, b, c; Biering *et al.*, 2013; Madhun *et al.*, 2014a, b). Studies on the occurrence of viral infections in early life stages of salmonids are only fragmentary (e.g. Plarre *et al.*, 2005).

Due to the limited data available, the disease status (outbreak statistics) in Norwegian fish farming is used as a proxy of the infection pressure from farmed salmon to wild salmonids. Information regarding disease outbreaks on Norwegian fish farms is gathered by the Norwegian Veterinary Institute (NVI) and published annually in their Fish Health Reports (e.g. Johansen, 2013). These data record official diagnoses from NVI, as well as information from the local fish health services. This information is likely to be biased towards the more serious diseases, particularly those that are required by law to be reported to government authorities. Subclinical infections may be common and may also contribute to the spread of pathogens. However, these infections are usually not detected. Despite shortcomings, these data are the best currently available information and give a reasonably good indication of the disease status of the majority of farmed fish in Norway.

Most diseases in Norwegian salmon and rainbow trout farms are represented by only a few outbreaks, often representing geographically separate cases (Johansen, 2013). However, some diseases have a large number of outbreaks/diagnoses, and are those most likely to cause elevated infection pressures that may affect wild populations. At present the most common diseases in Norwegian salmon farming are the viral diseases PD, IPN, CMS and HSMI (Table 7; Johansen, 2013). In addition, AGD due to *Paramoeba perurans* is an emerging problem (Hjeltnes, 2014). The listed viral diseases have caused some 400 or more outbreaks each year since 2005. Outbreaks are often more frequent in certain regions and at certain times of the year, leading to a consideration also of spatial and temporal variation in the potential infection pressure (Table 8).

Table 7. Number of disease outbreaks for the most important diseases in Norwegian salmon farming (Johansen, 2013).

	2005	2006	2007	2008	2009	2010	2011	2012
PD	45	58	98	108	75	88	89	137
HSMB	83	94	162	144	139	131	162	142
IPN	208	207	165	158	223	198	154	119
CMS	71	80	68	66	62	49	74	89

Table 8. Overview of the main periods where salmon and sea trout reside in coastal areas.

	Coastal area	J	F	M	A	M	J	J	A	S	O	N	D
Smolt migration	South Norway					xx	xx	x					
	Central Norway					x	xx	x					
	Northern Norway						x	xx					
Return	South Norway					x	xx	xx	x				
	Central Norway					x	xx	xx	xx	x			
	Northern Norway					x	xx	xx	xx	x			
Sea Trout (sea)	South Norway				x	xx	xx	xx	x	x	x		
	Central Norway				x	xx	xx	xx	x	x	x		
	Northern Norway				x	x	xx	xx	x	x			

Southern Norway: south Norway up to Sogn and Fjordane, Central Norway: Møre and Romsdal-Trøndelag; Northern Norway: Nordland-Finnmark. x ¼ a few fish in coastal areas, xx ¼ large numbers of fish in coastal areas.

We have considered the following elements of risk assessment regarding the viral agents salmonid alphavirus (SAV), infectious pancreatic necrosis virus (IPNV), piscine myocarditis virus (PMCV), and piscine reovirus (PRV); release, exposure, and consequences as follow. The release assessment is the assumed infection pressure as proxied by the outbreak statistics. The exposure assessment is a consideration of the spatial and temporal concurrence of wild salmonids with release. A consequence assessment should consider two aspects: (i) evidence for virus transmission and (ii) impact of viral infections. However, the impact is in all the considered cases are unknown, and only evidence for virus transmission can be discussed.

Pancreas Disease: Salmonid Alphavirus

PD in Atlantic salmon and rainbow trout is caused by SAV. In Norway, there are currently two regionalized PD epidemics caused by SAV3 (south of Hustadvika, 63° N) and SAV2 (north of 63° N). Experimental studies show transmission of SAV via water, and epidemiological studies provide evidence for horizontal farm to farm spread (Nelson *et al.*, 1995; McLoughlin *et al.*, 1996; Kristoffersen *et al.*, 2009; Stene, 2013; Stene *et al.*, 2014). The virus has been shown to survive for several weeks in the environment (Graham *et al.*, 2007) and thus may be carried long distances with currents (Stene, 2013; Stene *et al.*, 2014). SAV2 may have a different outbreak pattern than SAV3, since outbreaks tend to occur later in the year (Johansen, 2013). For the southern region (SAV3), the period in which smolts migrate and adult salmon return coincides with many SAV3 outbreaks (Table 8; Johansen, 2013). In the northern region (SAV2), most outbreaks occur later in the year. This may signify that most of the smolt migration precedes peak virus spread in the SAV2 region. On the other hand, returning salmon and sea trout are likely more exposed, but screening indicates that very few wild fish are infected (Biering *et al.*, 2013). Infected escaped salmon can enter rivers in fall, possibly exposing wild fish including naïve juveniles to the virus (Madhun *et al.*, 2014a).

Screening of sea trout (Biering *et al.*, 2013; Madhun *et al.*, 2014b) indicates that sea trout in areas with high frequency of PD outbreaks are not infected with SAV. This is in accordance with injection experiments, which suggest that sea trout is more resistant to SAV than salmon (Boucher *et al.*, 1995). SAV infections have been detected in wild salmonids and wild flatfish (Nylund, 2007; Snow *et al.*, 2010; Biering *et al.*, 2013), but PD has not been observed in wild fish.

Altogether, there are yet no data that confirm SAV transmission from farmed salmon to wild fish, but transmission of virus to wild fish is considered likely due to the large number of outbreaks and the documented efficient horizontal transmission of SAV. The probability of transmission of SAV to wild salmon is considered to be moderate for migrating smolts in the southern PD-region due to the temporal overlap between outbreaks and migration, whereas it is considered to be low in the northern PD-region as most of the migration is finished before the major outbreak period. For returning salmon, the probability of infection is considered to be low in both PD-regions based on the available screening results. The probability of SAV transmission to wild salmon is considered to be low in areas with no or few outbreaks. The probability of infection of sea trout during the marine phase is also considered to be low.

Infectious Pancreatic Necrosis: Infectious Pancreatic Necrosis Virus

IPNV is a robust, long-lived birnavirus that infect many different fish species in both fresh water and seawater (e.g. Reno, 1999). The virus is enzootic in Norway. IPN cause significant losses in fish farming in most areas in Norway (Johansen, 2013). However, there are indications of a downwards trend in outbreaks and losses, which might be caused by the increased use of IPN resistant fish (Johansen, 2013). The virus is shed into the water by infected fish, and is spread to other farms by water currents (see, e.g. Mortensen, 1993; Wallace *et al.*, 2005; Raynard *et al.*, 2007;Johansen *et al.*, 2011). A higher prevalence of IPNV has been found in wild fish near salmon farms with clinical outbreaks of IPN, compared with fish at distant sites (Wallace *et al.*, 2005, 2008). Fish surviving an IPNV infection often become persistent carriers of the virus, but viral shedding from carriers has not been demonstrated (Johansen *et al.*, 2011). The prevalence of IPNV in wild fish is low (Brun, 2003; Wallace *et al.*, 2005), and farmed fish are probably a major source of virus in the marine environment. Disease outbreaks in wild salmon have not been described, but mortality in wild marine fish due to IPN has been reported elsewhere (Stephens *et al.*, 1980; Mcallister *et al.*, 1984).

Due to the large number of annual outbreaks, the demonstrated robustness and infectivity of the virus, as well as the wide range of hosts, the probability of IPNV transmission from farmed to wild fish is considered as moderate. The extent and consequences of such transmission are unknown.

Heart and Skeletal Muscle Inflammation: Piscine Reovirus

HMSI affects farmed salmon along the entire coast of Norway. HSMI outbreaks mainly occur 5–9 months after sea-transfer. The causative agent is an *Orthoreovirus*,PRV (Palacios *et al.*, 2010; Løvoll *et al.*, 2012). The disease can be produced experimentally using infected tissue, infected cell culture, or by cohabitation (Kongtorp *et al.*, 2004; Martinez-Rubio *et al.*, 2012, 2013). PRV is present in high densities in salmon with HSMI, but high infection intensity can be found also in clinically healthy salmon. PRV infections have been detected in wild salmon along the entire coast of Norway and have been detected in sea trout (Biering *et al.*, 2013; Garseth *et al.*, 2013b). Analyses of PRV genotypes in wild Atlantic salmon, farmed salmon, and sea trout have suggested an extensive spread of the virus along the coast, and establishment in wild populations. This spread is probably due to extensive transportation of fish between areas over a long period (Garseth *et al.*, 2013c). However, there are no reports of HSMI in PRV infected wild salmonids (e.g. Garseth *et al.*, 2013b). In Norway sea trout are only rarely (1.4–3%) infected (Garseth *et al.*, 2012; Biering *et al.*, 2013) with PRV. PRV infections have been detected in some marine fish species, but the virus genotype is unknown (Wiik-Nielsen *et al.*, 2012). It is not known how long and in what quantities PRV is shed from infected fish nor viral survival in seawater. However, modelling suggests that the virus can be transported over longer distances than SAV (Aldrin *et al.*, 2010; Kristoffersen *et al.*, 2013). The latter findings suggest that the virus is relatively stable and may spread over large areas.

Since PRV infections are widespread in farmed salmon, may readily be transmitted, and is detectable in >10% of the wild salmon examined, it is considered likely that PRV is transmitted from farmed to wild salmon. HSMI occurrence in wild salmon and other wild fish are unknown (Garseth*et al.*, 2013b, Madhun *et al.*, 2014b).

Cardiomyopathy Syndrome: Piscine Myocarditis Virus

CMS is a serious disease in salmon, and is caused by a Totivirus, PMCV (Løvoll *et al.*, 2010; Haugland *et al.*, 2011). CMS can be transmitted experimentally by injecting heart tissue homogenates from diseased fish, PMCV from cell culture, and by cohabitation (Haugland *et al.*, 2011). Infections are long-lasting, with a gradual development of cardiac pathology.

PMCV has been detected in farmed Atlantic salmon along the entire Norwegian coast, but is not as widespread as PRV. PMCV infection has also been detected in a few wild salmon in Norway (Garseth *et al.*, 2012), and CMS-like lesions have been observed in the hearts of wild salmon before the discovery of the virus (Poppe and Seierstad, 2003). Large-sized wild salmon represent the only known natural reservoir for PMCV. There are no studies on shedding of PMCV from diseased fish or carriers, or on virus survival in water.

Due to the large number of hosts, prolonged infections, and the documented virus spread in cohabitation experiments, we assume that the virus is present in the environment and that the infection pressure around farms harbouring the virus is elevated. Examination of wild returning salmon detected only very few (3/1350) infected with PMCV (Garseth *et al.*, 2012; Biering *et al.*, 2013). Such a low prevalence, which may represent natural rather than fish farming-related infections, suggests that PMCV transmission from farmed to wild salmon is infrequent. Therefore, the probability of infection in wild salmon due to virus released from farms is considered low. Due to a general lack of data, particularly regarding young fish, it is not possible to assess the impact of PMCV infection in wild salmon.

Limitations of the Analysis

There is a scarcity of data on infections and lack of evidence for disease in wild salmonids for the four viral agents considered. A large number of outbreaks suggest extensive spread of virus, and consideration of timing also often substantiates an exposure of wild salmonids to the viral agents. A serious limitation in the risk assessment is a lack of information on infections due to these agents in wild fish, particularly in salmon smolts. These may be less susceptible than their farmed peers, but may also suffer mortality due to the infections. Such mortality would likely occur through predation (i.e. virus induced). In either case, the returning 1 or 2 sea-winter salmon could be found to be virus free, as is indeed generally the case regarding SAV. Juvenile fish is often particularly susceptible to viral infections. Escaped virus infected farmed salmon can enter rivers in fall (e.g. Madhun *et al.*, 2014a), where juvenile salmonids may be exposed to released virus. The impact of SAV, PRV, and PMCV infections on juvenile salmonids is unknown. The abundance of escaped salmon in Norwegian rivers (see above) indeed suggests a high potential for interaction at this stage. Hence, due to lack of data, consequence assessments cannot be done for the four viral agents considered.

Here, we have focused on the four major diseases currently prevalent in Norwegian salmon farming. All are viral diseases, and a large number of outbreaks are reported annually. However, low prevalence does not necessarily mean low impact, and diseases that at present are under control might surge and become a threat to wild populations. Subclinical or apparently benign infections may also have unforeseen ecological effects in nature by affecting survival (i.e. predator avoidance) or recruitment.

We have not considered the possible impact of exotic pathogens on Norwegian salmon farming. There are a large number of potential pathogens (i.e. hazards) infecting salmonids elsewhere that could affect farmed and wild salmonids in Norway if introduced (Raynard *et al.*, 2007;Brun and Lillehaug, 2010). Such introductions may be irreversible, and difficult or impossible to contain, and the main risk factor is the movement of live fish or fertilized eggs.

Future Work

To evaluate the effects of salmon and rainbow trout farming on the infection status of wild salmonids, there is a need to increase our knowledge about the complete pathogen repertoire (viruses, bacteria, fungi, and parasites) present in wild fish in areas of high- and low-intensity fish farming. Long-term monitoring of selected wild populations would allow detection of changes in the infection status in the population. Experimental challenge experiments with SAV, PRV, and PMCV must be performed on juvenile Atlantic salmon and brown trout in freshwater. This would better allow a consideration of the threat posed by escaped salmon ascending rivers during fall. A more extensive genotyping of virus from wild and farmed fish would improve our understanding of both virus spread and genetic changes in pathogens that may occur in the high host density of farming areas.

ORGANIC LOAD AND NUTRIENTS FROM SALMON FARMS

The salmonid aquaculture industry has continuously restructured since 1999, with reductions in the number of farms, increased farm size and relocation of farms to deeper fjord (50–300 m) and current rich coastal aquaculture sites. During this period, the production has doubled (Gullestad *et al.*, 2011), with typical salmon farm produces between 3000 and 5000 tons in a 18 months period in sheltered coastal waters and as much as 14 000 tons at more dynamic coastal sites. This rapid development has led to increased concerns about the environmental impacts both at present and future predicted finfish production levels.

Increased awareness of elevated discharges of nutrients, excess feed, and faeces to the marine environment has resulted in greater scrutiny of the aquaculture industry

(Mente *et al.*, 2006; Taranger *et al.*, 2012a). To this end, the Ministry of Fisheries and Coastal Affairs have stated that "the environmental impact of aquaculture must be kept at an acceptable level and be within the assimilative capacity of the area" (Anon., 2009b).

Assessing the risk of organic enrichment and nutrient overloading of Norwegian finfish aquaculture at local and regional scales focus both on benthic and pelagic systems and is based on a combination of scientific knowledge and industry monitoring data, coupled with best professional judgment, and the precautionary principle.

RISK ASSESSMENT

Release Assessment

Intensive farming of finfish in open sea cages results in the release of organic and inorganic effluents (i.e. carbon, nitrogen, and phosphorus) in the form of waste feed, faeces, and metabolic by-products to the surrounding aquatic environments (Holmer *et al.*, 2005; Strain and Hargrave, 2005). At current production levels in Norway, salmonid farming (1.3 M tons cultivated fish in 2012) releases ~34 000 tons of N, 60 000 tons of C, and 9750 tons of P annually (Taranger *et al.*, 2013).

Expose Assessment

Accumulation of these effluents into the marine system can negatively impact the ecosystem by contributing to eutrophication of pelagic systems, fertilization of benthic macrophytes in the euphotic zone, and organic enrichment of benthic systems (Strain and Hargrave, 2005). However, the area of influence (local or regional locations) and degree of enrichment of the environment depends on a number of factors including the size of the farm (i.e. the biomass of fish), the ambient environmental conditions (i.e. hydrodynamics, water depth, wave exposure, topography and substrate type) and the husbandry practices at the individual fish farms (Holmer *et al.*, 2005). In Norway, detailed knowledge about the environmental effects of organic and nutrient enrichment from finfish aquaculture is mainly based on studies around sheltered and coastal fjord aquaculture sites.

At deep aquaculture sites, fish farming effluents can be traced into the wider environment and into benthic foodwebs up to at least 1 km from the farming site (Kutti *et al.*, 2007b; Olsen *et al.*, 2012). At low deposition levels, organic enrichment of benthic sediments (up to 500 m from the farming location) stimulates secondary production in soft bottom communities, resulting in shifts in benthic faunal community structure (Kutti *et al.*, 2007a; Kutti, 2008; Bannister *et al.*, 2014). In addition, excessive loading of organic effluents to sediments often leads to dramatic changes in biogeochemical processes leading to grossly anoxic conditions (Valdemarsen *et al.*, 2012). The emissions of dissolved nutrients from finfish farms are quickly diluted in the water column at dynamic sites and elevated nutrient levels are hardly detected 200 m away from the farm (H. Jansen, IMR, unpublished data; Sanderson *et al.*, 2008).

Consequence Assessment

Many studies have investigated benthic impacts of fish farming on soft sediment benthic systems, demonstrating that intensive fish farming modifies biogeochemical processes (Holmer and Kristensen, 1992; Holmer and Frederiksen, 2007; Norði *et al.*, 2011). Remineraliztion of the highly labile organic waste (i.e. fish feed and faeces) results in increased sediment oxygen demand and altered metabolic pathways, and a shift from aerobic (i.e. heterotrophic respiration) to anaerobic (i.e. sulphate reduction and methanogenesis) microbial degradation (Holmer and Kristensen, 1992; Holmer *et al.*, 2003; Valdemarsen *et al.*, 2009). Excessive organic enrichment can thus lead to highly modified sediment conditions (Valdemarsen *et al.*, 2012), impacting the structure and biomass of faunal communities (Kutti *et al.*, 2007b; Hargrave *et al.*, 2008;Valdemarsen *et al.*, 2010; Bannister *et al.*, 2014). Increased release of dissolved nutrients from fish farming activities may stimulate phytoplankton growth and plankton blooms (Gowen and Ezzi, 1994) and may change the composition of seaweed communities in the littoral zone (Rueness and Fredriksen, 1991; Bokn *et al.*, 1992; Munda, 1996; Pihl *et al.*, 1999; Worm and Sommer, 2000; Krause-Jensen *et al.*, 2007).

To a lesser extent, there are studies that have investigate the effects of intensive fish farming to other habitats and biota including maerl beds (Hall-Spencer *et al.*, 2006; Sanz-Lazaro *et al.*, 2011; Aguado-Giménez and Ruiz-Fernández, 2012), coral reefs (Bongiorni *et al.*, 2003; Villanueva *et al.*, 2006), seaweeds and seagrass beds (Worm and Sommer, 2000; Diaz-Almela *et al.*, 2008; Holmer *et al.*, 2008), megafaunal communities (Wilding *et al.*, 2012), and pelagic and dermersal fish (Tuya *et al.*, 2006;Fernandez-Jover *et al.*, 2007, 2011; Dempster *et al.*, 2011). A consensus of these studies is that if the assimilative capacities of these environments are exceeded, then impacts on individual species, habitats, and ecosystems will be pronounced.

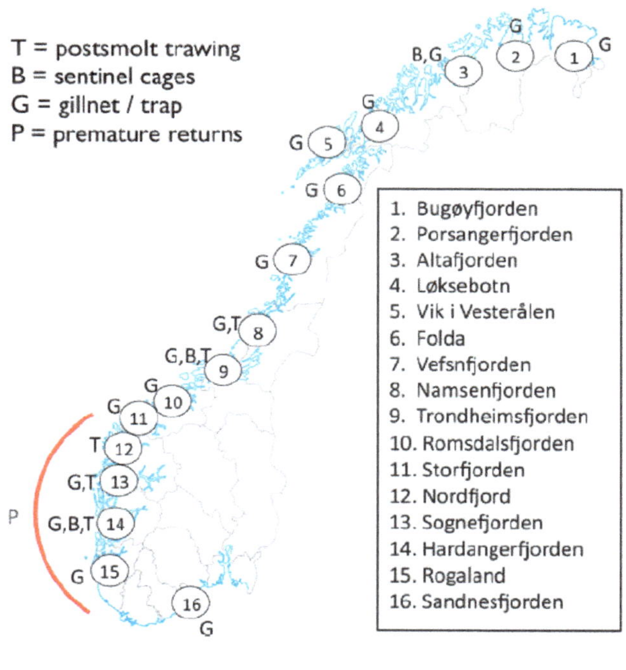

Figure 4. Sampling localities in Norwegian salmon lice monitoring programme applying various techniques such as gillnets and traps, post-smolt trawling, sentinel cages, and recording of premature return to rivers of sea trout during 2010–2013. The total number of sampling localities/sites with gillnets and traps has varied between 26 and 32 in the period.

Risk Estimation

Organic Loading on a Local Scale

On a local scale the endpoint: *unacceptable change in faunal communities and sediment chemistry in the production zone* is estimated. The criterion of unacceptable change is determined by Norwegian authorities and all salmon farms in Norway are monitored through mandatory investigations (MOM system; Hansen *et al.*, 2001). The MOM-B investigations are performed regularly under and in the closest vicinity of the fish cages and are based on qualitatively determined indicators such as chemical parameters (pH and redox potential), sensory parameters, and presence and/or absence of macro-infauna. The performance of these indicators against predefined thresholds categorizes the farming locations into different environmental conditions (1. low-, 2. medium-, 3. high-organic loading, and 4. organic overloading). The environmental condition 4 represents an unacceptable state when production cannot continue before the farming location has recovered. Data from the monitoring of Norwegian salmon farms are obtained from the Norwegian Directorate of Fisheries, including 2761 electronically reported MOM-B investigations undertaken beneath Norwegian fish farms between 2009 and 2013. The percentage of farms in an unacceptable ecological state (4) has been stable and <3% the last few years (Figure 5). This is probably a result of better localization of farms. The risk of unacceptable change in sediment chemistry and fauna communities in the production zone is low. However, according to national set thresholds for management of the production zone, low impact on this scale does not reflect pristine conditions, but merely that the farm is managed within acceptable conditions in regard to its local impact.

Several rigorous scientific examinations of benthic impacts have been conducted near salmon farms in western Norway (Hordaland) where benthic carbon loadings, benthic fauna responses, and sediment biogeochemical processes were studied (Kutti *et al.*, 2007a, b; Kutti, 2008; Valdemarsen *et al.*, 2012; Bannister *et al.*, 2014). In addition, there are generic scientific knowledge in respect to the flow of organic waste into benthic foodwebs along the Norwegian coast (Olsen *et al.*, 2012).

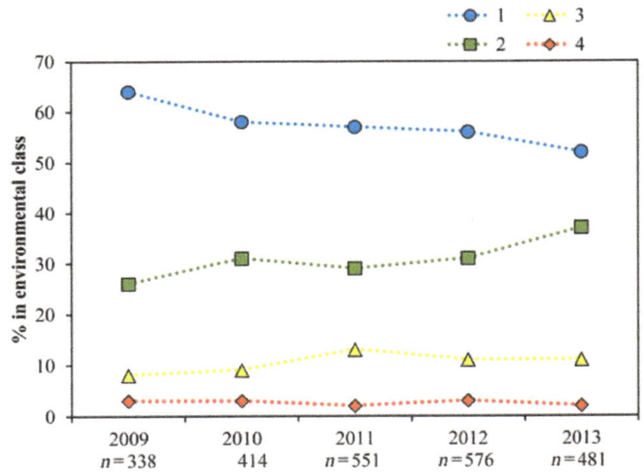

Figure 5. Impact of organic load from Norwegian marine finfish farms monitored by the mandatory MOM-B investigations (NS9410:2007) in the period 2009–2013. Data are given as percentage number of farms with ecological condition: 1 (blue), low organic loading; 2, (green), moderate organic loading; 3 (yellow), high organic loading (maximum allowed loading); 4 (red), overloading of the site, n = number of reported MOM-B investigations (data from Norwegian Directorate of Fisheries).

Organic Loading on a Regional Scale

The MOM-C system is an extended investigation of several sites (1–5) in the extended influence zone around farms and consists of quantitative measurements of the organic enrichment and the impact on biodiversity in infauna communities. The MOM-C investigations of fauna communities are following the Norwegian Standard (NS 9410) and farming sites are categorized into different environmental states (i.e. very good, good, moderate, poor, and very poor) according to nationally set thresholds (Molvær *et al.,* 1997). Based on the hydrographical conditions around the farm a distant point should be indentified in the most likely accumulation area beyond the production zone. The ecological condition at this distant site could be used as a proxy to estimate the risk of the endpoint;*significant change in bottom communities beyond the production zone (regional impact).*

Data from MOM-C investigations on 122 salmon farms, which represents ~10% of the farms currently operating in Norwegian waters, are compiled from the Norwegian Directorate of Fisheries. The data show that distant sites at 95% of the farms had a high or very high ecological classification according to national set thresholds for Shannon–Wiener diversity index (H′) (Molvær *et al.*, 1997) while 5% was classified in moderate conditions.

To provide an estimate of risk for impact on a regional scale, data have been retained from case studies in the Hardangerfjord (Husa *et al.*, 2014a) and regional monitoring in some other areas in Norwegian coastal waters (Vassdal *et al.*, 2012) according to parameters and thresholds defined in the Norwegian implementation of the European Water Framework Directive (Molvær *et al.*, 1997; WFD, 2000/60/EC; Anon., 2009a). These data show that the ecological conditions in fauna communities and oxygen values in deep regional basins are high to very high in fjords with high salmon farming activity. These findings were also supported by analysis of the relative importance of the extra contribution of organic farm waste to decomposing communities in the deep basins in the Hardangerfjord, estimating that current farming production increased oxygen consumption by 10% and decreased oxygen levels in bottom water with 0.09% (Aure, 2013). However, we do not have sufficient data from the entire Norwegian coastline to make a full risk estimation of the impact of organic loading on a regional scale.

Nutrient Emissions on a Local Scale

Local impact from nutrients and fine particulate material in the euphotic zone are currently not monitored around Norwegian fish farms, and we therefore have no data to estimate the endpoint; *nutrients from fish farms results in local eutrophication.*

Nutrient Emissions on a Regional Scale

To estimate the endpoint, *nutrients from fish farms results in regional eutrophication*, we do not have sufficient data from

Norwegian coastal waters to fulfil a complete risk estimation. However, three years monitoring of nutrient values and chlorophyll *a* in the Hardangerfjord area (Husa *et al.*, 2014b) and in Rogaland County, a sensitive area for fish farming due to lower water exchange (Vassdal *et al.*, 2012), show that ecological conditions for these parameters are within national acceptances thresholds (Molvær *et al.*, 1997) suggesting high or very high water quality.

These data coupled with modelling estimations on potential increase in phytoplankton production (Figure 6; Skogen *et al.*, 2009) suggest low risk of regional impacts from aquaculture in Norway. The potential increase in phytoplankton production is based on knowledge about the water transport mechanisms, coupled with typical natural values of nitrogen and phosphorous in the Norwegian Coastal Current and the calculated extra contribution to nutrient concentrations from fish farms in each Norwegian county. Assuming that theoretically all the nitrogen released from fish farms is assimilated in phytoplankton growth, an increase in the natural phytoplankton biomass were calculated and compared with the threshold of a 50% increase in phytoplankton biomass that is defined as eutrophication by OSPAR (Anon., 2010).

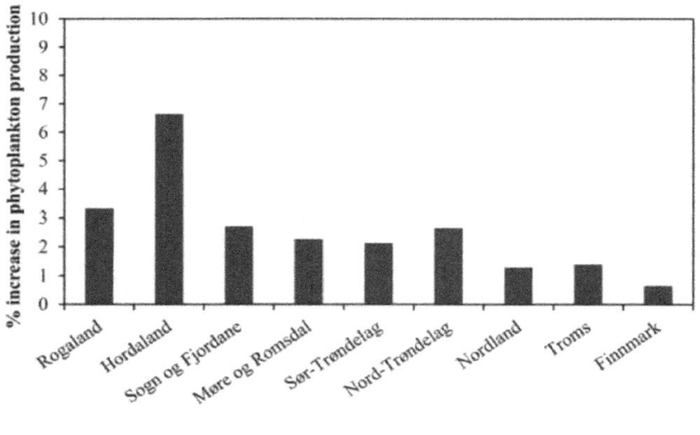

Figure 6. Estimated percentage increase in phytoplankton production due to emissions of dissolved nitrogen from finfish farms in 2012 in each Norwegian county, based on 100% exploitation of the nitrogen to carbon fixation.

Limitations

The risk assessment in the period 2010–2013 is based on a limited dataset, restricted scientific knowledge, and national monitoring methods that require upgrading/revising. On a local scale, the use of the MOM-B monitoring dataset for assessing local impacts of organic enrichment should be used cautiously. The MOM-B system is built on a qualitative assessment limited in its efficacy outside of soft sediment habitats. Using the "MOM system" to monitor the environmental effects to other benthic habitats such as hard, mixed and sandy bottom habitats, seaweed and kelp habitats, or other sensitive habitats including sponge aggregations and cold water corals reefs will lead to uncertain monitoring results. Therefore, reported benthic conditions underneath fish farms should be used cautiously in the risk assessment approach. The number of MOM-C investigations available for this risk assessment was limited to 122 fish farming sites.

Considering there are more or less 1000 fish farming locations along the Norwegian Coastline, the use of these investigations to provide an overview of the impact of organic enrichment from aquaculture within the influence area, should be approached cautiously, given they only represent ~10% of current production sites. Scientific data are restricted to fjord habitats in western Norway, there is a dearth of scientific studies representing the different geographical settings (i.e. southern, mid, and northern Norway) and the different benthic habitat types (coastal sandy sediments, hard bottom habitats, and sensitive and vulnerable habitats).

Furthermore, at regional scales national monitoring programmes have only started to begin (from 2008); therefore, existing knowledge and data of regional impacts from fish farming are mainly restricted to two counties and limited time. These limitations result in a precautionary approach used in the above risk assessment.

Future Work

To achieve the goal of the Ministry of Fisheries and Coastal Affairs that the environmental impact of aquaculture must be kept at an acceptable level and be within the assimilative capacity of the area, further scientific research and coastal monitoring efforts are needed. It is crucial that further knowledge is developed on understanding the interaction of organic and nutrient waste release on different habitat types (ca. hard bottom habitats, coastal sandy sediment habitats, and benthic boreal systems). In addition, habitats of ecological significance and sensitive species (ca. coral reefs, sponge aggregations, maerl beds, seagrass meadows, and spawning areas) require detailed investigations to understand their responses to organic and nutrient loadings, thus allowing more informed decisions on ecological impacts to be made. Furthermore, given the heterogeneity of benthic substrates along the Norwegian coastline, and the existing limitations of the MOM-B monitoring standard, improved monitoring tools need to be established to enable monitoring of local impacts from fish farms on non-soft sediment substrates. MOM-C investigations should be performed more often and at a greater number of stations along a gradient from enrichment. Furthermore, sampling for therapeutants and fatty acids should be incorporated in the MOM-C investigations to detect the pressure of drugs on the environment and also to determine if the impacts detected are in fact related to fish farming activities.

Moreover, there is also a need to develop new modelling tools that can predict the dispersal of organic and nutrient wastes from fish farms, which will enable better placement of monitoring stations for both MOM-C and regional monitoring to increase the likelyhood of detecting impacts. Finally, regional monitoring programmes should incorporate greater sampling coverage (i.e. sampling locations) and frequency along the entire Norwegian coastline. Monitoring programmes should also identify possible risk areas for regional impacts and place further emphasis on monitoring in these habitats.

DISCUSSION

The main approach in the risk assessment on the environmental impact of fish farming in Norway was to review the state-of-the-art on various hazards and potential risk factors, review national monitoring data with relevance for the risk assessment, and score qualitatively the risk of impact in broad categories; low, moderate, and high risk of impact when monitoring data were available (Taranger *et al.*, 2011a, b, 2012b, 2013).

When the first risk assessment of the environmental impact of Norwegian fish farming was initiated in 2010 (Taranger *et al.*, 2011a), only very broad definitions and goals of aquaculture sustainability were presented in a strategy document by the Norwegian government (Anon., 2009b). The goals for sustainability put forward in this document are listed in Table 1, and covers the topics; diseases/parasites, genetic interactions with wild populations, pollution and discharges, marine site structure and zoning, as well as sustainability of feed and feed resources. These goal are very generic and do not specify sustainability indicators and related thresholds for societal/political acceptance of the environmental impact of fish farming in Norway.

The Institute of Marine Research in Norway and the Norwegian Veterinarian Institute were requested by the Norwegian Ministry of Fisheries and Coastal Affairs in 2011 to propose scientifically based sustainability indicators and related thresholds to score the severity of the potential environmental impacts of aquaculture. This included indicators for the risk of genetic introgression of escaped farmed salmon as well impact of salmon lice from fish farming on wild salmonid populations. As a result, indicators and suggested thresholds to score the environmental impact as low, moderate, or high were proposed in 2012 (Taranger *et al.*, 2012a). These recommendations were in part approved and implemented into a governmental report on Norwegian seafood policy in 2013 (Anon., 2013c), and in a newly established Quality Norm for the management and protection of wild Atlantic salmon in Norway approved by the Norwegian government in late 2013 (Anon., 2013a).

As a result, the suggested indicators for environmental impact as the associated suggested thresholds for scoring of the impacts have been adopted in the most recent version of the Risk assessment of Norwegian fish farming (Taranger *et al.*, 2014). They are the basis of the current assessment of risk of genetic introgression and impact on salmon lice on wild salmonid populations. The assessments are based on monitoring programmes on the numbers of escaped salmon in Norwegian rivers over the last 3 years, and the level of salmon lice infections on wild salmonids in Norwegian coastal waters during the last 4 years.

New scientific findings have recently emerged, such as the measured level of introgression of escaped farmed salmon into wild salmon populations in 20 Norwegian rivers (Glover *et al.*, 2013a), and new estimates on the additional marine mortality caused by salmon lice on Atlantic salmon from the rivers Dale and Vosso, in Hordaland, Norway (Skilbrei *et al.*, 2013;Vollset *et al.*, 2014). Such findings assist in developing more quantitative assessments of the impact of number of escaped salmon in rivers, and the impact of salmon lice from farming on wild salmonids, respectively.

In contrast to the situation for genetic introgression and impact of salmon lice, there are very limited monitoring data on potential transfer of other diseases and parasites from salmon farming to wild fish in Norway. Hence, the assessment on the risk of disease transfers to wild fish is mainly based on an analysis of the frequency of disease outbreaks in the ~1000 sites for salmon farming along the Norwegian coast. This is supported with a review on the knowledge of risk of disease transmission for the most relevant pathogens, and some available data on prevalence of pathogens and/or any disease outbreaks in wild Atlantic salmon and sea trout populations in Norway (Taranger *et al.*, 2014).

Regarding the local and regional impact of organic load and release of nutrient from marine salmon farming, such environmental impact indicators and associated monitoring programmes are defined and adapted by the Norwegian fish farming authorities regarding the local zone under and close to the farms, whereas the indicators and monitoring programmes for regional effects are being implemented in some counties in Norway (Vassdal *et al.*, 2012; Husa *et al.*, 2014a). The local zone under and close to the farm is monitored with a risk-based frequency using the relatively simple MOM-B method, while the more sensitive MOM-C method with detailed analysis of the species compositions in soft bottom samples near the farms is only applied occasionally (Hansen *et al.*, 2001). Both these methods have limitations, e.g. they require soft bottom, and are currently under revisions.

Data from regional monitoring has only become available in a few counties in the last years, but new programmes are starting up in several counties. The regional monitoring will to a large degree be based upon environmental indexes and environmental quality elements and related threshold for scoring of quality according to the Norwegian implementation of EUs Water Framework Directive (Anon., 2009a).

The current risk assessment of local of organic matter and nutrient release are based on the mandatory MOM-B monitoring from all farms and MOM-C analyses from a limited number of fish farms (Taranger *et al.*, 2014). The regional impact is evaluated based on available models and some investigations in the counties of Hordaland and Rogaland which both have high salmon farming activity compared with the area of available coastal water. As discussed above under the section on impact of organic matter and nutrients, the analyses have limitations, and a more extensive and improved monitoring programme is needed both on local and regional scale.

Moreover, a range of other potential risk factors, such as use of various pharmaceuticals, transfers of xenobiotics with the feed, use of copper as antifouling agent on sea cages, interactions with fisheries and other ecological impacts of sea cage farming, as well as ecological impacts of catch, transport and use of wrasses as cleaner fish against salmon lice are discussed in the risk assessment of Norwegian fish farming (Taranger *et al.*, 2014), but are not included in the current analysis.

CONCLUSIONS AND SUMMARY OF MAIN FINDINGS

This represents the first risk assessment of cage-based salmonid aquaculture in Norway, which is world's largest producer of farmed Atlantic salmon. While there are several limitations in the approaches used to estimate the risks, as has been discussed in the sections above, this work has provided the Norwegian authorities with a framework upon which to evaluate the most important identified hazards against environmental goals for sustainability. The primary findings from the present risk assessment can be summarized as follows:

- Based upon the observed frequency of farmed escaped salmon on the spawning grounds of wild populations in the period 2010–2012, 21 of the 34 populations included in the risk assessment were in moderate–to-high risk of experiencing genetic changes due to introgression of farmed salmon. However, a recent study of 20 Norwegian rivers has demonstrated that there is only a moderate correlation between the observed frequency of escapees and introgression of farmed salmon (Glover *et al.*, 2013a); therefore, validation of the level of introgression in a higher number of native populations will be required in the future.

- During the period 2010–2013, salmon lice infections mainly resulting from salmon farming were estimated at a total of 109 stations covering relevant areas of the Norwegian coastline using wild sea trout as a proxy for local infection pressure on wild salmonids. Twenty-seven of these stations indicated moderate or high likelihood of mortality for wild migrating salmon smolts. For sea trout later in the season, 67 of the stations indicated moderate or high likelihood of mortality on wild sea trout.
- The high frequency of the viral disease outbreaks for PD, IPN, heart and skeletal muscle inflammation, and CMS in Norwegian salmon farming suggests extensive release of the causal pathogens for these diseases in many areas. Migrating wild salmon and local sea trout are likely to be exposed to these pathogens. However, the extent of this exposure and consequences remains largely unknown. Screening of wild salmonids has revealed low to very low prevalence of the viruses SAV, IPNV, PMCV, and low prevalence of PRV in salmon. Furthermore, these viruses have never been documented to cause disease in wild Norwegian salmonids. Thus, a general lack of data prohibits complete risk estimation for these diseases.
- From a total of ~500 yearly investigations of local organic loading under fish farms, 2% of them displayed unacceptable conditions in the benthic sediments and faunal composition in 2013, whereas 11% classified with a high organic loading but still within the threshold. The remaining 87% of the farms had a moderate-to-high ecological conditions. The risk of eutrophication and organic over loading in the benthic communities beyond the production area of the farm is considered low based upon case studies and monitoring data from a limited area of the Norwegian coast.

Given the rapid expansion of open sea cage farming in Norway, and internationally, and the range of ecological impacts that either are demonstrated or suspected, there is an urgent need for better knowledge about such impacts, to implement improved monitoring programmes for the most important hazards, and also to improve procedures for risk assessments including useful environmental risk indicators and to facilitate processes that involves definitions on the societal acceptance levels of the various impacts.

ACKNOWLEDGEMENTS

We appreciate the excellent work of all those participating in the work with the Norwegian Risk assessment (Risikorapporten). This includes in alphabetical order Jon Albretsen, Jan Aure, Britt Bang-Jensen, Geir Dahle, Arne Ervik, Per-Gunnar Fjelldal, Bengt Finstad, Bjørn Einar Grøsvik, Pia Kupka Hansen, Tom Hansen, Mikko Heino, Peder Andreas Jansen, Daniel Jimenez, Ingrid Askeland Johnsen, Knut Jørstad, Tore Kristiansen, Anja Kristoffersen, Rosa Maria Serra Llinnares, Stein Mortensen, Rune Nilsen, Håkon Otterå, Sonal Patel, Bjørn-Steinar Sæther, Ole B. Samuelsen, Maria Q. Sanchez, Anne Sandvik, Jofrid Skardhamar, Anne Berit Skiftesvik, Ove Skilbrei, Øystein Skaala, Cecilie Skår, Lars H. Stien, Ingebrigt Uglem Terje van der Meeren, and Vidar Wennevik.

REFERENCES

1. Aguado-Gime´nez F., and Ruiz-Ferna´ndez J.M. 2012. Influence of an experimental fish farm on the spatio-temporal dynamic of a Mediterranean mae¨rl algae community. Marine Environmental Research, 74: 47–55.
2. Aldrin M., Storvik B., Frigessi A., Viljugrein H., and Jansen P. A. 2010. A stochastic model for the assessment of the transmission pathways of heart and skeleton muscle inflammation, pancreas disease and infectious salmon anaemia in marine fish farms in Norway. Preventive Veterinary Medicine, 93: 51–61.
3. Anon. 2006. Report of the Study Group on Risk Assessment and Management Advice (SGRAMA). ICES Resource Management Committee, ICES CM 2006/RMC: 04, Ref LRC ACFM, ACE ACME 71. ICES, Copenhagen.
4. Anon. 2009a. Classification of environmental condition of water. Directorate group for implementation of the water framework (in Norwegian). Veileder 01:2009. 180 pp.
5. Anon. 2009b. Strategy for an Environmentally Sustainable Norwegian Aquaculture Industry (in Norwegian). Ministry of Fisheries and Coastal Affairs, Oslo, Norway. 34 pp.
6. Anon. 2010. Report of the Working Group on Environmental Interactions of Mariculture (WGEIM). ICES CM 2010SSGHIE:08. 58 pp.
7. Anon. 2011. Status for the Norwegian salmon stocks in 2011 (in Norwegian). Report from Vitenskapelig ra˚d for lakseforvaltning, nr. 3. 285 pp.
8. Anon. 2012. Salmon lice and effects on wild salmonids-from individual response to effects on stocks (in Norwegian). Thematic Report from Vitenskapelig ra˚d for lakseforvaltning no. 3. 56 pp.
9. Anon. 2013a. Quality norm from wild populations of Atlantic salmon (Salmo salar) (in Norwegian). Ministry of Climate and Environment, Oslo, Norway. 26 pp.
10. Anon. 2013b. Status for the Norwegian salmon stocks in 2013 (in Norwegian). Report from Vitenskapelig ra˚d for lakseforvaltning no 5. 136 pp.
11. Anon. 2013c. The worlds leading seafood nation (in Norwegian). White papers no. 22. Ministry of Fisheries and Coastal Affairs, Oslo, Norway. 139 pp.
12. Araki H., Berejikian B. A., Ford M. J., and Blouin M. S. 2008. Fitness of hatchery-reared salmonids in thewild. Evolutionary Applications, 1: 342–355.
13. Araki H., and Schmid C. 2010. Is hatchery stocking a help or harm? Evidence, limitations and future directions

in ecological and genetic surveys. Aquaculture, 308: S2 –S11.

14. Asplin L., Boxaspen K. K., and Sandvik A. D. 2011. Modeling the distribution and abundance of planktonic larval stages of Lepeophtheirus salmonis in Norway. In Salmon lice: an integrated approach to understanding parasite abundance and distribution. Ed. by S. Jones , and R. Beamish . Wiley-Blackwell, Oxford, UK.
15. Aure J. 2013. Farming and oxygen in the Hardangerfjord (in Norwegian). Havforskningsrapporten, Fisken og Havet Særnummer. pp. 42–44.
16. Austin B., and Austin D. A. 2007. Bacterial fish pathogens. Diseases of farmed and wild fish. Praxis Publishing Ltd., Chichester, UK.
17. Bakke T. A. 1991. A review of the inter- and intraspecific variability in salmonid hosts to laboratory infections with Gyrodactylus salaris Malmberg. Aquaculture, 98: 303 –310.
18. Bakke T. A., and Harris P. D. 1998. Diseases and parasitesinwild Atlantic salmon (Salmo salar) populations. Canadian Journal of Fisheries and Aquatic Sciences, 55: 247 –266.
19. Bakke T. A., Harris P. D., Jansen P. A., and Hansen L. P. 1992. Host specificity and dispersal strategy in gyrodactylid monogeneans, with particular reference to Gyrodactylus salaris (Platyhelminthes, Monogenea). Diseases of Aquatic Organisms, 13: 63–74.
20. Bakke T. A., and MacKenzie K. 1993. Comparative susceptibility of native Scottish and Norwegian stocks of Atlantic salmon, Salmo salar L., to Gyrodactylus salaris Malmberg: Laboratory experiments. Fisheries Research, 17: 69 –85.
21. Bannister R. J., Valdemarsen T., Hansen P. K., Holmer M., and Ervik A. 2014. Changes in benthic sediment conditions under an Atlantic salmon farm at a deep, well-flushed coastal site. Aquaculture Environment Interactions, 5: 29–47.
22. Berglund Andreassen K. 2013. Effects of infections with salmon lice (Lepeophtheirus salmonis) on wild smolts of salmon (Salmo salar L.) and trout (Salmo trutta L.) (in Norwegian). Master thesis, University of Tromsø, Norway. 61 pp.
23. Besnier F.,Glover K.A., and Skaala Ø. 2011. Investigating genetic change in wild populations: modelling gene flow from farm escapees. Aquaculture Environment Interactions, 2: 75–86.
24. Biering E., Madhun A. S., Isachsen C. H., Omdal L. M., Einen A. C. B., Garseth A. H., Bjørn P. A.,et al. 2013. Annual Report on health monitoring of wild anadromous salmonids in Norway. Report from the Norwegian Veterinary Institute and the Institute of Marine Research, Bergen, Norway.
25. Bjørn P. A., and Finstad B. 1997. The physiological effects of salmon lice infection on sea trout postsmolts. Nordic Journal of Freshwater Research, 73: 60–72.
26. Bjørn P. A., Finstad B., Asplin L., Skilbrei O., Nilsen R., Serra Llinares R. M., and Boxaspen K. K. 2011a. Method development for surveillance and counting of salmon lice on wild salmonids. Rapport fra Havforskningen 8-2011. 58 pp.
27. Bjørn P. A., Finstad B., Kristoffersen R., McKinley R. S., and Rikardsen A. H. 2007a. Differences in risks and consequences of salmon louse, Lepeophtheirus salmonis(Krøyer), infestation on sympatric populations of Atlantic salmon, brown trout, and Arctic charr within northern fjords. ICES Journal of Marine Science, 64: 386–393.
28. Bjørn P. A., Finstad B., Nilsen R., Asplin L., Uglem I., Skaala Ø., and Boxaspen K. K., et al. 2007b. Norwegian national surveillance of salmon lice epidemics on wild Artlantic salmon, sea trout and Arctic charr in connection with Norwegian national salmon rivers and fjords. NINA report 377. 33 pp.
29. Bjørn P. A., Finstad B., Nilsen R., Uglem I., Asplin L., Skaala Ø., and Hvidsten N. A. 2010. Norwegian national surveillance 2009 of salmon lice epidemics on wild Atlantic salmon, sea trout and Arctic charr in connection with Norwegian national salmon rivers and fjords . NINA Report 547. 50 pp.
30. Bjørn P. A., Nilsen R., Serra Llinares R. M., Asplin L., Askeland Johnsen I., Karlsen Ø., Finstad B., et al. 2013. Salmon lice infections on wild salmonids along the Norwegian coast in 2013. Final report to Norwegian Food Safety Authority. Report from Havforskningsinstituttet, no. 32–2013.
31. Bjørn P. A., Nilsen R., Serra Llinares R. M., Asplin L., Boxaspen K. K., Finstad B., Uglem I., et al. 2012. Salmon lice infections on wild salmonids along the Norwegian coast in 2012. Final report to Norwegian Food Safety Authority. Report from Havforskningen, 31 –2012.
32. Bjørn P. A., Nilsen R., Serra Llinares R. M., Asplin L., Boxaspen K. K., Finstad B., Uglem I., et al. 2011b. Final report to the Norwegian Food Safety Authority of the salmon lice infections along the Norwegian coast in 2011. Report From Havforskningen, 19-2011: 34.
33. Bjørn P. A., Sivertsga°rd R., Finstad B., Nilsen R., Serra-Llinares R. M., and Kristoffersen R. 2011c. Area protection may reduce salmon louse infection risk to wild salmonids. Aquaculture Environment Interactions, 1: 233 –244.
34. Bokn T., Murray S. N., Moy F. E., and Magnusson J. B. 1992. Changes in fucoid distribution and abundances in the inner Oslofjord, Norway: 1974–80 versus 1988–90. Acta Phytogeographica Suecia, 78: 117–124.
35. Bongiorni L., Shafir S., and Rinkevich B. 2003. Effects of particulate matter released by a fish farm (Eilat, Red Sea) on survival and growth of Stylophora pistillata coral nubbins. Marine Pollution Bulletin, 46: 1120–1124.
36. Boucher P., Raynard R. S., Houghton G., and Laurencin F. B. 1995. Comparative experimental transmission of pancreas disease in Atlantic salmon, rainbow trout and brown trout. Diseases of Aquatic Organisms, 22: 19 –24.
37. Bourret V., O'Reilly P. T., Carr J. W., Berg P. R., and Bernatchez L. 2011. Temporal change in genetic integrity suggests loss of local adaptation in a wild Atlantic salmon (Salmo salar) population following introgression by farmed escapees. Heredity, 106: 500 –510.

38. Brun E. 2003. Chapter 4: Epidemiology.In IPN in salmonids – a review, pp. 51 –67. Ed. by B. Skjelstad , E. Brun , and I. Jensen . VESO, Oslo, Norway.

39. Brun E., and Lillehaug A. 2010. Risk profile of the diseases in Norwegian aquaculture. Report 9 –2010 from National Veterinary Institute, Oslo, Norway.

40. Buschmann A. H., Riquelme V. A., Herna´ndez-Gonza´lez M. C., Varela D., Jime´nez J. E., Henrı´quez L. A., Vergara P. A., et al. 2006. A review of the impacts of salmonid farming on marine coastal ecosystems in the southeast Pacific. ICES Journal of Marine Science, 63: 1338–1345.

41. Clifford S. L., McGinnity P., and Ferguson A. 1998a. Genetic changes in an Atlantic salmon population resulting from escaped juvenile farm salmon. Journal of Fish Biology, 52: 118 –127.

42. Clifford S. L., McGinnity P., and Ferguson A. 1998b. Genetic changes in Atlantic salmon (Salmo salar) populations of northwest Irish rivers resulting from escapes of adult farm salmon. Canadian Journal of Fisheries and Aquatic Sciences, 55: 358–363.

43. Combes C. 2001. Parasitism: The ecology and evolution of intimate interactions. University of Chicago Press, Chicago, USA.

44. Crozier W. W. 1993. Evidence of genetic interaction between escaped farmed salmon and wild Atlantic salmon (Salmo salar L.) in a Northern Irish river. Aquaculture, 113: 19 –29.

45. Crozier W. W. 2000. Escaped farmed salmon, Salmo salar L., in the Glenarm River, Northern Ireland: genetic status of the wild population 7 years on. Fisheries Management and Ecology, 7: 437–446.

46. Dempster T., Sanchez-Jerez P., Fernandez-Jover D., Bayle-Sempere J., Nilsen R., Bjørn P. A., and Uglem I. 2011. Proxy measures of fitness suggest coastal fish farms can act as population sources and not ecological traps for wild gadoid fish. PLoS ONE, 6: e15646, 9 pp.

47. Diaz-Almela E., Marba N., Alvarez E., Santiago R., Holmer M., Grau A., Mirto S., et al. 2008. Benthic input rates predict seagrass (Posidonia oceanica) fish farm-induced decline. Marine Pollution Bulletin, 56: 1332–1342.

48. Dillane E., McGinnity P., Coughlan J. P., Cross M. C., de Eyto E., Kenchington E., and Prodohl P.,et al. 2008. Demographics and landscape features determine intrariver population structure in Atlantic salmon (Salmo salar L.): the case of the River Moy in Ireland. Molecular Ecology, 17: 4786–4800.

49. Diserud O. H., Fiske P., and Hindar K. 2010. Regional impact of escaped farm salmon on wild salmon populations in Norway. NINA Report 622. 40 pp.

50. Dobson A. P., and Hudson P. J. 1986. Parasites, disease and the structure of ecological communities. Trendsin Ecology& Evolution, 1: 11–15.

51. EFSA. 2012. EFSA Panel on Animal Health and Welfare: guidance on risk assessment for animal welfare. EFSA Journal, 10: 2513, 30 pp. doi:10.2903/j.efsa.2012.2513.

52. Einum S., and Fleming I. A. 1997. Genetic divergence and interactions in the wild among native, farmed and hybrid Atlantic salmon. Journal of Fish Biology, 50: 634–651.

53. Fernandez-Jover D., Jimenez J. A. L., Sanchez-Jerez P., Bayle-Sempere J., Casalduero F. G., Lopez F. J. M., and Dempster T. 2007. Changes in body condition and fatty acid composition of wild Mediterranean horse mackerel (Trachurus mediterraneus, Steindachner, 1868) associated to sea cage fish farms. Marine Environmental Research, 63: 1 –18.

54. Fernandez-Jover D., Martinez-Rubio L., Sanchez-Jerez P., Bayle-Sempere J. T., Jimenez J. A. L., Lopez F. J. M., Bjørn P. A., et al. 2011. Waste feed from coastal fish farms: a trophic subsidy with compositional side-effects for wild gadoids. Estuarine Coastal and Shelf Science, 91: 559 –568.

55. Finstad B., and Bjørn P. A. 2011. Present status and implications of salmon lice on wild salmonids in Norwegian coastal zones. In Salmon lice: an integrated approach to understanding parasite abundance and distribution, pp. 281–305. Ed. by S. Jones , and R. Beamish . Wiley-Blackwell, Oxford, UK.

56. Finstad B., Bjørn P. A.,Grimnes A., and Hvidsten N. A. 2000. Laboratory and field investigations of salmon lice [Lepeophtheirus salmonis (Krøyer)] infestation on Atlantic salmon (Salmo salar L.) postsmolts. Aquaculture Research, 31: 795 –803.

57. Finstad B., Bjørn P. A., Todd C. D.,Whoriskey F., Gargan P. G., Førde G., and Revie C. 2010. The effect of sea lice on Atlantic salmon and other salmonid species. InAtlantic salmon ecology, pp. 253–276. Ed. by Ø. Aas , S. Einum , A. Klemetsen , and J. Skurdal . Wiley-Blackwell, Oxford, UK.

58. Finstad B., and Jonsson N. 2001. Factors influencing the yield of smolt releases in Norway. Nordic Journal of Freshwater Research, 75: 37 –55.

59. Fiske P. 2013. Surveillance of escaped farmed salmon in rivers in the autumn 2010–2012. NINA Report 989. 33 pp.

60. Fiske P., Lund R. A., and Hansen L. P. 2006. Relationships between the frequency of farmed Atlantic salmon, Salmo salar L., in wild salmon populations and fish farming activity in Norway, 1989– 2004. ICES Journal of Marine Science, 63: 1182–1189.

61. Fleming I. A., Hindar K., Mjølnerød I. B., Jonsson B., Balstad T., and Lamberg A. 2000. Lifetime success and interactions of farm salmon invading a native population. Proceedings of the Royal Society of London Series B Biological Sciences, 267: 1517–1523.

62. Fleming I. A., Jonsson B., Gross M. R., and Lamberg A. 1996. An experimental study of the reproductive behaviour and success of farmed and wild Atlantic salmon (Salmo salar). Journal of Applied Ecology, 33: 893 –905.

63. Fraser D. J., Weir L. K., Bernatchez L., Hansen M. M., and Taylor E. B. 2011. Extent and scale of local adaptation in salmonid fishes: review and meta-analysis. Heredity, 106: 404–420.

64. Garcia de Leaniz C., Fleming I. A., Einum S., Verspoor E., Jordan W. C., Consuegra S., Aubin-Horth N.,et al. 2007. A critical review of adaptive genetic variation in Atlantic salmon: implications for conservation. Biological Reviews, 82: 173 –211.

65. Gargan P. G., Forde G., Hazon N., Russell D. J. F., and Todd C. D. 2012. Evidence for sea lice-induced marine mortality of Atlantic salmon (Salmo salar) in western Ireland from experimental releases of ranched smolts treated with emamectin benzoate. Canadian Journal of Fisheries and Aquatic Sciences, 69: 343 –353.

66. Garseth A. H., Biering E., and Aunsmo A. 2013a. Associations between piscine reovirus infection history traits in wild-caught Atlantic salmon Salmo salar L. in and life Norway. Preventive Veterinary Medicine, 112: 138 –146.

67. Garseth A. H., Biering E., and Tengs T. 2012. Piscine myocarditis virus (PMCV) in wild Atlantic salmon Salmo salar. Diseases of Aquatic Organisms, 102: 157–161.

68. Garseth A° . H., Ekrem T., and Biering E. 2013c. Phylogenetic evidence of long distance dispersal and transmission of piscine reovirus (PRV) between farmed and wild Atlantic salmon. PLoS ONE, 8: e82202.

69. Garseth A. H., Fritsvold C., OpheimM., Skjerve E., and Biering E. 2013b. Piscine reovirus (PRV) in wild Atlantic salmon, Salmo salar L., and sea-trout, Salmo trutta L., in Norway. Journal of Fish Diseases, 36: 483–493.

70. Gaughan D. J., Mitchell R.W., and Blight S. J. 2000. Impact of mortality, possibly due to herpesvirus, on pilchard Sardinops sagax stocks along the south coast of Western Australia in 1998–99. Marine and Freshwater Research, 51: 601 –612.

71. GESAMP. 2008. Assessment and communication of environmental risks in coastal aquaculture. IMO/ FAO/UNESCO-IOC/UNIDO/ WMO/IAEA/UN/ UNEP Joint Group of Experts on Scientific Aspects of Marine Environmental Protection. Reports and Studies GESAMP No. 76. FAO, Rome. 198 pp.

72. Gjedrem T. 2010. The first family-based breeding program in aquaculture. Reviews in Aquaculture, 2: 2–15

73. Glover K. A. 2010. Forensic identification of fish farm escapees: the Norwegian experience. Aquaculture Environment Interactions, 1: 1 –10. Glover K. A., Ottera H., Olsen R. E., Slinde E., Taranger G. L., and Skaala Ø. 2009. A comparison offarmed,wild and hybrid Atlantic salmon (Salmo salar L.) reared under farming conditions. Aquaculture, 286: 203–210. Glover K. A., Pertoldi C., Besnier F., Wennevik V., Kent M., and Skaala Ø. 2013a. Atlantic salmon populations invaded by farmed escapees: quantifying genetic introgressionwith a Bayesian approach and SNPs. BMC Genetics, 14: 4. Glover K. A., Quintela M., Wennevik V., Besnier F., Sørvik A. G. E., and Skaala Ø. 2012. Three decades of farmed escapees in the wild: A spatio-temporal analysis of population genetic structure throughout Norway. PLoS ONE, 7: e43129. Glover K. A., Skilbrei O., and Skaala Ø. 2008. Genetic assignment identifies farm of origin for Atlantic salmon Salmo salar escapees in a Norwegian fjord. ICES Journal of Marine Science, 65: 912–920. Glover K. A., Sørvik A. G. E., Karlsbakk E., Zhang Z., and Skaala Ø. 2013b. Molecular genetic analysis of stomach contents reveals wild Atlantic cod feeding on piscine reovirus (PRV) infected Atlantic salmon originating from a commercial fish farm. PLoS ONE, 8: e60924. Gowen R. J., and Ezzi I. A. 1994. Assessment and prediction of the potential for hypernutrification and eutrophication associated with cageculture of salmonids in Scottish waters. Dunstaffnage Marine Laboratory, Oban, Scotland. 137 pp. Graham D. A., Staples C., Wilson C. J., Jewhurst H., Cherry K., Gordon A., and Rowley H. M. 2007. Biophysical properties of salmonid alphaviruses: influence of temperature and pH on virus survival. Journal of Fish Diseases, 30: 533–543. Grimnes A., and Jakobsen P. J. 1996. The physiological effects of salmon lice infection on post-smolt of Atlantic salmon (Salmo salar). Journal of Fish Biology, 48: 1179 –1194. Gullestad P., Bjørgo S., Eithun I., Ervik A., Gudding R., Hansen H., Røsvik I. O., et al. 2011. Effective and sustainable use of areas for aquaculture – land of desire (in Norwegian). Report from an expert panel appointed by Ministry of Fisheries and Coastal Affairs, Oslo, Norway. 198 pp. Hall-Spencer J., White N., Gillespie E., Gillham K., and Foggo A. 2006. Impact of fish farms on maerl beds in strongly tidal areas. Marine Ecology Progress Series, 326: 1 –9. Hansen H., Bachmann L., and Bakke T. A. 2003. Mitochondrial DNA variation of Gyrodactylus spp. (Monogenea, Gyrodactylidae) populations infecting Atlantic salmon, grayling, and rainbow trout in Norway and Sweden. International Journal for Parasitology, 33: 1471–1478. Hansen L. P. 2006. Migration and survival of farmed Atlantic salmon (Salmo salar L.) released from two Norwegian fish farms. ICES Journal of Marine Science, 63: 1211–1217. Hansen P. K., Ervik A., Schanning M. T., Johannsen P., Aure J., Jahnsen T., and Stigebrandt A. 2001. Regulating the local environmental impact of intensive marine fish farming. II. The monitoring programme of the MOM system (Modelling – Ongrowing fish farms – Monitoring). Aquaculture, 194: 75 –92. Hargrave B. T., Holmer M., and Newcombe C. P. 2008. Towards a classification of organic enrichment in marine sediments based on biogeochemical indicators. Marine Pollution Bulletin, 56: 810 –824. Haugland O., Mikalsen A. B., Nilsen P., Lindmo K., Thu B. J., Eliassen T. M., Roos N., et al. 2011. Cardiomyopathy syndrome of Atlantic salmon (Salmo salar L.) is caused by a double-stranded RNA virus of the totiviridae family. Journal of Virology, 85: 5275–5286. Hazon N., Todd C. D., Whelan B., Gargan P. G., Finstad B., Bjørn P. A., and Wendelar Bonga S. E., et al. 2006. Sustainable management of interactions between aquaculture and wild salmonid fish. Final report for the SUMBAWS EU project. pp. 1 –293

74. Glover K. A. 2010. Forensic identification of fish farm escapees: the Norwegian experience. Aquaculture Environment Interactions, 1: 1 –10.

75. Glover K. A., Ottera H., Olsen R. E., Slinde E., Taranger G. L., and Skaala Ø. 2009. A comparison offarmed,wild

and hybrid Atlantic salmon (Salmo salar L.) reared under farming conditions. Aquaculture, 286: 203–210.

76. Glover K. A., Pertoldi C., Besnier F., Wennevik V., Kent M., and Skaala Ø. 2013a. Atlantic salmon populations invaded by farmed escapees: quantifying genetic introgressionwith a Bayesian approach and SNPs. BMC Genetics, 14: 4.

77. Glover K. A., Quintela M., Wennevik V., Besnier F., Sørvik A. G. E., and Skaala Ø. 2012. Three decades of farmed escapees in the wild: A spatio-temporal analysis of population genetic structure throughout Norway. PLoS ONE, 7: e43129.

78. Glover K. A., Skilbrei O., and Skaala Ø. 2008. Genetic assignment identifies farm of origin for Atlantic salmon Salmo salar escapees in a Norwegian fjord. ICES Journal of Marine Science, 65: 912–920.

79. Glover K. A., Sørvik A. G. E., Karlsbakk E., Zhang Z., and Skaala Ø. 2013b. Molecular genetic analysis of stomach contents reveals wild Atlantic cod feeding on piscine reovirus (PRV) infected Atlantic salmon originating from a commercial fish farm. PLoS ONE, 8: e60924.

80. Gowen R. J., and Ezzi I. A. 1994. Assessment and prediction of the potential for hypernutrification and eutrophication associated with cageculture of salmonids in Scottish waters. Dunstaffnage Marine Laboratory, Oban, Scotland. 137 pp.

81. Graham D. A., Staples C., Wilson C. J., Jewhurst H., Cherry K., Gordon A., and Rowley H. M. 2007. Biophysical properties of salmonid alphaviruses: influence of temperature and pH on virus survival. Journal of Fish Diseases, 30: 533–543.

82. Grimnes A., and Jakobsen P. J. 1996. The physiological effects of salmon lice infection on post-smolt of Atlantic salmon (Salmo salar). Journal of Fish Biology, 48: 1179 –1194.

83. Gullestad P., Bjørgo S., Eithun I., Ervik A., Gudding R., Hansen H., Røsvik I. O., et al. 2011. Effective and sustainable use of areas for aquaculture – land of desire (in Norwegian). Report from an expert panel appointed by Ministry of Fisheries and Coastal Affairs, Oslo, Norway. 198 pp.

84. Hall-Spencer J., White N., Gillespie E., Gillham K., and Foggo A. 2006. Impact of fish farms on maerl beds in strongly tidal areas. Marine Ecology Progress Series, 326: 1 –9.

85. Hansen H., Bachmann L., and Bakke T. A. 2003. Mitochondrial DNA variation of Gyrodactylus spp. (Monogenea, Gyrodactylidae) populations infecting Atlantic salmon, grayling, and rainbow trout in Norway and Sweden. International Journal for Parasitology, 33: 1471–1478.

86. Hansen L. P. 2006. Migration and survival of farmed Atlantic salmon (Salmo salar L.) released from two Norwegian fish farms. ICES Journal of Marine Science, 63: 1211–1217.

87. Hansen P. K., Ervik A., Schanning M. T., Johannsen P., Aure J., Jahnsen T., and Stigebrandt A. 2001. Regulating the local environmental impact of intensive marine fish farming. II. The monitoring programme of the MOM system (Modelling – Ongrowing fish farms – Monitoring). Aquaculture, 194: 75 –92.

88. Hargrave B. T., Holmer M., and Newcombe C. P. 2008. Towards a classification of organic enrichment in marine sediments based on biogeochemical indicators. Marine Pollution Bulletin, 56: 810 –824.

89. Haugland O., Mikalsen A. B., Nilsen P., Lindmo K., Thu B. J., Eliassen T. M., Roos N., et al. 2011. Cardiomyopathy syndrome of Atlantic salmon (Salmo salar L.) is caused by a double-stranded RNA virus of the totiviridae family. Journal of Virology, 85: 5275–5286.

90. Hazon N., Todd C. D., Whelan B., Gargan P. G., Finstad B., Bjørn P. A., and Wendelar Bonga S. E., et al. 2006. Sustainable management of interactions between aquaculture and wild salmonid fish. Final report for the SUMBAWS EU project. pp. 1 –293

91. Helland I. P., Finstad B., Uglem I., Diserud O. H., Foldvik A., Hanssen F., Bjørn P. A., et al. 2012. What determines salmon lice infections on wild salmonids? Statistical calculations of data from the national salmon lice surveillance program 2004–2010 (in Norwegian). NINA Rapport, 891. 51 pp.

92. Heuch P. A., Bjørn P. A., Finstad B., Holst J. C., Asplin L., and Nilsen F. 2005. A review of the Norwegian "National Action Plan Against Salmon Lice on Salmonids": The effect on wild salmonids. Aquaculture, 246: 79 –92.

93. Hindar K., Fleming I. A., McGinnity P., and Diserud A. 2006. Genetic and ecological effects of salmon farming on wild salmon: modelling from experimental results. ICES Journal of Marine Science, 63: 1234–1247.

94. Hjeltnes B. (Ed.) 2014. Report on fish health—2013 (in Norwegian). Veterinærinstituttet, Oslo, Norway. 44 pp.

95. Holm M., Holst J. C., and Hansen L. P. 2000. Spatial and temporal distribution of post-smolts of Atlantic salmon (Salmo salar L.) in the Norwegian Sea and adjacent areas. ICES Journal of Marine Science, 57: 955–964.

96. Holmer M., Argyrou M., Dalsgaard T., Danovaro R., Diaz-Almela E., Carlos M. D. E., Frederiksen M., et al. 2008. Effects of fish farm waste on Posidonia oceanica meadows: Synthesis and provision of monitoring and management tools. Marine Pollution Bulletin, 56: 1618–1629.

97. Holmer M., Duarte C. M., Heilskov A., Olesen B., and Terrados J. 2003. Biogeochemical conditions in sediments enriched by organic matter from net-pen fish farms in the Bolinao area, Philippines. Marine Pollution Bulletin, 46: 1470–1479.

98. Holmer M., and Frederiksen M. S. 2007. Stimulation of sulfate reduction rates inMediterranean fishfarm sediments inhabited by the seagrass Posidonia oceanica. Biogeochemistry, 85: 169–184.

99. Holmer M., and Kristensen E. 1992. Impact of marine fish cage farming on metabolism and sulfate reduction of underlying sediments. Marine Ecology Progress Series, 80: 191 –201.

100. Holmer M., Wildish D., and Hargrave B. 2005. Environmental effects of marine finfish aquaculture. In

Handbook of environmental chemistry, pp. 181–206. Ed. by B. Hargrave. Springer-Verlag, Berlin.

101. Holst J.C., Jakobsen P. J.,Nilsen F., HolmM., Asplin L., and Aure J. 2007. Mortality of seaward-migrating post-smolts of Atlantic salmon due to salmon lice infection in Norwegian salmon stocks. In Salmon at the edge, pp. 136 –137. Blackwell Science Ltd, Bodmin, Cornwall.

102. Husa V., Kutti T., Ervik A., Sjøtun K., Hansen P. K., and Aure J. 2014a. Regional impact from fin-fish farming in an intensive production area (Hardangerfjord, Norway). Marine Biology Research, 10: 241–252.

103. Husa V., Steen H., and Sjøtun K. 2014b. Historical changes in macroalgal communities in Hardangerfjord (Norway). Marine Biology Research, 10: 226–240.

104. Hvidsten N. A., Finstad B., Kroglund F., Johnsen B. O., Strand R., Arnekleiv J. V., and Bjørn P. A. 2007. Does increased abundance of sea lice influence survival of wild Atlantic salmon post-smolt? Journal of Fish Biology, 71: 1639 –1648.

105. Hyatt A. D., Hine P. M., Jones J. B., Whittington R. J., Kearns C., Wise T. G., and Crane M. S., et al. 1997. Epizootic mortality in the pilchard Sardinops sagax neopilchardus in Australia and New Zealand in 1995. II. Identification of a herpesvirus within the gill epithelium. Diseases of Aquatic Organisms, 28: 17–29.

106. Hytterød S., Lie Linaker M., Hansen H., Mo T. A., and Tavornpanich S. 2014. The surveillance programme for Gyrodactylus salaris in Atlantic salmon and rainbow trout in Norway. Surveillance programmes for terrestrial and aquatic animals in Norway. Annual report 2013. Norwegian Veterinary Institute, Oslo. 6 pp.

107. Jackson D., Cotter D., Newell J., McEvoy S., O'Donohoe P., Kane F., McDermott T., et al. 2013. Impact of Lepeophtheirus salmonis infestations on migrating Atlantic salmon, Salmo salar L., smolts at eight locations in Ireland with an analysis of lice-induced marine mortality. Journal of Fish Diseases, 36: 273 –281.

108. Jackson D., Cotter D., O′ Maoile′idigh N., O'Donohoe P., White J., Kane F., Kelly S., et al. 2011. An evaluation of the impact of early infestation with the salmon louse Lepeophtheirus salmonis on the subsequent survival of outwardly migrating Atlantic salmon, Salmo salar L., smolts. Aquaculture, 320: 159 –163.

109. Jansen P. A., Kristoffersen A. B., Viljugrein H., Jimenez D., Aldrin M., and Stien A. 2012. Sea lice as a density-dependent constraint to salmonid farming. Proceedings of the Royal Society B: Biological Sciences, 279: 2330–2338.

110. Johansen L. H., Jensen I., Mikkelsen H., Bjørn P. A., Jansen P. A., and Bergh Ø. 2011. Disease interaction and pathogens exchange between wild and farmed fish populations with special reference to Norway. Aquaculture, 315: 167–186.

111. Johansen R. (Ed.) 2013. Report on fish health—2012 (in Norwegian). Veterinærinstituttet, Oslo, Norway. 44 pp.

112. Johnsen B. O., and Jensen A. J. 1994. The spread of furunculosis in salmonids in Norwegian rivers. Journal of Fish Biology, 45: 47–55.

113. Johnsen B. O., Møkkelgjerd P. I., and Jensen A. J. 1999. The parasite Gyrodactylus salaris on salmon in Norwegian rivers, status report at start of the year 2000 (in Norwegian). NINA Oppdragsmelding, pp. 1 –129.

114. Jones S., and Beamish R. 2011. Salmon lice: an integrated approach to understanding parasite abundance and distribution. WileyBlackwell, Oxford, UK.

115. Jonsson B., and Jonsson N. 2006. Cultured Atlantic salmon in nature: a review of their ecology andinteractionwithwild fish. ICES Journal of Marine Science, 63: 1162–1181.

116. Karlsson S., Moen T., Lien S., Glover K. A., and Hindar K. 2011. Generic genetic differences between farmed and wild Atlantic salmon identified from a 7 K SNP-chip. Molecular Ecology Resources, 11: 247–253.

117. Kileng Ø., Johansen L-H., and Jensen I. 2011. ILA-virus in wild and farmed fish in Troms (in Norwegian). Nofima report 28/2011, Tromsø, Norway.

118. Kongtorp R. T., Kjerstad A., Taksdal T., Guttvik A., and Falk K. 2004. Heart and skeletal muscle inflammation in Atlantic salmon, Salmo salar L.: a new infectious disease. Journal of Fish Diseases, 27: 351–358.

119. Krause-Jensen D., Middleboe A. L., Carstensen J., and Dahl K. 2007. Spatial patterns of macroalgal abundance in relation to eutrophication. Marine Biology, 152: 25 –36.

120. Kristoffersen A. B., Bang Jensen B., and Jansen P. A. 2013. Risk mapping of heart and skeletal muscle inflammation in salmon farming. Preventive Veterinary Medicine, 109: 136–143.

121. Kristoffersen A. B., Viljugrein H., Kongtorp R. T., Brun E., and Jansen P. A. 2009. Risk factors for pancreas disease (PD) outbreaks in farmed Atlantic salmon and rainbow trout in Norway during 2003–2007. Preventive Veterinary Medicine, 90: 127–136.

122. Krkosˇek M., LewisM. A., andVolpe J. P. 2005. Transmission dynamics of parasitic sea lice from farm to wild salmon. Proceedings of the Royal Society B-Biological Sciences, 272: 689–696.

123. Krkosˇek M., Revie C., Gargan P. G., Finstad B., and Todd C. D. 2013a. Comment on Jackson et al. «Impact of Lepeophtheirus salmonisinfestations on migrating Atlantic salmon, Salmo salar L., smolts at eight locations in Ireland with an analysis of lice-induced marine mortality». Journal of Fish Diseases, 37: 415 –417.

124. Krkosˇek M., Revie C. W., Gargan P. G., Skilbrei O. T., Finstad B., and Todd C. D. 2013b. Impact of parasites on salmon recruitment in the Northeast Atlantic Ocean. Proceedings of the Royal Society B: Biological Sciences, 280: 20122359.

125. Kutti T. 2008. Regional impact of organic loading from a salomonid farm – dispersal, sedimentation rates and benthic fauna response. PhD thesis, University of Bergen.

126. Kutti T., Ervik A., and Hansen P. K. 2007a. Effects of organic effluents from a salmon farm on afjord system. I. Vertical export and dispersal processes. Aquaculture, 262: 367–381.

127. Kutti T., Ervik A., and Høisæther T. 2008. Effects of organic effluents from a salmon farm on a fjord system. III. Linking deposition rates of organic matter and benthic productivity. Aquaculture, 282: 47 –53.

128. Kutti T., Hansen P. K., Ervik A., Høisæther T., and Johannessen P. J. 2007b. Effects of organic effluents from a salmon farm on a fjord system. II. Temporal and spatial patterns in infauna community composition. Aquaculture, 262: 355 –366.

129. Løvoll M., Alarcon M., Jensen B. B., Taksdal T., Kristoffersen A. B., and Tengs T. 2012. Quantification of piscine reovirus (PRV) at different stages of Atlantic salmon Salmo salar production. Diseases of Aquatic Organisms, 99: 7 –12.

130. Løvoll M., Wiik-Nielsen J., Grove S., Wiik-Nielsen C. R., Kristoffersen A. B., Faller R., Poppe T. et al. 2010. A novel totivirus and piscine reovirus (PRV) in Atlantic salmon (Salmo salar) with cardiomyopathy syndrome (CMS). Virology Journal, 7: 309.

131. Madhun A. S., Biering E., Isachsen C. H., Omdal L. M., Einen A. C. B., Wennevik V., Svaˀsand T.,et al. 2014a. Annual report on health monitoring of wild anadromous salmonids in Norway. Report from the Norwegian Veterinary Institute and the Institute of Marine Research, Bergen, Norway.

132. Madhun A. S., Karlsbakk E., Isachsen C. H., Omdal L. M., Eide Sørvik A. G., Skaala Ø., and Barlaup B. T., et al. 2014b. Potential disease interaction reinforced: double-virus-infected escaped farmed Atlantic salmon, Salmo salar L., recaptured in a nearby river. Journal of Fish Diseases: in press, DOI: 10.1111/jfd.12228.

133. Martinez-Rubio L., Morais S., Evensen O.,Wadsworth S., Ruohonen K., Vecino J. L. G., and Bell J. G., et al. 2012. Functional feeds reduce heart inflammation and pathology in Atlantic salmon (Salmo salar L.) following experimental challenge with Atlantic salmon reovirus (ASRV). PLoS ONE, 7: e40266.

134. Martinez-Rubio L., Morais S., Evensen O., Wadsworth S., Vecino J. G., Ruohonen K., and Bell J. G.,et al. 2013. Effect of functional feeds on fatty acid and eicosanoid metabolism in liver and head kidney of Atlantic salmon (Salmo salar L.) with experimentally induced Heart and Skeletal Muscle Inflammation. Fish & Shellfish Immunology, 34: 1533–1545.

135. May R.M. 1983. Parasitic infections as regulators of animal interactions. American Scientist, 71: 36 –45.

136. May R. M., and Anderson R. M. 1979. Population biology of infectious diseases: Part II. Nature, 280: 455–461.

137. Mcallister P. E., Newman M. W., Sauber J. H., and Owens W. J. 1984. Isolation of infectious pancreatic necrosis virus (serotype-Ab) from diverse species of estuarine fish. Helgolander Meeresuntersuchungen, 37: 317–328.

138. McGinnity P., Prodohl P., Ferguson K., Hynes R., O'Maoileidigh N., Baker N., Cotter D., et al. 2003. Fitness reduction and potential extinction of wild populations of Atlantic salmon, Salmo salar, as a result of interactions with escaped farm salmon. Proceedings of the Royal Society of London Series B Biological Sciences, 270: 2443–2450.

139. McGinnity P., Stone C., Taggart J. B., Cooke D., Cotter D., Hynes R., McCamley C., et al. 1997. Genetic impact of escaped farmed Atlantic salmon (Salmo salar L.) on native populations: use of DNA profiling to assess freshwater performance of wild, farmed, and hybrid progeny in a natural river environment. ICES Journal of Marine Science, 54: 998 –1008.

140. McLoughlin M. F., Nelson R. T., Rowley H. M., Cox D. I., and Grant A. N. 1996. Experimental pancreas disease in Atlantic salmon Salmo salar post-smolts induced by salmon pancreas disease virus (SPDV). Diseases of Aquatic Organisms, 26: 117 –124.

141. Mente E., Pierce G. J., Santos M. B., and Neofitou C. 2006. Effect of feed and feeding in the culture of salmonids on the marine aquatic environment: a synthesis for European aquaculture. Aquaculture International, 14: 499 –522.

142. Molvær J., Knutzen J., Magnusson J. B., Rugg B., Skei J., and Sørensen J. 1997. Classification of environmental quality in fjords and coastal waters (in Norwegian). Norwegian Environment Agency, 3. 35 pp.

143. Mork J. 1991. One-generation effects of farmed fish immigration on the genetic differentiation of wild Atlantic salmon in Norway. Aquaculture, 98: 267–276.

144. Mørk T., and Hellberg H. 2005. The surveillance and control programme for Gyrodactylus salaris in Atlantic salmon and rainbow trout in Norway. In Surveillance and control programmes for terrestrial and aquatic animals in Norway, pp. 137–139. Annual report 2004.

145. Mortensen S. H. 1993. Passage of infectious pancreatic necrosis virus (IPNV) through invertebrates in an aquatic food-chain. Diseases of Aquatic Organisms, 16: 41–45.

146. Munda I. M. 1996. The northern Adriatic Sea. In Marine benthic vegetation, pp. 369–402. Ed. byW. Schramm , and P. Nienhuis . Springer, Berlin, Heidelberg.

147. Nash C. 2007. Guidelines for ecological risk assessment of marine aquaculture. Aquaculture Europe, 32: 5 –11.

148. Nelson R. T., Mcloughlin M. F., Rowley H. M., Platten M. A., and Mccormick J. I. 1995. Isolation of a toga-like virus from farmed Atlantic salmon Salmo salar with pancreas disease. Diseases of Aquatic Organisms, 22: 25 –32.

149. Nolan D. T., Reilly P., andWendelar Bonga S. E. 1999. Infection with low numbers of the sea louse Lepeophtheirus salmonis induces stressrelated effects in postsmolt Atlantic salmon (Salmo salar). Canadian Journal of Fisheries and Aquatic Sciences, 56: 947–959.

150. Norði G. A., Glud R. N., Gaard E., and Simonsen K. 2011. Environmental impacts of coastal fish farming: carbon and nitrogen budgets for trout farming in Kaldbaksfjorour (Faroe Islands). Marine Ecology Progress Series, 431: 223–241.

151. Norris A. T., Bradley D. G., and Cunningham E. P. 1999. Microsatellite genetic variation between and within farmed and wild Atlantic salmon (Salmo salar) populations. Aquaculture, 180: 247 –264.

152. Nylund A. 2007. Salmonid Alphavirus (SAV).InReview of disease interactions and pathogen exchange between farmed and wild finfish and shellfish in Europe, pp. 41–45. Ed. by R. Raynard , T. Wahli , I. Vatsos , and S. Mortensen . Veterinærmedisinsk Oppdragssenter AS.

153. Olsen S. A., Ervik A., and Grahl-Nielsen O. 2012. Tracing fish farm waste in the northern shrimp Pandalus borealis (Krøyer, 1838) using lipid biomarkers. Aquaculture Environment Interactions, 2: 133–144.

154. OSPAR. 2010. OSPAR commission 2010. Quality status report 2010. http://qsr2010.ospar.org/en/index.html.

155. Palacios G., Lovoll M., Tengs T., Hornig M., Hutchison S., Hui J., Kongtorp R. T., et al. 2010. Heart and skeletal muscle inflammation of farmed salmon is associated with infection with a novel reovirus. PLoS ONE, 5: e11487.

156. Parrish D. L., Behnke R. J., Gephard S. R., McCormick S. D., and Reeves G. H. 1998. Why aren't there more Atlantic salmon (Salmo salar)? Canadian Journal of Fisheries and Aquatic Sciences, 55: 281–287.

157. Pihl L., Svenson A., Moksnes P. O., and Wennehage H. 1999. Distribution of green algal mats throughout shallow soft bottoms of the Swedish Skagerrak archipelago in relation to nutrient sources and wave exposure. Journal of Sea Research, 41: 281–295.

158. Plarre H., Devold M., Snow M., and Nylund A. 2005. Prevalence of infectious salmon anaemia virus (ISAV) in wild salmonids in western Norway. Diseases of Aquatic Organisms, 66: 71 –79.

159. Poppe T. T., and Seierstad S. L. 2003. First description of cardiomyopathy syndrome (CMS)-related lesions in wild Atlantic salmon Salmo salar in Norway. Diseases of Aquatic Organisms, 56: 87–88.

160. Raynard R., Wahli T., Vatsos I., and Mortensen S. 2007. DIPNET – review of disease interactions and pathogen exchange between farmed and wild finfish and shellfish in Europe. p. 452. European Commission/ Veterinærmedisinsk Oppdragssenter.

161. Reno P. W. 1999. Infectious pancreatic necrosis virus and its prevalence. In Fish diseases and disorders. Vol 3: viral, bacterial and fungal infections, pp. 1 –55. Ed. by P. T. K. Woo , and D. W. Bruno . CABI Publishing, Walingford, UK.

162. Roberge C., Einum S., Guderley H., and Bernatchez L. 2006. Rapid parallel evolutionary changes of gene transcription profiles in farmed Atlantic salmon. Molecular Ecology, 15: 9–20.

163. Rueness J., and Fredriksen S. 1991. An assessment of possible pollution effects on the benthic algae of the outer Oslofjord, Norway. Oebalia, 17: 223 –235.

164. Samuel-Fitwi B., Wuertz S., Schroeder J. P., and Schulz C. 2012. Sustainability assessment tools to support aquaculture development. Journal of Cleaner Production, 32: 183 –192.

165. Sanderson J. C., Cromey C. J., Dring M. J., and Kelly M. S. 2008. Distribution of nutrients for seaweed cultivation around salmon cages atfarm sites in north–west Scotland. Aquaculture, 278: 60–68.

166. Sanz-Lazaro C., Belando M. D., Marin-Guirao L., Navarrete-Mier F., and Marin A. 2011. Relationship between sedimentation rates and benthic impact on Maerl beds derived from fish farming in the Mediterranean. Marine Environmental Research, 71: 22 –30.

167. Serra-Llinares R. M., Bjørn P. A., Finstad B., Nilsen R., Harbitz A., BergM., and Asplin L. 2014. Salmon lice infection on wild salmonids in marine protected areas: an evaluation of the Norwegian "National Salmon Fjords". Aquaculture Environment Interactions, 5: 1–16.

168. Skaala Ø., Glover K. A., Barlaup B. T., Sva°sand T., Besnier F., Hansen M. M., and Borgstrøm R. 2012. Performance of farmed, hybrid, and wild Atlantic salmon (Salmo salar) families in a natural river environment. Canadian Journal of Fisheries and Aquatic Sciences, 69: 1994 –2006.

169. Skaala Ø., Høyheim B., Glover K. A., and Dahle G. 2004. Microsatellite analysis in domesticated and wild Atlantic salmon (Salmo salar L.): allelic diversity and identification of individuals. Aquaculture, 240: 131–143.

170. Skaala Ø., Wennevik V., and Glover K. A. 2006. Evidence of temporal genetic change in wild Atlantic salmon, Salmo salar L., populations affected by farm escapees. ICES Journal of Marine Science, 63: 1224–1233.

171. Skilbrei O. T. 2010a. Adult recaptures of farmed Atlantic salmon postsmolts allowed to escape during summer. Aquaculture Environment Interactions, 1: 147–153.

172. Skilbrei O. T. 2010b. Reduced migratory performance of farmed Atlantic salmon post-smolts from a simulated escape during autumn. Aquaculture Environment Interactions, 1: 117 –125.

173. Skilbrei O. T. 2013. Migratory behaviour and ocean survival of escaped out-of-season smolts of farmed Atlantic salmon Salmo salar. Aquaculture Environment Interactions, 3: 213–221.

174. Skilbrei O. T., Finstad B., Urdal K., Bakke G., Kroglund F., and Strand R. 2013. Impact of early salmon louse, Lepeophtheirus salmonis, infestation and differences in survival and marine growth of sea-ranched Atlantic salmon, Salmo salar L., smolts 1997–2009. Journal of Fish Diseases, 36: 249 –260.

175. Skilbrei O. T., Glover K. A., Samuelsen O. B., and Lunestad B. T. 2008. A laboratory study to evaluate the use of emamectin benzoate in the control of sea lice in sea-ranched Atlantic salmon (Salmo salar L.). Aquaculture, 285: 2 –7.

176. Skilbrei O. T., Heino M., and Sva°sand T. 2015. Using simulated escape events to assess the annual numbers and destinies of escaped farmed Atlantic salmon of different life stages from farms sites in Norway. ICES Journal of Marine Science, 72: 670–685.

177. Skilbrei O. T., and Wennevik V. 2006. Survival and growth of sea-ranched Atlantic salmon, Salmo salar L., treated against sea lice before release. ICES Journal of Marine Science: Journal du Conseil, 63: 1317–1325.

178. SkogenM. D., EknesM., Asplin L. C., and Sandvik A. D. 2009.Modelling the environmental effects of fish farming in a Norwegian fjord. Aquaculture, 298: 70 –75.

179. Snow M., Black J., Matejusova I., McIntosh R., Baretto E., Wallace I. S., and Bruno D. W. 2010. Detection of salmonid alphavirus RNA in wild marine fish: implications for the origins of salmon pancreas disease in aquaculture. Diseases of Aquatic Organisms, 91: 177 –188.

180. Solberg M. F., Glover K. A., Nilsen F., and Skaala Ø. 2013a. Does domestication cause changes in growth reaction norms? A study of farmed, wild and hybrid Atlantic salmon families exposed to environmental stress. PLoS ONE, 8: e54469.

181. Solberg M. F., Zhang Z. W., Nilsen F., and Glover K. A. 2013b. Growth reaction norms of domesticated, wild and hybrid Atlantic salmon families in response to differing social and physical environments. BMC Evolutionary Biology, 13: 234.

182. Sommerset I., Krossøy B., Biering E., and Frost P. 2005. Vaccines for fish in aquaculture. Expert Review of Vaccines, 4: 89 –101.

183. Stabell O. 1984. Homing and olfaction in salmonids: a critical review with special reference to the Atlantic salmon. Biological Reviews of the Cambridge Philosophical Society, 59: 333–388.

184. Sta°hl G. 1987. Genetic population structure of Atlantic salmon. In Population genetics and fishery management, pp. 121 –140. Ed. by N. Ryman , and F. Utter . University of Washington Press, Seattle.

185. Stene A. 2013. Transmission of pancreas disease in marine salmon farming in Norway. PhD thesis, Norwegian School of Veterinary Sciences, Oslo, Norway. 117 pp.

186. Stene A., Viljugrein H., Yndestad H., Tavornpanich S., and Skjerve E. 2014. Transmission dynamics of pancreas disease (PD) in a Norwegian fjord: aspects of water transport, contact networks and infection pressure among salmon farms. Journal of Fish Diseases, 37: 123–134.

187. Stephens E. B., Newman M.W., Zachary A. L., and Hetrick F. M. 1980. A viral etiology for the annual spring epizootics of Atlantic menhaden Brevoortia tyrannus (Latrobe) in Chesapeake Bay. Journal of Fish Diseases, 3: 387 –398.

188. Sterud E., Forseth T., Ugedal O., Poppe T. T., Jørgensen A., Bruheim T., and Fjeldstad H.,et al. 2007. Severe mortality in wild Atlantic salmon Salmo salar due to proliferative kidney disease (PKD) caused by Tetracapsuloides bryosalmonae (Myxozoa). Diseases of Aquatic Organisms, 77: 191–198.

189. Strain P., and Hargrave B. 2005. Salmon aquaculture, nutrient fluxes and ecosystem processes in Southwestern New Brunswick. In Handbook of environmental chemistry, pp. 29 –57. Ed. by B. Hargrave . Springer.

190. Taggart J. B., Verspoor E., Galvin P. T., Moran P., and Ferguson A. 1995. A minisatellite DNA marker for discriminating between European and North American Atlantic salmon (Salmo salar). Canadian Journal of Fisheries and Aquatic Sciences, 52: 2305 –2311.

191. Taranger G. L., Boxaspen K. K., Madhusn A. S., and Sva°sand T. (Eds) 2012b. Suggested first generation method for environmental impact indicators with respect to genetic influences from farmed salmon to wild salmon and the impact of sea lice from farmed fish on wild salmon populations (in Norwegian). Fisken og Havet 13-2012.

192. Taranger G. L., Sva°sand T., Kvamme B. O., Kristiansen T. S., and Boxaspen K. (Eds) 2014. Risk assessment of Norwegian aquaculture 2013 (in Norwegian). Fisken og Havet, Særnummer 2-2014.

193. Taranger G. L., Sva°sand T., Kvamme B. O., Kristiansen T. S., and Boxaspen K. K. 2012a. Risk assessment of Norwegian aquaculture [Risikovurdering norsk fiskeoppdrett] (In Norwegian). Fisken og havet, særnummer 2-2012. 131 pp.

194. Taranger G. L., Sva°sand T., Kvamme B. O., Kristiansen T. S., and Boxaspen K. K. 2013. Risk assessment of Norwegian aquaculture 2012 (In Norwegian). Fisken og Havet, særnummer 2-2013. 164 pp.

195. Taranger G. L., Sva°sand T., Madhun A. S., and Boxaspen K. K. 2011a. Risk assessment of environmental impact of Norwegian aquaculture 2010 [Risikovurdering miljøvirkninger av norsk fiskeoppdrett 2010] (in Norwegian). Fisken og Havet, særnr. 3-2010. 93 pp.

196. Taranger G. L., Sva°sand T., Madhun A. S., and Boxaspen K. K. 2011b. Risk assessment – environmental impact of Norwegian aquaculture. Extracts from Risk assessment of environmental impact of Norwegian aquaculture 2010 Fisken og Havet no. 3-2011. 52 pp.

197. Taylor E. B. 1991. A review of local adaptation in salmonidae, with particular reference to Pacific and Atlantic salmon. Aquaculture, 98: 185–207.

198. Torrissen O., Jones S., Asche F., Guttormsen A., Skilbrei O. T., Nilsen F., and Horsberg T. E., et al. 2013. Salmon lice – impact on wild salmonids and salmon aquaculture. Journal of Fish Diseases, 36: 171–194.

199. Tuya F., Sanchez-Jerez P., Dempster T., Boyra A., and Haroun R. J. 2006. Changes in demersal wild fish aggregations beneath a sea-cage fish farm after the cessation of farming. Journal of Fish Biology, 69: 682–697.

200. Tveiten H., Bjørn P. A., Johnsen H. K., Finstad B., and McKinley R. S. 2010. Effects of the sea louse Lepeophtheirus salmonis on temporal changes in cortisol, sex steroids, growth and reproductive investment in Arctic charr Salvelinus alpinus. Journal of Fish Biology, 76: 2318–2341.

201. Valdemarsen T., Bannister R. J., Hansen P. K., Holmer M., and Ervik A. 2012. Biogeochemical malfunctioning in sediments beneath a deepwater fish farm. Environmental Pollution, 170: 15 –25.

202. Valdemarsen T., Kristensen E., and Holmer M. 2009. Metabolic threshold and sulfide-buffering in diffusion controlled marine sediments impacted by continuous organic enrichment. Biogeochemistry, 95: 335–353.

203. Valdemarsen T., Kristensen E., and HolmerM. 2010. Sulfur, carbon, and nitrogen cycling in faunated marine sediments impacted by repeated organic enrichment. Marine Ecology Progress Series, 400: 37–53.

204. Vassdal T., Heggøy E., and Johansen P-O. 2012. Marine monitoring Rogaland (in Norwegian). Status report May 2012. UNI Miljø, SAM Marin. E-rapport, 26-2012.

205. Verspoor E., Beardmore J. A., Consuegra S., De Leaniz C. G., Hindar K., Jordan W. C., Koljonen M. L.et al. 2005. Population structure in the Atlantic salmon: insights from 40 years of research into genetic protein variation. Journal of Fish Biology, 67: 3 –54.

206. Villanueva R. D., Yap H. T., and Montano M. N. E. 2006. Intensive fish farming in the Philippines is detrimental to the reef-building coral Pocillopora damicornis. Marine Ecology Progress Series, 316: 165–174.

207. Vollset K. W., Skoglund H., Barlaup B. T., Pulg U., Gabrielsen S-E., Wiers T., and Ska˚r B., et al. 2014. Can the river location within a fjord explain the density of Atlantic salmon and sea trout? Marine Biology Research, 10: 268–278.

208. Wagner G. N., Fast M. D., and Johnson S. C. 2008. Physiology and immunology of Lepeophtheirus salmonis infections of salmonids. Trends in Parasitology, 24: 176–183.

209. Wagner G. N., McKinley R. S., Bjørn P. A., and Finstad B. 2003. Physiological impact of sea lice on swimming performance of Atlantic salmon. Journal of Fish Biology, 62: 1000–1009.

210. Wagner G. N., McKinley R. S., Bjørn P. A., and Finstad B. 2004. Short-term freshwater exposure benefits sea lice-infected Atlantic salmon. Journal of Fish Biology, 64: 1593 –1604.

211. Wallace I. S., Gregory A., Murray A. G., Munro E. S., and Raynard R. S. 2008. Distribution of infectious pancreatic necrosis virus (IPNV) in wild marine fish from Scottish waters with respect to clinically infected aquaculture sites producing Atlantic salmon, Salmo salar L. Journal of Fish Diseases, 31: 177 –186.

212. Wallace I. S., Murray A. G., Gregory A., and Raynard R. S. 2005. The prevalence of infectious pancreatic necrosis virus in wild marine fish from Scotland with reference to clinically infected salmon farms. Abstract in EAFP 12th international Convergence, 12.

213. Wells A., Grierson C. E., MacKenzie M., Russon I., Reinardy H., Middlemiss C., Bjørn P. A., et al. 2006. Physiological effects of simultaneous, abrupt seawater entry and sea lice (Lepeophtheirus salmonis) infestation of wild, sea-run brown trout (Salmo trutta) smolts. Canadian Journal of Fisheries and Aquatic Sciences, 64: 1360–1369.

214. Wells A., Grierson C. E., Marshall L., MacKenzie M., Russon I. J., Reinardy H., Sivertsga˚rd R., et al. 2007. Physiological consequences of prematurefreshwater return; for wild sea-run brown trout (Salmo trutta) postsmolts infested with sea lice (Lepeophtheirus salmonis). Canadian Journal of Fisheries and Aquatic Sciences, 64: 1360–1369.

215. Wiik-Nielsen C. R., Lovoll M., Sandlund N., Faller R., Wiik-Nielsen J., and Jensen B. B. 2012. First detection of piscine reovirus (PRV) in marine fish species. Diseases of Aquatic Organisms, 97: 255–258.

216. Wilding T. A., CromeyC. J., Nickell T. D., and Hughes D. J. 2012. Salmon farm impacts on muddy-sediment megabenthic assemblages on the west coast of Scotland. Aquaculture Environment Interactions, 2: 145–156.

217. Worm B., and Sommer U. 2000. Rapid direct and indirect effects of a single nutrient pulse in a seaweed-epiphyte-grazer system. Marine Ecology Progress Series, 202: 283–288.

218. Zhang Z.,Glover K. A.,Wennevik V., Sva˚sand T., Sørvik A.G. E., Fiske P., and Karlsson S.,et al. 2013. Genetic analysis of Atlantic salmon captured in a netting station reveals multiple escapement events from commercial fish farms. Fisheries Management and Ecology, 20: 42 –51.

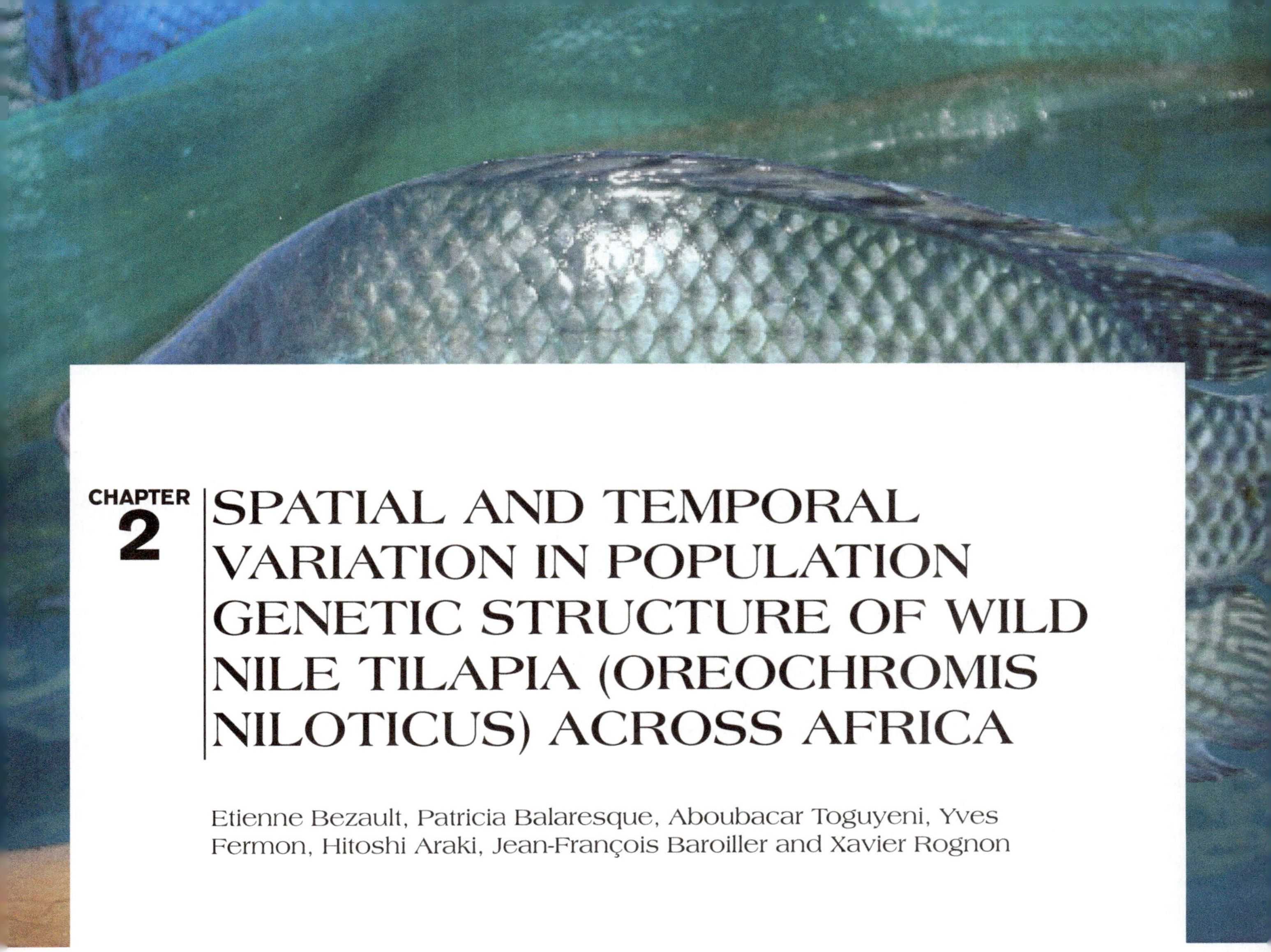

CHAPTER 2

SPATIAL AND TEMPORAL VARIATION IN POPULATION GENETIC STRUCTURE OF WILD NILE TILAPIA (OREOCHROMIS NILOTICUS) ACROSS AFRICA

Etienne Bezault, Patricia Balaresque, Aboubacar Toguyeni, Yves Fermon, Hitoshi Araki, Jean-François Baroiller and Xavier Rognon

BACKGROUND

The population genetic structure of living organisms is largely shaped by both historical and contemporary gene flow in the species range [1]. Furthermore, the factors structuring the genetic diversity are also able to play a key role in the process of diversification, adaptation and speciation [2]. Identifying the interactions between those factors is crucial to understand the evolutionary history of a species [3]. On the one hand, the extrinsic factors, such as climatic and geological events, shape the large-scale population differentiation pattern [4, 5, 6]. On the other hand, both extrinsic (*e.g.* habitat heterogeneity), and intrinsic factors (*e.g.* dispersal capability, mating system and habitat preference) have an impact on gene pool composition at intra-population level or over short time periods [4, 7, 8, 9]. Species with wide distribution and high dispersal capability are supposed to exhibit genetic differentiation at the paleo-biogeographic scales, with limited micro-geographic structure. On the contrary, species with limited distribution and restricted dispersal capability and/or specific mating behaviour are supposed to exhibit strong genetic structure both at micro-geographic and temporal scales. However, species with a wide distribution area and specific life history traits (*e.g.* restricted dispersal, limited population size) are expected to exhibit more complex genetic diversity pattern.

Cichlid fishes are well known examples of complex population genetic structure among African ichthyofauna. Lake-wide studies have revealed the impact of paleo-historical lake level fluctuations on genetic diversity (*e.g.*[10]). Small scale investigations have revealed the importance of both habitat heterogeneity, ecology and dispersal capability on population differentiation (*e.g.*[11]). To date, however, very few studies have investigated the factors responsible for the population genetic structures both at the micro- and the macro-geographic scales at the same time. They often represent only a relatively restricted paleo-biogeographic scale [12].

The Nile tilapia, *Oreochromis niloticus* (Linnaeus, 1758), is an interesting model-species to study the interactions

Citation: Bezault et al.: Spatial and temporal variation in population genetic structure of wild Nile tilapia (Oreochromis niloticus) across Africa 2011 12:102. doi:10.1186/1471-2156-12-102

between intrinsic and extrinsic factors on the structure of the natural populations from a local and temporal to a broader biogeographic scale. This economically important fish has one of the largest natural distributions among African freshwater fishes, covering the entire Nilo-Sudanian province (from Senegal to Nile basins), the Ethiopian Rift Valley province, the Kivu province, north Tanganyika province (Ruzizi) and the Northern part of the East African Rift Valley. This species shows an exceptional capacity of adaptation, which allowed its colonisation of a wide range of habitats from small forest rivers to large drainage and lakes, as well as alkaline pools with hot springs [13, 14]. The description of seven sub-species based on eco-morphology [13] largely reflects their adaptive divergences.

Due to its great interest for aquaculture and fisheries, the Nile tilapia and a few other tilapia species have been introduced outside their natural distributions [14]. Introduced tilapias have become invasive especially in areas originally not containing any tilapiine cichlid, within as well as outside Africa. In area inhabited by congeneric tilapias, on the other hand, they have often led to hybridisation with the local allopatric species [15]. However, between sympatric congeneric species of tilapias, signature of hybridisation has only been reported at evolutionary time scale, not at current ecological time scale. This is especially the case of the ancient mitochondrial introgression from *O. aureus* into *O. niloticus* populations restricted to West Africa, whereas the 2 species are also sympatric in the Nile River. This introgression has certainly happened during the drastic water level fluctuations of the Pleistocene [16]. But neither recent natural hybridisation event nor successful translocation of any allopatric tilapia (*i.e. Oreochromis spp.*) within the natural distribution of *O. niloticus* has been reported so far [15].

As most of the cichlid fish, the Nile tilapia exhibits interesting and complex life history traits, and especially well-developed social behaviour [13, 14, 17, 18, 19]. During reproduction, males show strong territoriality and females provide elaborated parental care (*i.e.* maternal mouth-brooding and guarding) [17]. Because of this reproduction behaviour and substrate affinity, the Nile tilapia is considered a rather sedentary species [14]. In addition, the lekking behaviour of males to attract females for reproduction suggests the existence of a certain level of sexual selection [17, 20]. These life history traits are expected to strongly affect the population dynamics of this species, via limited dispersal or reduced effective population size.

From the paleo-geographic point of view, Africa has experienced severe hydrogeographic modifications since the Pleistocene. The East African Rift valley has been subject to many tectonic disruptions of the water basins with inversion of the course of some rivers in the Nile basin (*sensu lato*), whereas the Sudano-Sahelian region experienced dramatic climatic fluctuations with alternating humid and dry phases [21, 22]. By their drastic modifications of the extension and connectivity of the different water-basins, all these paleo-geographic and climatic events have undoubtedly affected (1) the distribution of fish species in these ichthyofaunal provinces [23, 24] and (2) their population genetic structure.

Previous studies have suggested an influence of paleo-geographic events on the historical distribution of Nile tilapia, based either on morphological traits [13] or on moderately polymorphic molecular markers, *i.e.* allozymes and mtDNA, [16, 25, 26, 27, 28]. At the opposite, micro-geographic population structure has been reported in lacustrine populations of another tilapiine species, *Sarotherodon melanotheron*, with similar life history strategy, implying non-random mating [29]. However, to date, the relative importance of these paleo-geographic events and the factors influencing the micro-geographic structure of the current populations is not well understood.

In this study, we investigated the spatio-temporal genetic structure of ten natural populations of *O. niloticus*, which cover the main part of the species' natural distribution in Africa. By investigating both spatial and temporal genetic diversity based on nine microsatellite loci, we aimed (1) to understand the current population genetic structure of *O. niloticus* in its natural habitat range in Africa and (2) to evaluate to what extend the population genetic structure was shaped by the paleo-geographic events and the current geographic connectivity at the different spatio-temporal scales. We also discuss potential roles of life history of the species in the species-range population genetic differentiation.

RESULTS

Spatial and Temporal Sampling

Natural populations of *O. niloticus* were sampled from ten geographic sites across Africa to maximise the representation of the biodiversity of the species, naturally spread over the vast Nilo-Sudanian and the Ethiopian ichthyofaunal provinces [23, 30]. In order to complement the analyses of population genetic differentiation from macro-geographic to micro-geographic and temporal scales, further sampling was focused along the Volta basin.

Overall, this sample-set of 350 individuals covers six of the larger hydrographic basins naturally inhabited by*O. niloticus*, representing the Nilo-Sudanian ichthyo-province, including Sudano-Sahelian (Niger, Senegal and Volta basins) and Nilotic regions (Turkana and Nile basins) and the Ethiopian Rift Valley province (Awash basin) [23, 31]. According to taxonomy [13], this represents four sub-species of *O. niloticus*: *O. n. niloticus* (in West-Africa and Nile) *O. n. vulcani* (in Lake Turkana), *O. n. cancellatus* (in the main part of Awash basin) and *O. n. filoa* (in hot springs of the Awash basin) (Table 1 and Figure 1).

Table 1 Population samples information

Sub-species	Ichthyo. Regions	Basin *Lake or River*	Station	GPS location		*Sampling year*	Codes	Sample size
				Lat.				
	Ethiopian	**Awash**					**Aw**	
O. n. cancel-latus		*Lake Hora*	Debra Zeit	08°45' N	38°59' E	2002	**Hr**	30
O. n. cancel-latus		*Lake Koka*	Koka	08°24' N	39°01' E	2002	**Kk**	30
O. n. filoa		*Lake Metahara*	Abadir	08° 51' N	39°50' E	2002	**Me**	30
O. n. vulcani	**Nilotic**	**Lake Turkana**	El Molo Bay	03°37' N	36°02' E	1994	**Tu**	15
O. n. niloticus		**Nile**	Lake Manzala	31°17' N	31°59' E	1992	**Mz**	14
O. n. niloticus	**Sudano-Sahelian**	**Senegal**	Djouj Nat.Park	16°24' N	16°12' W	1991	**Se**	15
O. n. niloticus		**Niger**	Bamako	12°38' N	07°60' W	1990	**Nb**	10
		Volta					**Vo**	
O. n. niloticus		*Kou*	Bama	11°22' N	04°28' W	2004	**Ko**	16
O. n. niloticus		*Lake Volta*	Kpandu	06°48' N	00°18' E	Total	**Kp**	103
						Nov 2001	**KpN1**	*30*
						Mar 2002	**KpM2**	*13*
						Jul 2002	**KpJ2**	*30*
						Feb 2003	**KpF3**	*30*
O. n. niloticus		*River Volta*	Nyinuto	05°56' N	00°41' E	Total	**Ny**	87
						Nov 2001	**NyN1**	*28*
						Mar 2002	**NyM2**	*30*
						Feb 2003	**NyF3**	*29*
							Total:	**350**

Taxonomic, bio-geographic and sample size information of the ten populations of Nile tilapia, *Oreochromis niloticus*, studied at both spatial and temporal scales.

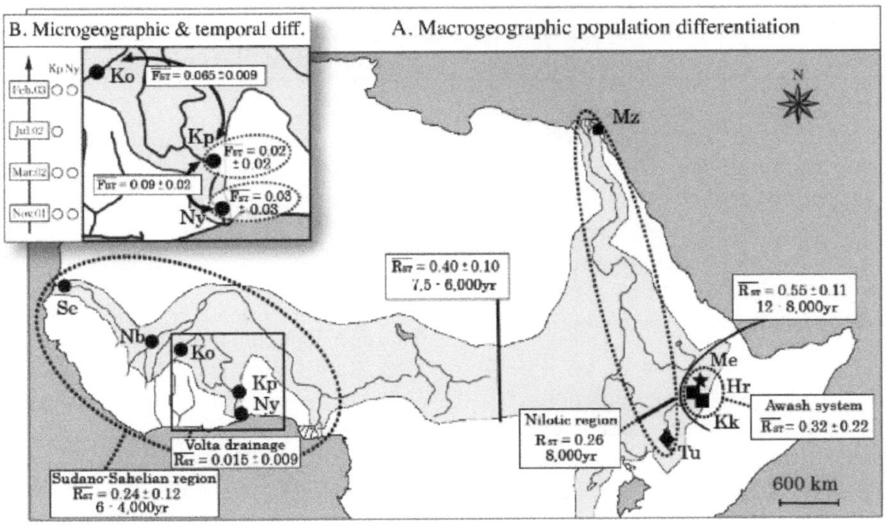

Figure 1 Population genetic differentiation of Nile tilapia, *Oreochromis niloticus* , across its natural distribution. Geographic representation of the 10 sampled populations of Nile tilapia within the natural distribution range of the species (shaded area) and summary of the genetic differentiation observed according to the clustering obtained from AMOVA analysis (see Table 4); (A) at macro-geographic level across Africa and (B) zoom on micro-geographic and temporal variability within the Volta basin. Separation times between hydrographical basins are based

on paleontological information (see References in the Discussion section). The genetic differentiations are the average pairwise population value observed within (dotted line) or between (solid line) basins or ichthyological provinces and regions; calculations are based on R_{ST} at Macro-geographic scale, while on F_{ST}-at micro-geographic and temporal scales, according to the test suggested by Hardy et al. [35] (see Table 3 & 5 respectively). The population symbols refer to the different morphological sub-species: *O. n. niloticus* (black dot), *O. n. vulcani* (black diamond), *O. n. cancellatus* (black square) and *O. n. filoa* (black star) - (see Table 1 for population codes).

Micro-geographic and temporal sampling on the Volta basin was conducted at three sampling localities along the hydrographic basin, of which two were sampled successively multiple times between 2001 and 2003 (Table1 and Figure 1). They represent 206 individual samples, which were used to estimate temporal variation of population genetic structure as well as effective size of these populations.

Intra-population Diversity

Using nine microsatellite loci, large variation in genetic diversity was observed among populations.

Table 2 Genetic diversity per population.

Populations		*A*	*A* r ± SD	H_e ± SD	H_{obs} ± SD	F_{IS}		**Relatedness**			
Spatial	***Temporal***							Mr_{xy}		Vr_{xy}	
Hr		2.33	2.15 ± 1.16	0.32 ± 0.28	0.29 ± 0.26	0.089					
Kk		2.11	1.68 ± 0.60	0.17 ± 0.21	0.17 ± 0.24	-0.027					
Me		2.78	2.31 ± 0.78	0.36 ± 0.23	0.34 ± 0.25	0.063					
Mz		5.00	4.75 ± 1.40	0.72 ± 0.10	0.66 ± 0.18	0.090					
Tu		5.33	4.94 ± 1.74	0.74 ± 0.11	0.69 ± 0.13	0.070					
Se		7.44	6.30 ± 3.41	0.67 ± 0.27	0.63 ± 0.29	0.058					
Nb		3.00	3.00 ± 1.41	0.42 ± 0.24	0.39 ± 0.28	0.076					
Ko		6.33	5.30 ± 1.77	0.65 ± 0.18	0.64 ± 0.19	0.020					
Kp		9.89	5.79 ± 1.94	0.69 ± 0.19	0.58 ± 0.15	0.158	***	-0.01		0.22	
	KpN1	*6.44*	*5.12 ± 2.08*	*0.64 ± 0.24*	*0.54 ± 0.21*	*1.154*	***	*-0.03*	**	*0.24*	
	KpM2	*4.56*	*4.49 ± 1.70*	*0.69 ± 0.17*	*0.63 ± 0.26*	*0.086*		*-0.08*		*0.29*	*
	KpJ2	*7.11*	*5.40 ± 1.79*	*0.66 ± 0.19*	*0.55 ± 0.15*	*0.180*	***	*-0.03*		*0.21*	
	KpF3	*7.56*	*5.90 ± 2.17*	*0.71 ± 0.17*	*0.65 ± 0.19*	*0.090*	*	*-0.04*		*0.21*	
Ny		8.56	5.37 ± 2.51	0.64 ± 0.19	0.55 ± 0.24	0.149	***	-0.04	***	0.35	
	NyN1	*6.57*	*5.35 ± 2.52*	*0.67 ± 0.16*	*0.52 ± 0.37*	*0.230*	***	*-0.07*		*0.40*	
	NyM2	*5.25*	*4.19 ± 2.10*	*0.55 ± 0.20*	*0.46 ± 0.27*	*0.156*	***	*-0.04*		*0.27*	**
	NyF3	*6.22*	*4.84 ± 2.73*	*0.60 ± 0.25*	*0.53 ± 0.26*	*0.111*	*	*-0.04*	*	*0.27*	**

Genetic diversity and allelic richness observed and estimated from rarefaction algorithm calculated over the nine loci for the ten study populations and their temporal samples (codes are given in Table 1), as well as level of relatedness r_{xy} [63], estimated based on mean (Mr_{xy}) and variance (Vr_{xy}) of the distribution of pairwise relatedness index, calculated for each temporal samples from each location and across temporal samples for each location.

A = number of alleles observed per population; *A* r = allelic richness per population based on rarefaction algorithm; H_e = non-biased expected heterozygosity; H_{obs} = observed heterozygosity (with standard deviations = SD); the F_{IS} = fixation index values with their significances after sequential Bonferroni correction; Test significance: *$P < 0.05$, ** $P < 0.01$, *** $P < 0.001$.

The populations of the Awash basin showed the lower gene diversity based on both allelic richness (*A* r = 1.68 - 2.31) and heterozygosity (H_o = 0.17 - 0.34). All other Nilotic and Sahelo-Sudanian populations presented an overall higher level of genetic diversity (*A* r = 4.75 - 6.30; H_o = 0.55 - 0.69), except for Niger revealing an intermediate level of diversity (*A* r = 3.00; H_o = 0.42).

Departure from Hardy-Weinberg equilibrium was detected for 5 of the 10 geographic populations (*i.e.* Hora, Metahara, Turkana, Kpandu and Nyinuto). The test remained significant after Bonferroni correction for the two populations from the Volta basin constituted of temporal sub-sampling (Kpandu, Kp, and Nyiinuto, Ny). Both populations showed significantly positive F_{IS} values, suggesting a deficit of observed heterozygotes ($P < 0.001$, Table 2). This result was consistent with a temporal Wahlund effect at the population level when pooling slightly differentiated temporal samples (F_{ST} = 0 - 0.053 - Table 3 and see below). However, significantly positive F_{IS} was also found in all but one temporal sample from these two populations (*i.e.* except KpM2 with the smallest sample size). The deficit of heterozygotes could be due to either i) null alleles, ii) a spatial Wahlund effect, or iii) non-random mating. After exclusion of the three loci potentially possessing null alleles (*see detail in Methods section*), however, four samples (*i.e.* two in Kpandu and two in Nynuto) still exhibited significant positive F_{IS} values. Thus, null alleles are not the main cause of the disequilibrium. The possibility of the spatial Wahlund effect, namely artificial substructure from the pool of different fishing catches within the temporal samples, was tested using AMOVA. No significant differentiation among seine catches was detected (P = 0.386;*data not shown*), suggesting the absence of Wahlund effect at this level. To investigate a possibility of non-random mating within population and cohorts, we analysed the level of relatedness among individuals at both i) site and ii) temporal levels, and compared the distribution of the observed pairwise relatedness index (r_{xy}) to simulated distribution expected under panmixia. At the site level, Nyinuto population presented an overall lower level of relatedness (Mr_{xy}, $P < 0.001$) than expected under panmixia, however the population from Kpandu did not differ from the null expectation. Furthermore, the majority of the temporal samples showed significant difference in level of relatedness than expected in a randomly mating population in both populations: two temporal samples showed higher average pairwise relatedness (*i.e.* KpN1 & NyF3) and one showed a higher variance of relatedness distribution (*i.e.* KpM2), while one sample showed a lower variance of relatedness distribution (*i.e.* NyM2 - Table 2). Thus, though the three possible causes of the deficit of heterozygotes are not mutually exclusive, non-random mating within each temporal sample is seemingly contributing to the observed pattern most.

Table 3 Matrix of macro-geographic population differentiation based on R_{ST} and F_{ST} estimators

	Hr	Kk	Me	Mz	Tu	Se	Nb	Ko	Kp	Ny
Hr		0.427***	0.072**	0.601***	0.472***	0.713***	0.636***	0.693***	0.587**	0.660***
Kk	0.296***		0.473***	0.696***	0.582***	0.833***	0.787***	0.816***	0.709***	0.770***
Me	0.059**	0.287***		0.553***	0.377***	0.692***	0.599***	0.671***	0.573***	0.632***
Mz	0.451***	0.553***	0.425***		0.259***	0.384***	0.247***	0.397***	0.425***	0.424***
Tu	0.442***	0.549***	0.416***	0.115***		0.431***	0.209***	0.469***	0.520***	0.517***
Se	0.489***	0.596***	0.467***	0.225***	0.229***		0.217***	0.152***	0.133**	0.101**
Nb	0.603***	0.723***	0.585***	0.326***	0.331***	0.243***		0.377***	0.374***	0.349***
Ko	0.507***	0.607***	0.481***	0.222***	0.214***	0.097***	0.306***		0.0249*ns*	0.0073*ns*
Kp	0.404***	0.456***	0.392***	0.216***	0.221***	0.054**	0.246***	0.054**		0.0142*ns*
Ny	0.455***	0.524***	0.440***	0.243***	0.224***	0.101***	0.261***	0.095**	0.078**	

Pairwise R_{ST} (above diagonal) and F_{ST} (below diagonal) for the ten spatial populations studied, with corresponding level of significance (based on 10,000 replicates). The significance of the test of Hardy [35] ($P < 0.001$) suggests that R_{ST} is more appropriate to describe genetic differentiation among these populations than F_{ST}.

A Bayesian-based population assignment analysis using STRUCTURE[32, 33] revealed that more than 91% of the individuals belong to a single genetic cluster with a posterior probability of > 95% (Figure 2a), further reflecting expected biogeographic affinities among populations (see below); only 3.7% of individuals (13 out of 350) showed with an assignment > 10% posterior probability to another genetic cluster, suggesting little impact of translocation or recent hybridisation in the studied populations.

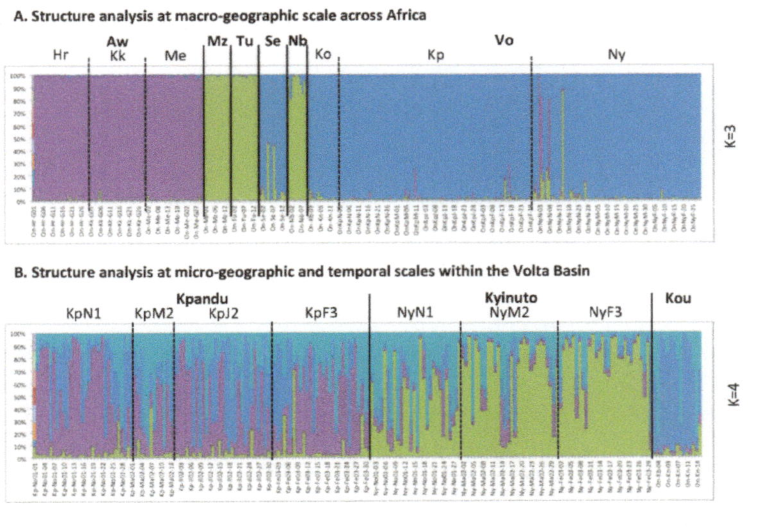

Figure 2 Population structure at macro-geographic, micro-geographic and temporal levels. Individual-based Bayesian clustering using STRUCTURE v.2.3.1[32, 33] to assess the genetic population structure at both **A)** macro-geographic and **B)** micro-geographic and temporal scales. For each scale, the number of genetic clusters (K), which was best supported by the procedure proposed by Evanno et al. (ΔK) [34] was represented (*i.e.* K = 3 for A and K = 4 for B). Different colours represent different genetic clusters in each figure. Code representing populations and hydrographic basins are given above graphs (see Table 1 for details).

Macro-geographic Populations Structure & Differentiation

At macro-geographic level, factorial correspondence analysis (FCA) discriminated three distinct groups of populations along the first two factorial axes (F1 & F2; Figure 3a): the first group corresponded to the three Ethiopian populations, Hora, Koka and Metahara, the second group to the five populations from the Sudano-Sahelian region, Niger, Senegal and Volta (Kou, Kpandu and Nyinuto), and the third group to the Nilotic populations of Nile and Turkana, which were further separated by the third factorial axis (F3; Figure 3b). This population clustering is reflecting the biogeographic separations rather than taxonomic description. The STRUCTURE analysis, for the best supported number of clusters (K = 3) according to the Evanno correction [34], revealed a similar clustering pattern between populations from the Ethiopian province and, Nilotic and Sudano-Sahelian regions (Figure 2a). However, the population from Niger was predominantly associated with the Nilotic group rather than the Sudano-Sahelian one. Nevertheless, the assignment of the Niger population become shared between Nilotic and Sudano-Sahelian groups and then predominantly associated with the Sudano-Sahelian group, at K = 4 and higher respectively. These results point out a slightly more central position of the Niger basin within the Sahelian region, compared to the Nilotic region (also supported by genetic differentiation - see below). Furthermore the separation of the two populations from the Nilotic region was also observed when K was higher than the optimal K = 4.

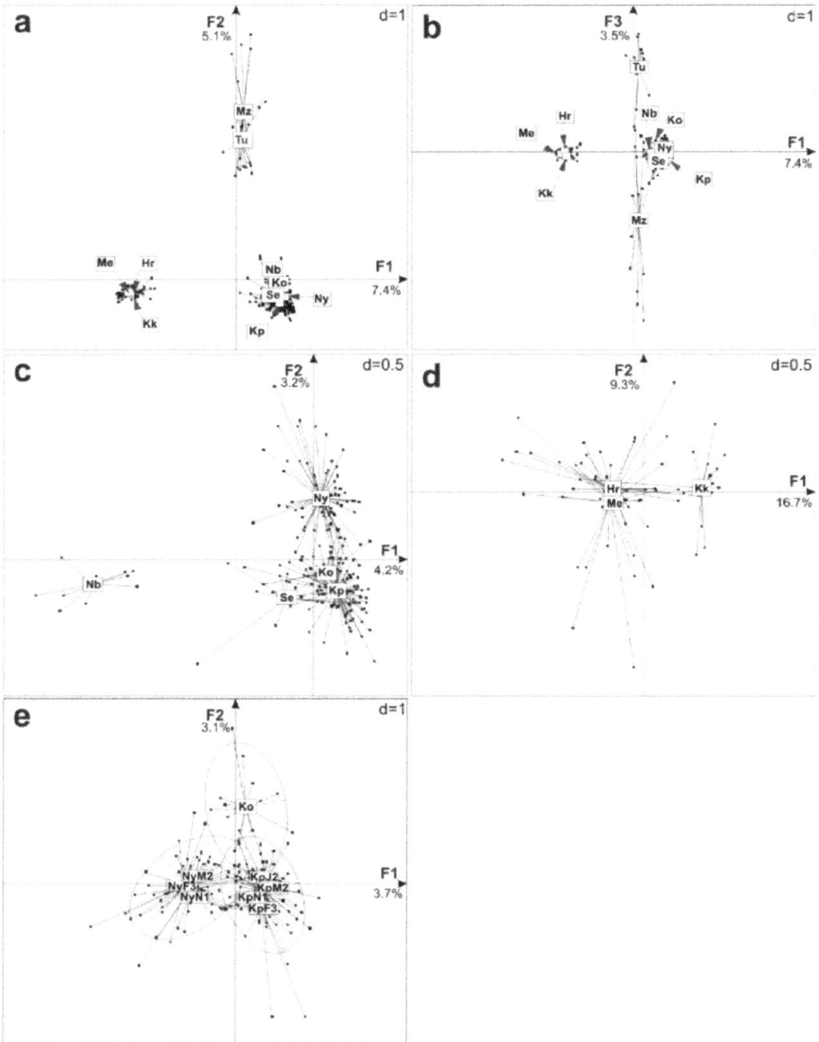

Figure 3 Hierarchical spatio-temporal population clustering. Factorial correspondence analyses based on genotype of *O. niloticus* samples at successive hierarchical spatial and temporal levels, depicting the best population clustering. For each population a star representation is used: every individual is linked with a line to the barycentre of its own population (labelled using sample code according to Table 1). At macro-geographic scale, we performed the projection of the 10 populations representing the overall biogeographic region studied (*i.e.* Ethiopian Rift Valley province, Nilotic, Sudano-Sahelian regions) on factorial planes F1 × F2 (**a**) and F1 × F3 (**b**), the projection of the Sudano-Sahelian region (*i.e.* West-African populations: Niger, Senegal and Volta basin populations) on factorial plane F1 × F2 (**c**), the projection of the Ethiopian Rift Valley province populations (*i.e.* Awash basin) on factorial plane F1 × F2(**d**). For the micro-geographic and temporal scales analysis, we performed the projection of the samples from the Volta basin including Kou, Kpandu and Nyinuto populations (temporal sample series are represented by stars and the area of each geographic population is represented by a 95% confidence ellipse), on factorial plane F1 × F2 (**e**). For each correspondence analysis, the percentage of variation represented by each factorial axis is given aside of its label.

The significance ($P < 0.001$) of the test suggested by Hardy et al. [35] revealed that SMM-based estimator (R_{ST}, see Methods) is the most pertinent to describe genetic differentiation at macro-geographic scale The highest population differentiations were found between regions (Table 3), especially between Ethiopian and Sudano-Sahelian populations ($\overline{R_{ST}} = 0.695 \pm 0.08$), , while the Nilotic populations exhibit a high degree of genetic differentiation from either Ethiopian or Sudano-Sahelian populations ($\overline{R_{ST}} = 0.55 \pm 0.11$ and $\overline{R_{ST}} = 0.40 \pm 0.10$, respectively; Figure 1). The hierarchical AMOVA conducted with populations clustered by biogeographic sub-regions revealed that 49.4% of the genetic variance was actually distributed between Ethiopian province and, Nilotic and Sudano-Sahelian regions, whereas 6.7% was distributed among population within regions, both with very high significance ($P < 0.001$). The alternative AMOVA model, with populations clustered according to taxonomy, showed comparatively less fitted variance partitioning, with lower percentage of variance explained by the clustering factor (47.3%). Although this result was highly significant, a larger part of the genetic variance was attributed to the between-populations within group component (9.1%). When samples were clustered by hydrographic basins, 49.8% of the genetic variability was partitioned among groups, and 3.5% was detected among populations within basins ($R_{SC} = 0.069$, $P < 0.001$), revealing intra-basin heterogeneity.

Analyses conducted at the biogeographic region level reveal variable level of differentiation, as 61.2% of the genetic variance of the system is contained within populations of the Ethiopian Rift Valley province (n = 3), while 90.6% for the Sudano-Sahelian region (n = 5). Within Ethiopian Rift valley province, the multivariate analysis widely separates the three populations across the first factorial plane (F1 × F2; Figure 3d), confirmed by an average differentiation of $\overline{R_{ST}} = 0.32 \pm 0.22$ (Figure 1): Hora and Metahara populations being clearly overlapping, with the lowest pairwise differentiation ($R_{ST} = 0.072$, $P < 0.01$), while Koka appeared more isolated, especially from Metahara (0.473, $P < 0.001$). Within the Sudano-Sahelian region, 24.8% of the variance was observed between hydrographical basins (Table 4). Multivariate analysis and pairwise genetic differentiation reveal that a large part of this differentiation is due to the separation of the Niger from the remaining basins (F1; Figure 3c), $\overline{R_{ST}} = 0.264 \pm 0.001$ (Figure 1). The among-population, within-group differentiation was low and not significant ($R_{SC} = 0.002$) indicating certain homogeneity within basins. Within the Volta basin, almost all of the genetic variance (~100%) was found within populations with very little differentiation between sample sites ($\overline{R_{ST}} = 0.015 \pm 0.009$; Figure 1). This result was concordant with the fact that populations within the Volta basin are still connected.

Table 4 Hierarchical partitioning of the genetic variance.

Stat.		**Variance components (%)**			**Φ Statistics**		
Estim.	**Grouping**	**Within pops**	**Among groups**	**Among pops within groups**	Φ_{ST}	Φ_{CT}	Φ_{SC}
	Macro-geographic scale						
	All populations in one group	54.69		45.31	**0.453** ***		
	All populations grouped by region [0]	43.90	49.42	6.68	**0.561** ***	**0.494** ***	**0.132** ***
	All populations grouped by basin [1]	46.74	49.78	3.48	**0.533** ***	**0.498** ***	**0.069** ***
	All populations grouped by taxonomy [2]	43.72	47.14	9.14	**0.563** ***	**0.471** ***	**0.173** ***
R_{ST}	within Sudano-Sahelian region	90.61		9.39	**0.090** ***		
	within Sudano-Sahel. grouped by basin[3]	75.03	24.79	0.18	**0.250** ***	0.248^{ns}	0.002^{ns}
	within Ethiopian region (Awash)	61.16		38.84	**0.388** ***		
	within Nilotic region (Turkana & Nile)	73.05		26.95	**0.269** ***		
	within Volta basin	100.45		-0.45	0.004^{ns}		
	Micro-geographic & temporal scale						
	within Volta basin	93.67		6.33	**0.063** ***		
F_{ST}	within Volta grouped by site [4]	91.56	6.48	1.96	**0.084** ***	**0.065** **	**0.021** ***
	within Kpandu (all temporal samples)	98.71		1.29	**0.013** *		
	within Nyinuto (all temporal samples)	92.61		7.39	**0.074** **		

Hierarchical analysis of molecular variance (AMOVA) calculated based on (1) *R*-Statistics for the macro-geographic scale and on (2) *F*-Statistics for the micro-geographic and temporal scales, considering the results of the test suggested by Hardy *et al.* [35];

the populations grouping reflects ichthyological provinces and regions, hybrographic basins, and taxonomy status (cf. Table 1).

[0] 3 groups: (Ethiopian: Hr, Kk, Me), (Nilotic: Mz, Tu), (Sudano-Sahelian: Se, Nb, Ko, Kp, Ny)

[1] 6 groups: (Awash: Hr, Kk, Me), (Nile: Mz), (Turkana: Tu), (Senegal: Se), (Niger: Nb), (Volta: Ko, Kp, Ny)

[2] 4 groups: (*O.n.cancellatus*: Hr, Kk), (*O.n.filoa*: Me), (*O.n.vulcani*: Tu), (*O.n.niloticus*: Mz, Se, Nb, Ko, Kp, Ny)

[3] 3 groups: (Senegal: Se), (Niger: Nb), (Volta: Ko, Kp, Ny)

[4] 3 groups: (Ko), (Kp: KpN1, KpM2, KpJ2, KpF3), (Ny: NyN1, NyM2, NyF3)

Test significance: ns = non-significant, * $P < 0.05$, ** $P < 0.01$, *** $P < 0.001$.

Overall, these results demonstrate that ichthyofaunal provinces definition and hydrographical basins boundaries are the two major structuring factors of *O. niloticus* populations at large geographic scales.

Micro-geographic and Temporal Comparisons

The analysis of micro-geographic and temporal variations has been conducted along the Volta basin. Based on factorial correspondence analysis (Figure 3e), three groups visibly emerged in the first factorial plan (F1 × F2), clustering the individuals according to their different geographic origins, Kou, Kpandu and Nyinuto. Temporal samples from Kpandu and Nyinuto were very close to each other within each location. Similarly, the STRUCTURE analysis reveals an optimal number of four clusters, representing preferentially the three different geographic locations, without strict discrimination (Figure 2b). The fourth genetic cluster was relatively equally distributed across all sites. No clear differentiation among temporal samples was observed at K = 4 or higher number of cluster.

This pattern was confirmed by the partition of the genetic variance and pairwise population differentiations based on F_{ST} (*i.e.* the non-significance of the test proposed by Hardy et al [35] suggesting that IAM-based estimator is the most suitable to depict genetic differentiation at this scale - Table 5, see also Methods).

Table 5 Matrix of micro-geographic and temporal population differentiation based on F_{ST} estimators.

	Ko	KpN1	KpM2	KpJ2	KpF3	NyN1	NyM2	NyF3
Ko								
KpN1	0.068**							
KpM2	0.069***	0.029*						
KpJ2	0.052**	0.016*ns*	-0.015*ns*					
KpF3	0.072***	0.024*	0.038**	0.030*				
NyN1	0.101***	0.103***	0.060***	0.092***	0.053**			
NyM2	0.116***	0.130***	0.118***	0.106***	0.105***	0.045**		
NyF3	0.106***	0.099***	0.100***	0.103***	0.080***	0.053***	-0.0037*ns*	

$\mathbf{F}_{ST}$ (below diagonal) for the eight spatio-temporal samples from the 3 geographic populations studied along the Volta basin, with corresponding level of significance (based on 10,000 replicates). The non-significance of the test of Hardy [35] suggests that $\mathbf{F}_{ST}$ is more appropriate to describe genetic differentiation among these populations than $\mathbf{R}_{ST}$.

Test significance: ns = non-significant, * $P < 0.05$, ** $P < 0.01$, *** $P < 0.001$.

S; below diagonal) and associated level of significance (above diagonal).

The highest genetic differentiation was found between geographic sites in the Volta basin (F_{ST} = 0.05 - 0.13, $\overline{F_{ST}}$ = 0.09 ± 0.02; Figure 1), representing 6.5% of the genetic variance of the system (Table 4). The genetic differentiations within temporal series in Kpandu (F = 0 - 0.04, $\overline{F}$ = 0.02 ± 0.02) and Nyinuto (F = 0 - 0.05, $\overline{F}$ = 0.03 ± 0.03) were lower than observed between sample sites, but yet significant (Table 5). Partition of the genetic variance revealed that 2.1% of the genetic variance was due to genetic differentiation in time (across both Kpandu and Nyinuto; Table 4). To examine whether the genetic differentiation between temporal series was due to the effect of one sample site or both, we performed separate AMOVA on Kpandu and Nyinuto samples, respectively. Interestingly, we found that genetic variance due to among temporal series variability was significant for both sites: 1.3% for Kpandu (F_{ST} = 0.013, $P < 0.05$) and 7.4% for Nyinuto (F_{ST} = 0.074, $P < 0.01$).

This demonstrates that subtle temporal changes of gene pool composition need to be considered in population structuring of *O. niloticus* at micro-geographic scale.

Effective Population Size and Immigration Rate

The effective population size (N_e) was estimated based on two focal populations in the Volta basin (Kpandu & Nyinuto). We used three different methods, based either on temporal variations of allele frequency within a population, compared the global gene pool of the basin [36], or on the level of linkage disequilibrium (LD-*Ne*) within population [37]. Two temporal methods were used in this study. The LD-*Ne* and the first temporal method (Ne_{CLOSED}) assume absence of immigration to estimate N_e [38]. The second temporal method (Ne_{open}) relaxes this assumption and assume connectivity between populations to jointly estimate N_e and immigration rate (*m*) [36] (Table 6). Point estimates of N_e ranged between 31-143 for Kp and 23-34 for Ny. While different methods provided different confidence intervals (N_e: from 21-53 to 70-817 for Kp, and from 17-33 to 23-55 for Ny), real N_e would likely be in or around the above ranges (see discussion in [39]). Estimates of the immigration rate (*m*) were similar for both populations ($m \sim 0.097$ for Kp, and $m \sim 0.094$ for Ny).

Table 6 Effective population size and dispersal rate.

Estimator	Ne_{OPEN} **&*m***					Ne_{CLOSED}		**LD-*Ne***	
Population	**Source**	***Ne***	**95%CI**	***m***	**95%CI**	***Ne***	**95%CI**	***Ne***	**95%CI**
Kpandu	*Ko+Ny*	31	21-53	0.097	0.053-0.160	143	70-817	61	37-142
Nyinuto	*Ko+Kp*	27	19-44	0.094	0.052-0.157	34	23-55	23	17-33

Estimates of effective population size (*Ne*) and dispersal rate (*m*) from temporal samples of two populations from the Volta basin using three different methods: (1) maximum likelihood temporal method to jointly estimate Ne_{open} and migration *m*[36], (2) pseudo-likelihood temporal method to estimate Ne_{closed} [38], and (3) point-estimate Linkage-Disequilibrium based method LD-*Ne* [37] averaged over the different temporal samples (KpM2 (n = 13) was excluded from the calculations).

DISCUSSION

Phylogeography of the Nile tilapia: Past History and Current Genetic Structure at Macro-Geographic Scale

From the Miocene (*i.e.* 20-5 My BP) to the Pleistocene (*i.e.* 2 My to 10,000 yrs BP), numerous paleo-geographic (*e.g.* causing disruption of river drainages) and paleo-climatic events (*e.g.* humid and dry phase cycles), by changing the connectivity between the different hydrographical systems, have deeply affected African ichthyofauna [22, 23, 24, 40]. Biogeographic data and fossil records suggest a Nilotic origin of the Nile tilapia (Upper Pliocene ~ 5-2 My BP [13]) - compatible with allozyme data [25]. The present study identified three genetic clusters of Nile tilapia populations corresponding to the major biogeographic subdivisions: (1) the Ethiopian Rift Valley ichthyofaunal province (*i.e.* the Awash system), (2) the Nilotic region (*sensu lato*, comprising the presently separated but previously connected Nile River and Turkana basin) and (3) the Sudano-Sahelian region (*i.e.* West-African basins) (Figure 1), the two later belonging to the Nilo-Sudanian ichthyofaunal province [23]. The comparison of the AMOVA's results revealed a first differentiation of population between ichthyofaunal regions (49%), as also depicted by STRUCTURE analysis. Therein genetic variance was further partitioned between hydrographic basins (1-3% of genetic variance between basins independently of ichthyological regions), and further remaining significant molecular variance among populations within basins (3.5%). This pattern is consistent with those observed in the few other fish species widespread across the Nilo-Sudanian area and studied by genetic markers, *Tilapia zillii*[25], *Oreochromis aureus*[16] and *Clarias gariepinus*[41]. In particular, these studies also emphasise the deep split between the Nile and the Sudano-Sahelian populations. The westward spread of this species from the Nile to Sudano-Sahelian rivers has probably occurred during the Pleistocene [13, 14], presumably facing the raise of mega-paleo-lakes and drainage connectivity in the Saharan region [21, 24, 40].

The Ethiopian populations exhibited the lowest levels of polymorphism within each population and the largest genetic divergence from the other ones (Tables 2). The isolation of the Awash from the Nile system occurred by tectonic disruption of this river basin about 12,000-8,000 years ago [23], which can explain the substantial genetic divergence between these two basins ($\overline{R_{ST}} = 0.55 \pm 0.011$ - Figure 1). The low genetic diversity in all three Awash populations (Table 2) and rather high differentiation ($\overline{R_{ST}} = 0.32 \pm 0.22$ - Figure 1) among them suggest strong effects of genetic drift within this «islands system», which exhibits an extensive level of fragmentation between the Awash River and several relatively isolated highland lakes. Following Trewavas [13], two different morphological sub-species coexist in the Awash basin: *O. n. cancellatus* in the river and cold-water lakes (*e.g.* Lakes Hora and Koka), and *O. n. filoa* endemic to the hot spring areas (*e.g.* Lake Metahara). The lower genetic distance between Hora and Metahara ($R_{ST} = 0.07$) compared to Koka ($R_{ST} = 0.43$ and 0.47 respectively) does not reflect their taxonomic status (*i.e.* paraphyly of *O. n. cancellatus*). In addition, the description of *O. n. filoa* is mostly based on the reduced number of some meristic characters (*i.e.* number of vertebra, scales), which might be subject to developmental plasticity [13]. These results, as well as the similar reproduction pattern and the absence of post-zygotic barrier [42, 43] support the hypothesis that *O. n. filoa* and *O. n. cancellatus* are two ecotypes, rather than distinct subspecies. However, further specific experiments, especially according to environment-induced developmental plasticity and mate choice preference, should be conducted to confirm this hypothesis.

In the Nilotic system *sensu lato*, disconnection occurred between the Nile river and the currently endorheic Lake Turkana basin about 8,000 years ago [21, 44]. Individuals from these two geographically distant populations have been previously identified as two distinct subspecies, *O. n. niloticus* and *O. n. vulcani*respectively [13]. However, compared to the large geographic distance and separation time existing between these two populations, the genetic differentiation was expected to be higher than observed (R_{ST} = 0.26 - Figure1) - with respect to other values observed in the present study. Connections between Lake Turkana and the Nile River have probably existed at least periodically until a few thousand years ago [44], as it seems to be reflected by nuclear and mtDNA data [16, 25, 26, 27].

The Sudano-Sahelian basins were isolated from the Nile River by multiple paleo-geographic events between 7,500 and 6,000 years ago [21]. Although morphological analyses have clustered populations from the Nile River with those of the Sudano-Sahelian basins [13], previous allozyme data [25, 27] and this study ($\overline{R_{ST}}$ = 0.38 ± 0.07) revealed a clear disconnection between these two biogeographic entities. This independent history is reinforced by the differential pattern of ancient introgression of mtDNA from *O. aureus* to *O. niloticus*involving the whole West African area, but not reported in any Nilotic population [16].

The Chad-Chari system appears to have played a key role between the Nilotic and Sudanian biogeographic entities, providing an unidirectional connectivity for the ichthyofauna from the Nile to the West African drainages, via the Niger, during the periods of growth of the mega-paleo-Lake Chad [21, 24, 40, 45]. The West African populations are the most geographically widespread, but also the least morphologically diversified [13]. They also exhibited the lowest level of genetic differentiation between basins ($\overline{R_{ST}}$ = 0.24 ± 0.12 - Figure1). Such a pattern might be explained by the persistence of episodic connections among Senegal, Niger and Volta headwater between 6,000 and 4,500 years ago [22]. Despite these recent gene flow events, all the Sudano-Sahelian populations could be strictly distinguished using microsatellites, even for populations at an intra-basin level, such as for the Volta River, contrary to the previous results obtained using allozyme or mitochondrial data [25].

As expected, no signature of anthropogenic impact on population differentiation was identified in this study. Paleo-geographic and climatic events appear to be important factors for the current pattern of the genetic structure of Nile tilapia populations within its natural distribution across Africa.

Microgeography and Temporal Variation

Within the Volta basin, most of the genetic variability was distributed within populations (93.7%), from the upstream Kou River, to the central Lake Volta (Kpandu) and downstream Volta River (Nyinuto). This suggests their recent differentiation and/ or non-interrupted gene flow. The genetic differentiation within temporal series in Kpandu ($\overline{F}$ = 0.02 ± 0.02) and Nyinuto ($\overline{F}$ = 0.03 ± 0.03) was lower than between sample sites ($\overline{F_{ST}}$ = 0.09 ± 0.02; Figure 1), but still significant and representing more than 2% of the genetic variance of the system. These results suggest a geographic differentiation between populations, and a quick turnover of the gene pools within population within only a few generations.

Our estimates of the effective population size from three different approaches consistently showed larger $N_e$in Kpandu than in Nyinuto (Table 6). The estimated migration rate (*m*), obtained jointly with Ne_{open}[36], provided similar values for both populations. Although the size of each sample for these estimates was limited (n = 28 - 30), the congruence of the different N_e estimates together with relatively small confidence intervals indicates a good coherence and reliability of our estimations [39, 46]. The population of Kpandu, located at the shore of Lake Volta (one of the largest artificial lakes in the world), is believed to have a large census size and/or well connected to neighbouring populations. In contrast, the population of Nyinuto, located at the downstream limit of Nile tilapia distribution along the Volta River, is likely to have a more limited size and to be easily subject to fluctuations (*e.g.* variation of water level and salinity - E. Bezault *pers. obs.*), and/or more isolated, than a central basin population. Despite the potential limit associated to our sampling strategy, our N_e estimates consistently showed relatively small effective population size of the natural Nile tilapia populations relative to expected census population size (*i.e.* fisheries records expressed in millions of tonnes per year for the Volta basin; E. K. Abban, *pers. comm.*) and a non-negligible dispersal rate.

The hypothesis of the effect of specific social and reproductive behaviour of the Nile tilapia emerges from the positive F_{IS} observed in both populations and the significant relatedness detected in some temporal samples (*i.e.* 3 out of the 4 temporal samples with significant relatedness showing either a higher mean or variance relatedness distribution, revealing respectively overall related individuals or subgroups of related individuals). A shoaling behaviour involving relatives could for instance explain such a pattern. This behaviour has been observed in other populations of *O. niloticus*[47, 48] as well as in another tilapia species, *Sarotherodon melanotheron*[29, 47]. These authors suggested this pattern to be due to direct inbreeding, implying preferential reproduction between individuals related among themselves, as observed in other cichlid species [49]. Furthermore, it has been claimed that this behaviour appears preferentially in open-water habitats, and especially in lagoon systems [29]. In the present study, the pattern of inbreeding observed in both lake and river habitats cannot be totally explained by shoaling behaviour among relatives, due also to the absence of sub-structure related to seine net catches (*i.e.* potential shoal) within temporal samples. A more direct effect of a non-panmictic reproductive system should be considered. Polygynandric mating system, generally exhibited, by maternal mouthbrooding cichlids, as *O. niloticus*, is generally associated with sexual selection and pronounced between-individual variations in reproduction success [17, 20]. Such variance of fitness has been reported for male Nile tilapia but only in captivity [50]. The rapid turnover in gene pool and the restricted effective population sizes, observed in our study, could be explained by the differential contribution of breeders to within and between cohorts over generations.

Furthermore the non-negligible dispersal rate detected in the present study opens up new research perspectives on cichlids. While sophisticated reproduction strategy and extended parental care are expected to increase territoriality and consequently prevent adult migration, our results suggest that the exceptional adaptive plasticity and riverine affinity of the tilapia could also be seen in its dispersal leading to larger migration rate than more specialised (stenotypic) cichlid species from East African Great Lakes [11, 20, 51]. This hypothesis is corroborated by the population genetic analysis of the riverine haplochromines generalist,*Pseudocrenilabrus multicolour victoriae*, presenting relatively large distribution area with geographic structure being the main driver of population structure and gene-flow [12]. It has also been demonstrated in this species important change of gene pool composition at short temporal scale (*i.e.* 2 years), which has been associated with drastic fluctuation of hydrographic regime [52]. Larger estimates of effective population size (including higher confidence interval reaching infinity) have been obtained for *P. m. victoriae* in this study compared to effective population size estimated here for *O. niloticus*. Unfortunately no census population size estimate for any of these species is available to effectively compare the N_e/N ratio, however, in the case of *P. m. victoriae* it cannot be ruled out that the larger effective population size observed is influenced by flooding regime.

CONCLUSIONS

The distribution and pattern of genetic diversity of a species over its natural range is the result of multiple and interlinked factors, which acted through space and time. For the Nile tilapia, *O. niloticus*, the genetic pattern observed from large geographic to temporal scales first reflects the hierarchical impact of paleo-geographic and climatic events as well as the impact of hydrographic connectivity on population genetic structure. Furthermore, the hierarchical pattern of genetic differentiation seems to reflect only bio-geographic events, without revealing any signature of anthropogenic impact. Compared to the most studied, highly radiating cichlid lineages, the morphologically and ecologically generalist species, as especially the tilapiines, [13, 19], might have distinct population dynamic characteristics. The relatively small effective population size associated with a moderate dispersal and the rapid change in gene pool composition in only a few generations tend to reinforce our current knowledge about the biology of this species, and they might reflect a potential impact of natural history strategy on population dynamics. Further investigations are needed to confirm this hypothesis. In addition, more research emphasis should be put on micro-geographic population structures to evaluate a more precise relationship between census and effective population size, the extent of dispersal among the natural population of Nile tilapia, as well as the balance between selection and gene-flow across different habitats. Furthermore, the possibility that Nile tilapia populations may be organised in meta-populations within larger and more connected systems (like within the Lake Volta) should be further investigated, from a genetic resource as well as from a conservation point of view.

METHODS

Sampling and Analysis Strategy

We collected 350 samples of *O. niloticus* from ten geographic sites across Africa (Table 1 and Figure 1). Our samplings aimed 1) to cover the macro-geographic area in the natural range of this species as extensive as possible and include different ichthyofaunal provinces and regions and therein hybrographic basins [30], and 2) to include successive temporal sampling through the short period of time in local populations, which allow us to survey the temporal genetic stability at the micro-geographic level. In both cases, a special attention was paid to avoid a potential influence of anthropogenic fish translocations and introductions, which could cause introgression and hybridisation. For that purpose, sampling sites without any record of *Oreochromis spp.* introduction were selected. In addition, all sampled individuals (considering only fully mature specimens, > 120 mm) were visually checked according to specific morphological characters [13] and confirmed the absence of sign of hybridisation for specimens analysed [16, 42, 43]. Sampling of Volta and Awash basins were done between 2001 and 2004, whereas samples from Lake Turkana, Lake Manzala, Niger and Senegal were a subset of populations already investigated for enzyme polymorphism and mtDNA [16, 25, 26], which have been considered pure *O. niloticus* at the nuclear genome level [16].

The sampling locations covered six hydrographic basins in Africa, belonging to the Sudano-Sahelian (Niger, Senegal and Volta) and Nilotic (Turkana, Nile) regions of the Nilo-Sudanian province and Ethiopian Rift valley (Awash basin) ichthyofaunal province [23, 30, 31] and represented four sub-species (*O. n. niloticus*, *O. n. vulcani*, *O. n. cancellatus* and *O. n. filoa*) according to their eco-morphological characters [13].

Analysis of micro-geographic and temporal variation was focussed on the Volta basin, where temporal sample series were collected repeatedly over a period of 18 months in two of the three geographic populations: Kpandu and Nyinuto (Table 1). These temporal samples were collected precisely at the same geographic location (verified by GPS coordinates) with the same standardised fishing protocol using seine nets. To minimise the potential impact of shoals' capture (*i.e.* biased sampling among relatives), each temporal sample was composed of individuals taken from multiple seine catches.

For our analyses, we considered two different datasets. (1) The macro-geographic analyses were conducted including all samples from the ten geographic populations considered (n = 350; temporal sample series from the two sites of the Volta basin pooled respectively into a single population-sample per site, *i.e.* Kpandu and Nyinuto). (2) The micro-geographic and temporal analyses were conducted at the scale of the entire Volta basin, including the samples from the 3 sites along the basin (n = 206): one sample from the Kou River (upstream Volta basin), four temporal samples from Kpandu (on the Lake Volta) and three temporal samples from Nyinuto (downstream on the Volta River) (Table 1 and Figure 1).

Microsatellite Typing

Nine unlinked microsatellites were selected from the *O. niloticus* genomic DNA library [53]. Genomic DNA was extracted from fin clips or muscle fragments stored in 95% ethanol using a phenol-chloroform procedure [54]. DNA samples were amplified by PCR using the indirect fluorescent tagging procedure described by Schuelke [55]. The amplification was performed using standard PCR-mix concentration and a touch-down PCR procedure, allowing standardised conditions over the loci analysed, using a 384-well-plate technology (Genotyping Platform, Genopole Montpellier Languedoc-Roussillon, CIRAD). PCR products were detected on an automated Li-Cor gel sequencer (IR2, Lincoln, Neb.), and genotyped were accessed by eyes.

Genetic Diversity Analysis

The total number of alleles (A), observed (H_o) and non-biased expected (H_{nb}) heterozygosities were calculated for each locus separately and across all loci using GENETIX 4.03 [56] and FSTAT 2.9.3 [57]. Allelic richness (Ar) was estimated for each population using rarefaction algorithm implemented in HP-rare [58], assuming an even number of genes per population, n = 20, based on the smallest population sample size in the study. For each sample, the fixation index, F_{IS}, was estimated across all loci, and statistical significance was assessed with permutation tests (n = 1000) using GENETIX. Corrections for multiple tests were performed using the sequential Bonferroni procedure [59].

Heterozygote deficits, as indicated by F_{IS} values significantly larger than 0, might reflect (1) sampling issues (*e.g.* Wahlund effect) or (2) impact of specific life history traits including a low dispersal or any social behaviour leading to a particular mating system. However, technical problems, such as null alleles, can also affect F_{IS}. Thus, we first ran MICRO-CHECKER[60] to estimate if null alleles might affect the fixation index. Potential null alleles were identified at three loci (UNH142, UNH162, and UNH211), however, excluding these three loci had very minor effects on F_{IS} (data not shown).

We used analysis of molecular variance, AMOVA [61], implemented in ARLEQUIN 2.0 [62] to test the potential population substructure due to different seine catches within the different temporal and local samples. We tested the possibility of non-random mating and inbreeding within population and cohorts as a potential cause of positive F_{IS} values by calculating the degree of relatedness within samples using the pairwise relatedness index r_{xy} of Queller & Goodnight [63] implemented in IDENTIX[64]. The observed arithmetic mean of r_{xy} (Mr_{xy}) in a sample was compared to that expected for a panmictic population. The observed variance of r_{xy} (Vr_{xy}) was also compared with the expectation from a panmictic population to test for the presence of multiple family groups within samples (which would increase the variance of r_{xy}) [64, 65]. For the sample pooled by station (Kp and Ny), the null distribution was generated by permuting genotypes within the population (1,000 permutations), following Castric *et al.*[65]. For the temporal samples, the null distribution was calculated based on 1,000 randomised sub-samples of the same size as the tested sample, taken from the total samples from the given geographic site (*i.e.* Kp or Ny) [66].

Population Clustering and Genetic Differentiation

Factorial correspondence analyses were performed using R [67] and the package ADE4 v1.4-9 [68] to investigate the relationship between populations from multi-locus genotypes without defining groups *a priori*. Successive correspondence analyses were performed to group populations from the largest to the finest geographic scales investigated.

Similarly, genetic population clustering was assessed using STRUCTURE v.2.3.1[32, 33] and conducted at i) macro-geographic scale and ii) micro-geographic and temporal scales. Simulations were conducted assuming an admixture model, with 10,000 burn-in steps and 100,000 MCMC steps. Ten independent runs were performed for each number of clusters assumed (K), with K varying between 2 and z+1, with z being the number of samples including in the analysis (*i.e.* respectively 11 for macro-geographic and 9 for micro-geographic and temporal analysis). The optimal number of clusters was assessed based on correction proposed by Evanno et al. [34].

Genetic parameters, providing a measure of the genetic differentiation existing between two populations, can be calculated using different estimators (for a discussion see [69]). The *F*-statistics [70], based on the infinite alleles model (IAM), relies solely on allele identity information. As it is an easy target for homoplasy, leading to the underestimation of genetic differentiation among highly structured populations (where $\mu > m$), this parameter generally is used to compare populations at small geographic scale. In contrast, the analogous *R*-statistics [71] based on the stepwise mutation model (SMM), takes into account allele size differences. As it typically suffers from high sampling variances, it is used for comparisons of large geographic scales. Here, as we were faced with distinct scales, we applied the test suggested by Hardy *et al.*[35] to choose the most suitable estimators. This test indicates whether or not allele sizes provide information on population differentiation, helping to choose between *F*- statistics and *R*- statistics in the most objective way. We applied this test implemented in SPAGeDi [72] for both datasets, and used the most pertinent statistics. Matrices of pairwise differentiation between populations (*R*-statistics or *F*- statistics) were computed using ARLEQUIN. The averaged genetic differentiation calculated over pairwise value between or within group of populations were noted $\overline{F_{ST}}$ or $\overline{R_{ST}}$.

Further, AMOVA [61] was also used to investigate how genetic variability was distributed in space and time. Considering macro-geographic scale, successive hierarchical levels were defined based alternatively on i) biogeographic regions and provinces (3 groups), ii) hydrographic basins (6 groups) and iii) sub-species taxonomy (4 groups - see Table 4 for details). Considering micro-geographic scales, hierarchical AMOVA was computed considering temporal samples within localities at i) the Volta basin and ii) the locality levels.

Estimating Effective Population Size and Migration Rate

We estimated effective population size (N_e) and migration rate (m) based on the spatio-temporal samples from the Volta basin. In order to reduce the sampling biases due to the small sample size, we excluded a temporal sample with the smallest sample size (KpM2, n = 13) from the estimation, and then retained only larger samples (n ≥ 28). We used three different approaches to estimate N_e (see discussion in [39, 46, 73]). Two of them are based on temporal variation of allele frequencies within populations. Both use likelihood methods and are implemented in MLNe 2.3 [36], but the first method (N_e-$_{CLOSED}$) assumes no immigration [38], assumption likely violated in the current context. The second method (N_e-$_{open}$), relaxes this assumption and allows the joint estimation of N_e and m[36]. For these approaches, the three temporal samples were considered to be from successive generations, assuming a generation time of about 6 months, as generally accepted for the Nile tilapia [14]. For the estimation of Ne_{open} of a given focal population, we pooled the samples from the other populations from the basin (*i.e.* Ko + Ny for Kp, and Ko + Kp for Ny) to mimic the allele frequencies in the source population of all potential immigrants. The third approach (LD-N_e), is based on the linkage disequilibrium method [37] and implemented in NE

ESTIMATOR[74]. In this method, no migration is assumed. LD-N_e was calculated for each temporal sample and then averaged over them. A point estimate and the 95% confidence intervals were obtained for each population by each method.

ACKNOWLEDGEMENTS

We would like to thank Eddie Abban and the team of the Water Research Institute (Ghana), Abebe Getahun (University of Addis Ababa), Pascal Bonnet & Géraud Laval (CIRAD/ILRI), the Office des Parcs Nationaux du Sénégal, Gabriele Hörstgen-Schwark (University of Göttingen), and Brendan McAndrew (University of Stirling) for their cooperation in obtaining samples. We also thank Stéphane Mauger, Gaëlle Chupot, René Guyomard (INRA), and Claire Billot (CIRAD) for their laboratory assistance; Agathe Bezault for the illustrations; Wendy Brand-Williams (INRA) for English improvement, and especially, Nicolas Poulet (ONEMA) for his help in statistical analyses and discussions about demography, Bernard Chevassus (INRA) and Mark Jobling (University of Leicester) for stimulating discussions, and Dylan Fraser (Dalhousie University), Irene Keller (EAWAG), Marta Barluenga (Museum of Natural History, Madrid), the editor Michael Hansen and the anonymous reviewers for their constructive comments. This study was supported by grants from the CIRAD, INRA & MENRT (France) to EB.

AUTHORS' CONTRIBUTIONS

EB, PB, XR, HA and JFB designed the research within the framework of the PhD project of EB; EB, YF, XR and AT collected the samples and data; EB and PB analysed the data; EB, PB, XR and HA led the writing. All authors read and approved the final manuscript.

REFERENCES

1. Holsinger KE, Weir BS: Genetics in geographically structured populations: defining, estimating and interpreting FST. Nature Review Genetics. 2009, 10 (9): 639-650. 10.1038/nrg2611.
2. Schluter D: The Ecology of Adaptive Radiation. Oxford: Oxford University Press. 2000
3. Weiss S, Persat H, Eppe R, Schlotterer C, Uiblein F: Complex patterns of colonization and refugia revealed for European grayling *Thymallus thymallus*, based on complete sequencing of the mitochondrial DNA control region. Mol Ecol. 2002, 11 (8): 1393-1407. 10.1046/j.1365-294X.2002.01544.x.
4. Tibbets CA, Dowling TE: Effects of intrinsic and extrinsic factors on population fragmentation in three species of North American minnows (Teleostei: Cyprinidae). Evolution. 1996, 50 (3): 1280-1292. 10.2307/2410668.
5. McClenaghan LR, Smith MH, Smith MW: Biochemical genetics of mosquitofish. IV. Changes of alleles frequencies through time and space. Evolution. 1985, 39: 451-160. 10.2307/2408377.
6. Bernatchez L, Wilson CC: Comparative phylogeography of nearctic and palearctic fishes. Mol Ecol. 1998, 7 (4): 431-452. 10.1046/j.1365-294x.1998.00319.x.
7. Laroche J, Durand JD, Bouvet Y, Guinand B, Brohon B: Genetic structure and differentiation among populations of two cyprinids, *Leuciscus cephalus* and *Rutilus rutilus*, in a large European river. Can J Fish Aquat Sci. 1999, 56 (9): 1659-1667.
8. Heithaus MR, Laushman RH: Genetic variation and conservation of stream fishes: influence of ecology, life history, and water quality. Can J Fish Aquat Sci. 1997, 54 (8): 1822-1836.
9. Salgueiro P, Carvalho G, Collares-Pereira MJ, Coelho MM: Microsatellite analysis of genetic population structure of the endangered cyprinid *Anaecypris hispanica* in Portugal: implications for conservation. Biol Conserv. 2003, 109 (1): 47-56. 10.1016/S0006-3207(02)00132-5.
10. Egger B, Koblmuller S, Sturmbauer C, Sefc KM: Nuclear and mitochondrial data reveal different evolutionary processes in the Lake Tanganyika cichlid genus *Tropheus*. BMC Evol Biol. 2007, 7: 1-14. 10.1186/1471-2148-7-1.
11. Wagner CE, McCune AR: Contrasting patterns of spatial genetic structure in sympatric rock-dwelling cichlid fishes. Evolution. 2009, 63 (5): 1312-1326. 10.1111/j.1558-5646.2009.00612.x.
12. Crispo E, Chapman LJ: Population genetic structure across dissolved oxygen regimes in an African cichlid fish. Mol Ecol. 2008, 17 (9): 2134-2148. 10.1111/j.1365-294X.2008.03729.x.
13. Trewavas E: Tilapiine fishes of the genera *Sarotherodon*, *Oreochromis* and *Danakilia*. 1983, London, UK: British Museum Natural History
14. Philippart J-C, Ruwet JC: Ecology and distribution of tilapias. The biology and culture of tilapias. Edited by: Pullin RSV, Lowe-McConnell RH. 1982, Manila, Philippines: ICLARM, 7: 15-60.
15. Eknath AE, Hulata G: Use and exchange of genetic resources of Nile tilapia (*Oreochromis niloticus*). Rev Aquac. 2009, 1 (3-4): 197-213. 10.1111/j.1753-5131.2009.01017.x.
16. Rognon X, Guyomard R: Large extent of mitochondrial DNA transfer from *Oreochromis aureus* to *O. niloticus* in West Africa. Mol Ecol. 2003, 12 (2): 435-445. 10.1046/j.1365-294X.2003.01739.x.
17. Keenleyside MHA: Parental care. Cichlid Fishes: Behaviour, ecology and evolution. Edited by: Keenleyside M. 1991, London, UK: Chapman & Hall, 191-208.
18. Lowe-McConnell RH: Ecology of cichlids in South American and African waters, excluding the African Great Lakes. Cichlid Fishes: Behaviour, ecology and evolution. Edited by: Keenleyside M. 1991, London, UK: Chapman & Hall, 60-83.
19. Klett V, Meyer A: What, if Anything, is a Tilapia? Mitochondrial ND2 Phylogeny of Tilapiines and the Evolution of Parental Care Systems in the African Cichlid Fishes. Mol Biol Evol. 2002, 19 (6): 865-883.
20. Kocher TD: Adaptative evolution and explosive speciation: the cichlid fish model. Nat Rev Genet. 2004, 5 (4): 288-298. 10.1038/nrg1316.

21. Beadle LG: The inland waters of tropical Africa. An introduction to tropical limnology. 1981, New York, USA.: Langman Inc, 2

22. Grove AT: The physical evolution of the river basins. The Niger and its neighbours Environmental history and hydrobiology, human use and health hazard of the major West African rivers. Edited by: Grove AT. 1985, Netherlands: A.A. Balkena Publishers, 21-60.

23. Roberts TR: Geographical distribution of African freshwater fishes. Zool J Linn Soc. 1975, 57: 249-319. 10.1111/j.1096-3642.1975.tb01893.x.

24. Lévêque C: Diversity dynamics and conservation: the freshwater fish of tropical Africa. 1997, Cambridge, UK: Cambridge University Press

25. Rognon X, Andriamanga M, McAndrew B, Guyomard R: Allozyme variation in natural and cultured populations in two tilapia species: *Oreochromis niloticus and Tilapia zillii*. Heredity. 1996, 76: 640-650. 10.1038/hdy.1996.91.

26. Rognon X, Guyomard R: Mitochondrial DNA differentiation among East and West African Nile tilapia populations. J Fish Biol. 1997, 51 (1): 204-207. 10.1111/j.1095-8649.1997.tb02525.x.

27. Agnèse JF, Adépo-Gourène B, Abban EK, Fermon Y: Genetic differentiation among natural populations of the Nile tilapia*Oreochromis niloticus* (Teleostei, cichlidae). Heredity. 1997, 79: 88-96. 10.1038/hdy.1997.126.

28. Seyoum S, Kornfield I: Taxonomic notes on the *Oreochromis niloticus* subspecies-complex (Pisces, Cichlidae), with a description of a new subspecies. Can J Zool. 1992, 70 (11): 2161-2165. 10.1139/z92-291.

29. Pouyaud L, Desmarais E, Chenuil A, Agnèse JF, Bonhomme F: Kin cohesiveness and possible inbreeding in the mouthbrooding tilapia *Sarotherodon melanotheron* (Pisces, Cichlidae). Mol Ecol. 1999, 8: 803-812. 10.1046/j.1365-294X.1999.00632.x.

30. Paugy D, Zaiss R, Troubat JJ: Faunafri - World Wide Web electronic publication. 2008, [http://www.poissons-afrique.ird.fr/faunafri/]

31. Paugy D: The Ethiopian subregion fish fauna: an original patchwork with various origins. Hydrobiologia. 2010, 649 (1): 301-315. 10.1007/s10750-010-0273-0.

32. Pritchard JK, Stephens M, Donnelly P: Inference of population structure using multilocus genotype data. Genetics. 2000, 155 (2): 945-959.

33. Falush D, Stephens M, Pritchard JK: Inference of population structure using multilocus genotype data: Linked loci and correlated allele frequencies. Genetics. 2003, 164 (4): 1567-1587.

34. Evanno G, Regnaut S, Goudet J: Detecting the number of clusters of individuals using the software structure: a simulation study. Mol Ecol. 2005, 14 (8): 2611-2620. 10.1111/j.1365-294X.2005.02553.x.

35. Hardy OJ, Charbonnel N, Freville H, Heuertz M: Microsatellite allele sizes: A simple test to assess their significance on genetic differentiation. Genetics. 2003, 163 (4): 1467-1482.

36. Wang JL, Whitlock MC: Estimating effective population size and migration rates from genetic samples over space and time. Genetics. 2003, 163 (1): 429-446.

37. Hill WG: Estimation of Effective Population-Size from Data on Linkage Disequilibrium. Genet Res. 1981, 38 (3): 209-216. 10.1017/S0016672300020553.

38. Wang JL: A pseudo-likelihood method for estimating effective population size from temporally spaced samples. Genet Res. 2001, 78 (3): 243-257. 10.1017/S0016672301005286.

39. Fraser DJ, Hansen MM, Ostergaard S, Tessier N, Legault M, Bernatchez L: Comparative estimation of effective population sizes and temporal gene flow in two contrasting population systems. Mol Ecol. 2007, 16 (18): 3866-3889. 10.1111/j.1365-294X.2007.03453.x.

40. Drake NA, Blench RM, Armitage SJ, Bristow CS, White KH: Ancient watercourses and biogeography of the Sahara explain the peopling of the desert. Proc Natl Acad Sci USA. 2011, 108 (2): 458-462. 10.1073/pnas.1012231108.

41. Rognon X, Teugels GG, Guyomard R, Galbusera P, Andriamanga M, Volckaert F, Agnese JF: Morphometric and allozyme variation in the African catfishes *Clarias gariepinus* and *C. anguillaris*. J Fish Biol. 1998, 53 (1): 192-207.

42. Bezault E: Etude du système de déterminisme du sexe au sein de populations naturelles de tilapia du Nil, *Oreochromis niloticus* : Importance des composantes génétiques et environnementales. PhD thesis. 2005, Orsay, France: Université Paris XI

43. Bezault E, Clota F, Derivaz M, Chevassus B, Baroiller JF: Sex determination and temperature-induced sex differentiation in three natural populations of Nile tilapia (Oreochromis niloticus) adapted to extreme temperature conditions. Aquaculture. 2007, 272: S3-S16.

44. Grove AT, Goudie AS: Late quaternary lake levels in the rift valley of southern Ethiopia and elsewhere in tropical Africa. Nature. 1971, 234: 403-405. 10.1038/234403a0.

45. Greenwood PH: Fish fauna of the Nile. The Nile, biology of the ancient river. Edited by: Rzóska J. 1976, The Hague, Netherlands, 127-141.

46. Palstra FP, Ruzzante DE: Genetic estimates of contemporary effective population size: what can they tell us about the importance of genetic stochasticity for wild population persistence?. Mol Ecol. 2008, 17: 3428-3447. 10.1111/j.1365-294X.2008.03842.x.

47. Agnèse JF, Adépo-Gourène B, Nyingi D: Functional microsatellite and possible selective sweep in natural populations of the black-chinned tilapia Sarotherodon melanotheron (Teleostei, Cichlidae). Marine Genomics. 2009, 1 (3-4): 103-107.

48. Nyingi D, De Vos L, Aman R, Agnèse J-F: Genetic characterization of an unknown and endangered native population of the Nile tilapia *Oreochromis niloticus* (Linnaeus, 1758) (Cichlidae; Teleostei) in the Loboi Swamp (Kenya). Aquaculture. 2009, 297 (1-4): 57-63. 10.1016/j.aquaculture.2009.09.017.

49. Thunken T, Bakker TCM, Baldauf SA, Kullmann H: Active inbreeding in a cichlid fish and its adaptive significance. Curr Biol. 2007, 17 (3): 225-229. 10.1016/j.cub.2006.11.053.

50. Fessehaye Y, El-Bialy Z, Rezk MA, Crooijmans R, Bovenhuis H, Komen H: Mating systems and male reproductive success in Nile tilapia (*Oreochromis niloticus*) in breeding hapas: A microsatellite analysis. Aquaculture. 2006, 256 (1-4): 148-158. 10.1016/j.aquaculture.2006.02.024.

51. Danley PD, Markert JA, Arnegard ME, Kocher TD: Divergence with gene flow in the rock-dwelling cichlids of Lake Malawi. Evolution. 2000, 54 (5): 1725-1737.

52. Crispo E, Chapman LJ: Temporal Variation in Population Genetic Structure of a Riverine African Cichlid Fish. J Hered. 2010, 101 (1): 97-106. 10.1093/jhered/esp078.

53. Lee WJ, Kocher TD: Microsatellites DNA markers for genetics mapping in *Oreochromis niloticus*. J Fish Biol. 1996, 49 (1): 169-171.

54. Estoup A, Martin O: Marqueurs microsatellites: isolement à l›aide de sondes non-radioactives, caractérisation et mise au point. 1996, [http://www.agroparistech.fr/svs/genere/microsat/microsat.htm]

55. Schuelke M: An economic method for the fluorescent labeling of PCR fragments. Nature Biotechnology. 2000, 18 (2): 233-234. 10.1038/72708.

56. Belkhir K, Borsa P, Chikhi L, Raufaste N, Bonhomme F: GENETIX 4.02, logiciel sous Windows pour la génétique des populations. 2001, Montpellier, France: Laboratoire Génome, Populations, Interactions, CNRS UMR 5000, Université de Montpellier II

57. Goudet J: FSTAT, a program to estimate and test gene diversities and fixation indices (version 2.9.3). Lausanne, Switzerland: University of Lausanne. 2001

58. Kalinowski ST: HP-RARE 1.0: a computer program for performing rarefaction on measures of allelic richness. Mol Ecol Notes. 2005, 5 (1): 187-189. 10.1111/j.1471-8286.2004.00845.x.

59. Rice WR: Anylizing tables of statistical tests. Evolution. 1989, 43 (1): 223-225. 10.2307/2409177.

60. van Oosterhout C, Hutchinson WF, Wills DPM, Shipley P: MICRO-CHECKER: software for identifying and correcting genotyping errors in microsatellite data. Mol Ecol Notes. 2004, 4 (3): 535-538. 10.1111/j.1471-8286.2004.00684.x.

61. Excoffier L, Smouse PE, Quattro JM: Analysis of molecular variance inferred from metric distances among DNA haplotypes - application to human mitochondrial-DNA restriction data. Genetics. 1992, 131 (2): 479-491.

62. Schneider S, Roessli D, Excoffier L: Arlequin Ver 2.000: A software for population genetics data analysis. 2000, Geneva, Switzerland

63. Queller DC, Goodnight KF: Estimating relatedness using genetic markers. Evolution. 1989, 43: 258-275. 10.2307/2409206.

64. Belkhir K, Castric V, Bonhomme F: IDENTIX, a software to test for relatedness in a population using permutation methods. Mol Ecol Notes. 2002, 2 (4): 611-614. 10.1046/j.1471-8286.2002.00273.x.

65. Castric V, Bernatchez L, Belkhir K, Bonhomme F: Heterozygote deficiencies in small lacustrine populations of brook charr*Salvelinus Fontinalis* Mitchill (Pisces, Salmonidae): a test of alternative hypotheses. Heredity. 2002, 89: 27-35.

66. Behrmann-Godel J, Gerlach G, Eckmann R: Kin and population recognition in sympatric Lake Constance perch (*Perca fluviatilis* L.): can assortative shoaling drive population divergence?. Behav Ecol Sociobiol. 2006, 59 (4): 461-468. 10.1007/s00265-005-0070-3.

67. R Development Core Team: R: A language and environment for statistical computing, reference index version 2.2.1. 2005, Vienna, Austria: R Foundation for Statistical Computing

68. Chessel D, Dufour AB, Thioulouse J: The ade4 package - I: One-table methods. R News. 2004, 4: 5-10.

69. Balloux F, Lugon-Moulin N: The estimation of population differentiation with microsatellite markers. Mol Ecol. 2002, 11 (2): 155-165. 10.1046/j.0962-1083.2001.01436.x.

70. Weir BS, Cockerham CC: Estimating F-statistics for the analysis of population structure. Evolution. 1984, 38: 1358-1370. 10.2307/2408641.

71. Slatkin M: A measure of population subdivision based on microsatellite allele frequencies. Genetics. 1995, 139 (1): 457-462.

72. Hardy OJ, Vekemans X: SPAGeDi: a versatile computer program to analyse spatial genetic structure at the individual or population levels. Mol Ecol Notes. 2002, 2 (4): 618-620. 10.1046/j.1471-8286.2002.00305.x.

73. Wang JL: Estimation of effective population sizes from data on genetic markers. Philosophical Transactions of the Royal Society B-Biological Sciences. 2005, 360 (1459): 1395-1409. 10.1098/rstb.2005.1682.

74. Peel D, Ovenden JR, Peel SL: Neestimator: Software for Estimating Effective Population Size (version 1.3). 2004, [http://www.dpi.qld.gov.au/cps/rde/dpi/hs.xsl/28_6908_ENA_HTML.htm]

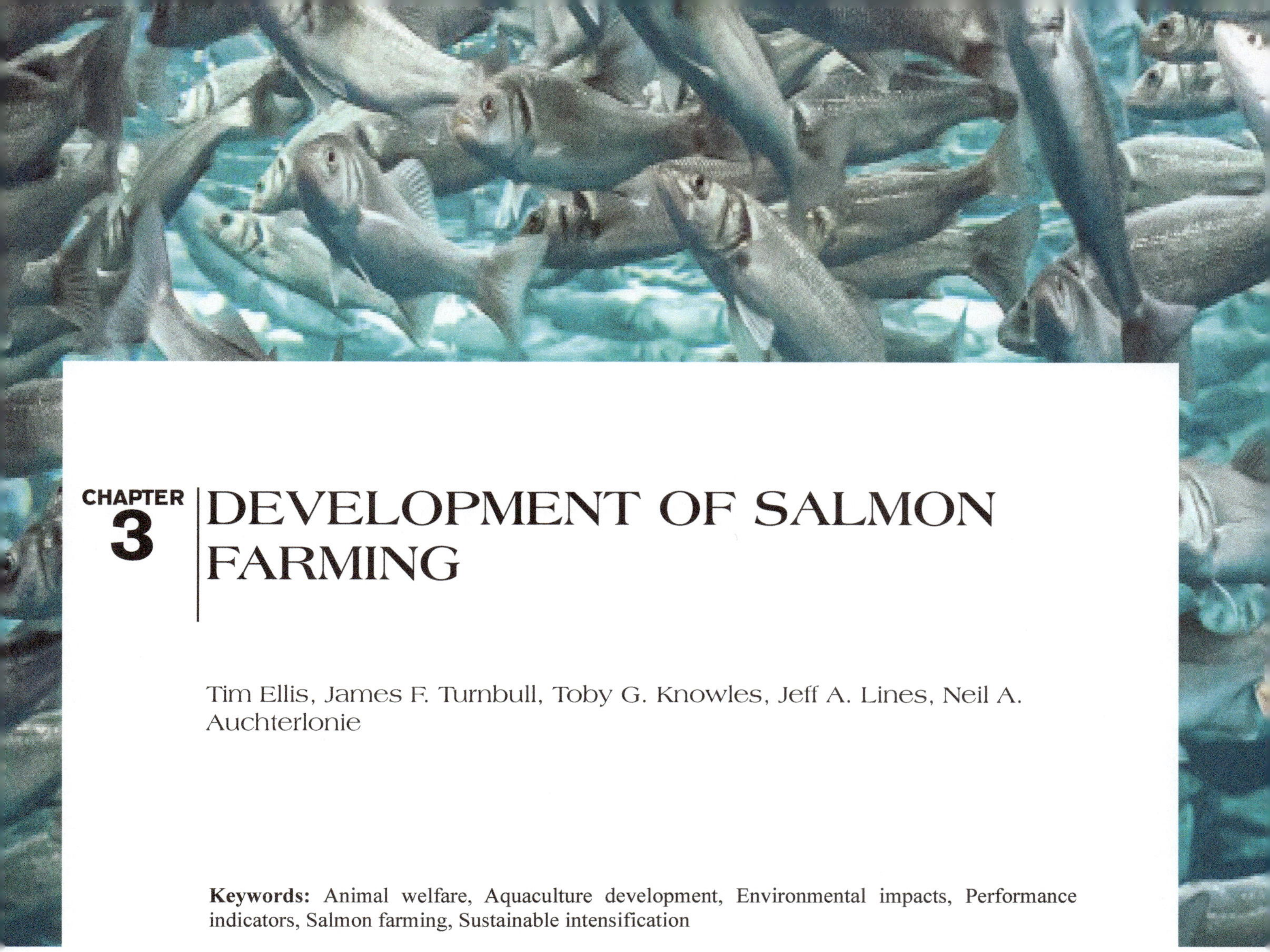

CHAPTER 3 DEVELOPMENT OF SALMON FARMING

Tim Ellis, James F. Turnbull, Toby G. Knowles, Jeff A. Lines, Neil A. Auchterlonie

Keywords: Animal welfare, Aquaculture development, Environmental impacts, Performance indicators, Salmon farming, Sustainable intensification

INTRODUCTION

Humans have traditionally relied upon capture fisheries for supplies of fish and shellfish. However, while production from fisheries has remained static over the last three decades, production from aquaculture has increased 12-fold (8.6% $year^{-1}$), providing 42% of seafood in 2012 (FAO, 2014). Continued expansion of aquaculture is viewed as a key strategy to ensure global food and nutrition security (Godfray et al., 2010 and Beveridge et al., 2013) and close the "fish-gap", i.e. the disparity between seafood supply and demand (FAO, 2007). It is noteworthy that the production growth of fed farmed species (i.e. reliant on external sources of feed) has outstripped that of non-fed species (i.e. feeding on in situ food sources) (Shepherd, 2012).

The global production of farmed Atlantic salmon (*Salmo salar*), a fed aquaculture species, has followed a similar trajectory to that of global aquaculture and is the fastest growing food production system in the world (Shepherd and Little, 2014). Reported production has increased from just 1 t in 1964 in a single country (Norway) to > 2 million t in 2012, across 11 different countries (Fig. 1). The apparent success of Atlantic salmon aquaculture (hereafter termed salmon farming) has been attributed to i) its ease of culture, ii) development in areas encouraged by governments, iii) development at a time coinciding with the rise of supermarkets and consumer interest in healthy eating, and iv) product attributes, i.e. a high fillet yield and a product that can be sold in diverse forms, e.g. fresh, sushi, cured, ready-meals, frozen (Forster, 2010, Ashe and Bjørndal, 2011 and Seafish, 2012). Farming has changed salmon from a luxury product to a global commodity that is an affordable staple seafood product for consumers in the industrialised world (Pelletier and Tyedmers, 2007 and Forster, 2010).

Citation: Tim Ellis, James F. Turnbull, Toby G. Knowles, Jeff A. Lines, Neil A. Auchterlonie, Development Of Salmon Farming http://dx.doi.org/10.1016/j.aquaculture.2016.02.012

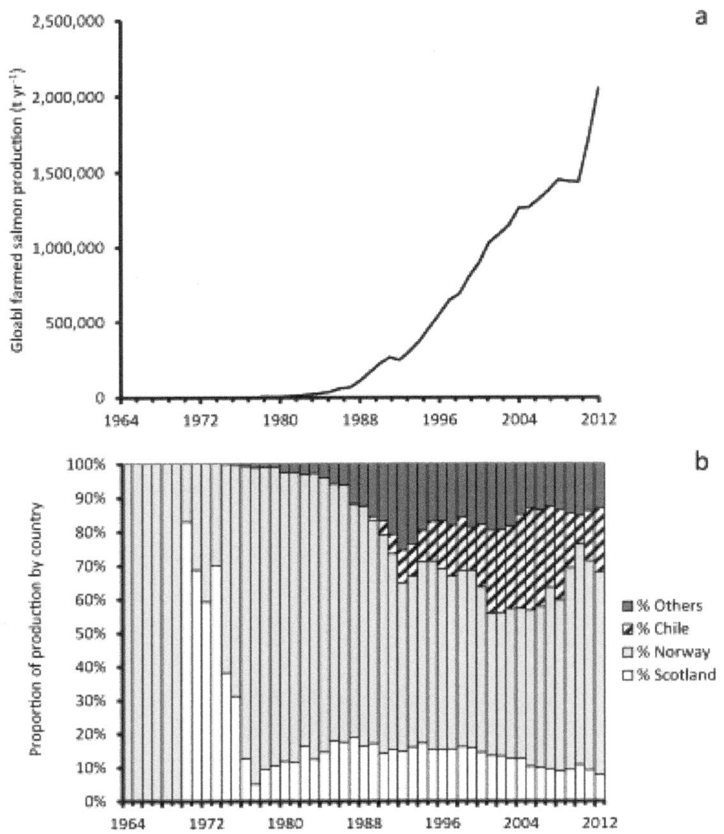

Figure. 1. Time series of global farmed Atlantic salmon production (data for 1950–2012 from http://data.fao.org(accessed 19/01/15) with nil production reported 1950–1963). a: harvest (t year^{-1}). b: Proportion contribution from Scotland, Norway, Chile and other countries (i.e. Canada, Faroe Islands, Australia, United States of America, Ireland, Russian Federation, France, Spain, Iceland, Sweden, Turkey and Denmark).

Salmon farming has spread from Norway across the natural species range of the northern Atlantic, and outside into both the northern and southern Pacific (west Canada, Chile, Australia). In 2012, the main producing nations were Norway, Chile and the United Kingdom (60%, 19%, and 8% respectively) (Fig. 1). Salmon farming is continuing to expand, with increases in global tonnage of 20% in both 2011 and 2012. In 2013, the production of farmed Atlantic salmon was 1500 times greater than the reported fishery catch (NASCO, 2014). With the recognised need to expand global aquaculture, salmon farming provides an ideal case study, to identify and discuss trends and issues pertinent to aquaculture development.

The Issues

In a prominent paper discussing strategies to meet the challenge of increasing demand for food caused by human population growth and higher consumption rates, Godfray et al. (2010) highlighted five key strategies. In addition to *expanding aquaculture*, the other strategies refer to increasing yields from food production chains by *adopting productive farming methods*, *selecting genotypes* best suited to farm conditions, *reducing waste*, and *changing human diets* to consumption of products from trophic levels and farming systems that maximise conversion efficiency. Godfray et al. (2010) also emphasised that implementing strategies of increasing agricultural (and aquacultural) food production should not be de-coupled from environmental considerations. The term "Sustainable Intensification" (SI) has been introduced to portray increasing the efficiency of agricultural food production through increases in yield relative to resource inputs (e.g. space, water, feed, energy) and outputs (e.g. greenhouse gas and eutrophication emissions, effects on biodiversity). SI therefore recognises that no food production chain is environmentally benign, and the focus should be on increasing production in conjunction with reducing pressures on the environment. Garnett et al. (2013) argue that SI also needs to safeguard animal welfare, the nutritional value of products, and rural economies.

Scottish Salmon Farming as a Case Study

Assessing the past history of aquaculture and trends will help to understand the present and plan for the future (sensu Hawkins et al., 2013). The Scottish salmon farming industry provides a suitable case-study of aquaculture development for a number of reasons:

- Extensive time series of annual statistics (up to 45 years) are available for the entire industry covering production, farm sites and systems, fish performance, and some socio-economic and environmental indicators. The statistics have been published within annual reports on the commercial Scottish finfish farming industry since 1979 by the competent authority (i.e. the body responsible for official control) for fish health in Scotland (www.scotland.gov.uk/Topics/marine/Publications/stats/FishFarmProductionSurveys; accessed 5 March 2015). [The Norwegian Directorate of Fisheries also publishes time series of statistics (from 8 to 20 years) for the Norwegian aquaculture industry (http://www.fiskeridir.no/english/statistics/norwegian-aquaculture/aquaculture-statistics/atlantic-salmon-and-rainbow-trout accessed 5 March 2015) with some earlier data available in the literature (Tilseth et al., 1991)].

- Scotland is a major producer of salmon. FAO reported production tonnages (Fig. 1) show that the UK (with production almost entirely in Scotland) was the leading salmon producing country in the early 1970s, then was second behind Norway until 1999, since when Chilean production has exceeded that in Scotland (apart from in 2010).
- The Scottish Government has aspirations for salmon harvest to continue to increase (by 28% between 2012 and 2020) and expansion of the industry is valued due to the income generated and employment in rural areas (Marine Scotland, 2014).
- Farmed salmon is the most popular fresh fish with UK consumers (Seafish, 2011), reflecting price accessibility (Naylor and Burke, 2005) and consumer preferences. Farms provide a reliable and consistent source of supply suited to processing and retailing (Naylor and Burke, 2005, Ashe and Bjørndal, 2011 and Shepherd and Little, 2014).
- Intensive aquaculture, and salmon farming in particular, have attracted criticisms and a preponderance of bad press (Naylor and Burke, 2005, Amberg and Hall, 2008 and Shepherd and Little, 2014). The various pressures that salmon farming methods place on the environment (through discharge of nutrients, organic particulates and chemicals; pathogen and escapee interactions with wild stocks; culling of predators; use of industrial fish in feed) and on fish welfare have been highlighted for Scotland (e.g. FAWC, 1996 and Scottish Executive, 2002).

Freshwater salmon culture started in the UK (and elsewhere) in the 19th century producing juvenile salmon for stocking rivers (www.fao.org/fishery/culturedspecies/Salmo_salar/en accessed 5 March 2015; Forster, 2010). Commercial salmon farming, i.e. on-growing in seawater in captivity until harvest, started in Norway and Scotland in the late 1960s with the introduction of floating net-pens (sea-cages) (Munro et al., 1980, Tilseth et al., 1991 and Forster, 2010). In 2014, there were 356 Scottish farms which employed 1,634 staff and produced 179,022 t wet weight at harvest (Munro et al., 2014) with a first sale value around £0.7 billion (http://news.scotland.gov.uk/News/Record-year-for-salmon-production-1cb9.aspx#downloads accessed 3 Nov. 2015). The value of the farmed salmon harvest therefore now exceeds that of the capture fishery landings in Scotland (Marine Scotland, 2014).

THE SCOTTISH FISH FARM PRODUCTION ANNUAL SURVEY REPORTS 1979–2014 AS A DATA SOURCE

The annual reports on the Scottish finfish farming industry provide a consistent, authoritative, time series of official data that document the development of the Scottish salmon farming industry and represent the source of information for the body of this review. The importance of annual reporting was recognised at the inception to a) give confidence in the statistics and b) provide insight into trends, which less frequent snapshot views would not (Munro et al., 1980 and Munro and Wadell, 1981). The reports provide one of the longest and most comprehensive time series of data available for any aquaculture sector globally, and the competent authority and Scottish industry must be commended for maintaining this output over 35 years.

The reports provide data on the Scottish salmon industry gathered largely by annual questionnaire surveys of aquaculture companies known by, or more recently compulsorily registered with, the competent authority for fish health. All companies actively engaged in freshwater and seawater salmon farming supply the requested information (e.g. Munro and Wallace, 2013); companies not returning the information are subject to additional requests (Munro and Wadell, 1981). The reports are therefore based on self-reporting by the industry, supplemented with additional information held by the competent authority (e.g. from health certificates of imports and exports; reports of escape events), and explanatory comments based upon the authors' knowledge of the industry. The data coverage is therefore of the entire industry operating in Scotland, rather than a sample (or extrapolation from a sample).

The questionnaire distributed to the industry has changed over time (e.g. DAFS, 1986), as has the extent, format and presentation of the data within the annual reports. The series of reports provide time series of variable length (up to 46 years) for various metrics of the industry. The reports frequently state that data published in previous years have been reassessed and updated where necessary (e.g. Munro and Gauld, 1996, Hastings and Smith, 2005 and Munro et al., 2014). As data are necessarily collected retrospectively, changes in company ownership may introduce uncertainty for particular years (Munro et al., 2014).

We selected metrics for various aspects of the industry from the most recent report, and time series were extended by successive extraction from earlier reports. Where occasional discrepancies were noted between reports, values from the most recent reports were assumed to represent corrected values. The annual reports tabulate data, with the number of years varying for each metric (e.g. range 1 to 22 (median 11) years in Munro and Wallace, 2015). In this review, we illustrate entire time series graphically to facilitate visualisation of long-term trends, in addition to inter-annual fluctuations.

The annual reports separate salmon farming into freshwater (ova to smolt) and seawater (smolt to harvest) production. Here we report metrics together to illustrate the parallel development of the two sub-sectors, and group indicators relating to 1) production, 2) sites and systems, 3) biological performance, 4) socio-economics, and 5) environmental pressures. Supplementary comments from the reports are cited to aid interpretation and discussion. Since 1985 (DAFS, 1986), some metrics have been sub-divided for different geographical regions within Scotland; regional information is not discussed within this review.

THE PRODUCTION CYCLE OF FARMED SALMON IN SCOTLAND

Wild Atlantic salmon are anadromous: the early life stages inhabit freshwater, the main growth phase occurs in seawater and the adults return to freshwater to reproduce. Salmon farming therefore occurs in both freshwater (hatchery and nursery) and seawater (on-growing to harvest). The production cycle of farmed salmon in Scotland can divided into successive stages:

Broodstock

Potential broodstock are selected in spring–summer (mostly 2nd sea winter fish, with some 3rd sea-winter or older fish) and held until autumn–winter for stripping (Munro and Gauld, 1996). Although some broodstock may be stripped at sea sites, it is common practice to move broodstock to freshwater sites for acclimation some weeks prior to stripping (Munro and Gauld, 1996). Ova production is related to age/size: 2nd sea-winter females of 8 kg produce around 12,000 ova; 3rd sea-winter females of 12 kg produce around 16,000 ova. Ova size is variable, with around 5000 ova L^{-1} (Munro and Gauld, 1995 and Munro and Gauld, 1996). Although the stripping season (winter) extends over two calendar years (October through to January), ova production is reported for the year in which it starts (Munro and Gauld, 1996).

Freshwater Hatchery Stage (ova, alevin and parr)

The stripped ova are fertilised and "laid down", hatch as alevin 12 weeks later and grow into parr (Munro and Gauld, 1996).

Freshwater Nursery Stage (parr to smolt)

The parr grow and undergo smoltification (physiological, morphological and behavioural changes that enable survival in seawater) in the Spring, cued by seasonal patterns of temperature and light. Through photoperiod control, smoltification can be advanced outside the natural spring timing, and growth can be manipulated by controlling water temperature and varying feeding regimes (Munro and Gauld, 1996). Historically, natural smolt were put to sea in Spring (April–June) after 1 or 2 years in freshwater, but photoperiod manipulation now enables smolt to be put to sea throughout the year (Munro and Gauld, 1995). Natural smolt are termed S1 and S2, and "out-of-season" smolt are described as S½ or S1½, depending upon age at smoltification (Munro and Gauld, 1997 and Munro and Wallace, 2013):

S½: < 12 months old, i.e. transferred to sea in calendar year of hatch. NB: The synonymous term "S0" is now used by the industry, but the term S½ is retained here for consistency with the annual reports;

S1: 12–18 months old, i.e. transferred to sea in January–June in year post hatch;

S1½: 19–24 months old, i.e. transferred to sea in July–December in year post hatch;

S2: > 24 months old when transferred to sea.

S½ are produced from the largest size grades and/or early spawned ova and are available for transfer to seawater as early as 6 months after first feeding (Munro and Gauld, 1996). S1½ and S2 tend to be "left-overs" or slow growers from previous batches (Munro and Gauld, 1997).

Seawater Stage (smolt to harvest)

Smolt are "put to sea" to be on-grown in seawater. The normal seawater production period is 18 months to 2 years (Stagg and Allan, 1999), with fish being harvested at various ages and times of year, depending upon growth. Stagg and Gauld (1998) noted a market demand for fish in the 3–4 kg range, and that larger fish tended to go for smoking (rather than as fresh fish). The annual reports provide details of harvest of fish at different ages/timings under terminology that has been dropped by industry (Munro and Gauld, 1995 and Munro et al., 2014), i.e.:

- input year fish: harvested within 1st calendar year of seawater transfer;
- Year 1 grilse: harvested January–August in the 2nd calendar year in seawater. NB: The term grilse as used in the reports (and here) reflects time of harvest, and therefore differs to use with wild salmon where it refers to maturation and return to freshwater after one winter at sea;
- Year 1 pre-salmon: harvested September–December in the 2nd calendar year in seawater;
- Year 2 "salmon": after 2 calendar years in seawater.

The annual reports do not suggest that harvest fish are held for > 2 years in seawater.

There is therefore a variety of age-classes in production in freshwater and seawater at any one time. In 2014, the Scottish salmon industry harvested 34.3 million fish (from 4 age groups), with inputs of 48 million smolt, 70.8 million ova and 0.003 million female broodstock salmon (Table 1). Having several year-classes in production at one time together with variable stage durations does compromise collation of certain metrics.

Table 1. Data on inputs and outputs of salmon for the Scottish salmon farming industry in 2014 — numbers of individuals, harvest volumes (t) and mean weights (kg) (data from Munro and Wallace, 2015).

Production stage	Life history stage/ origin	Number of individuals		Harvest (tonnes)		Mean weight at harvest (kg)
			Total		Total	
Broodstock in Scotland	Female fish stripped		2,711			
Ova production	Ova production — Scotland	33,450,000	92,312,000			
	Ova imported (outside GB)	58,862,000				

Ova laid down to hatch in Scotland	From Scottish production	14,418,000	70,827,000			
	Other GB production	2,725,000				
	Foreign ova	53,684,000				
Smolts produced in Scotland	S0.5	22,367,000	45,004,000			
	S1	22,473,000				
	S1.5	164,000				
Smolts put to sea in Scotland	Scottish	45,080,000	48,045,000			
	Smolts — English	893,000				
	Smolts — other	2,072,000				
Seawater harvest	Harvest in year 0: input year fish	286,000	34,314,000	720	179,022	2.5
	Harvest in year 1: grilse	9,048,000		46,686		5.2
	Harvest in year 1: pre-salmon	11,268,000		55,311		4.9
	Harvest in year 2: salmon	13,712,000		76,305		5.6

PRODUCTION METRICS

Ova

Data are available on numbers (and sources) of ova produced, exported, imported, and laid down to hatch in Scotland.

Data on ova production in Scotland (i.e. stripped from Scottish farmed broodstock) have been published since 1990. Ova production was at its peak of 224.4 M in 1990 and has since decreased by 85% to 33.5 M in 2014 ($r_s = -0.655$, $n = 25$, $p < 0.001$; Fig. 2a). Ova produced in Scotland can be marketed for human consumption (as salmon caviar; Munro and Gauld, 1996), exported to support salmon farming in other countries, or be laid down to hatch. There are no records of Scottish ova being sold for human consumption (Munro and Gauld, 1997 and Stagg and Gauld, 1998). The Scottish industry has traditionally exported ova, although numbers have decreased over time ($r_s = -0.627$, $n = 21$, $p < 0.005$; Fig. 2a): the salmon farming industry in Chile appears to have been the main recipient of exported ova in the mid to late 1990s (Munro and Gauld, 1997 and Stagg and Gauld, 1998). There was then an apparent hiatus, with export to Chile resuming in 2004 (Smith et al., 2005), but the trade ceased in 2010 (Walker and McAlister, 2011). Smaller numbers of ova from wild salmon have been exported to support restocking of European river systems (Munro and Gauld, 1997).

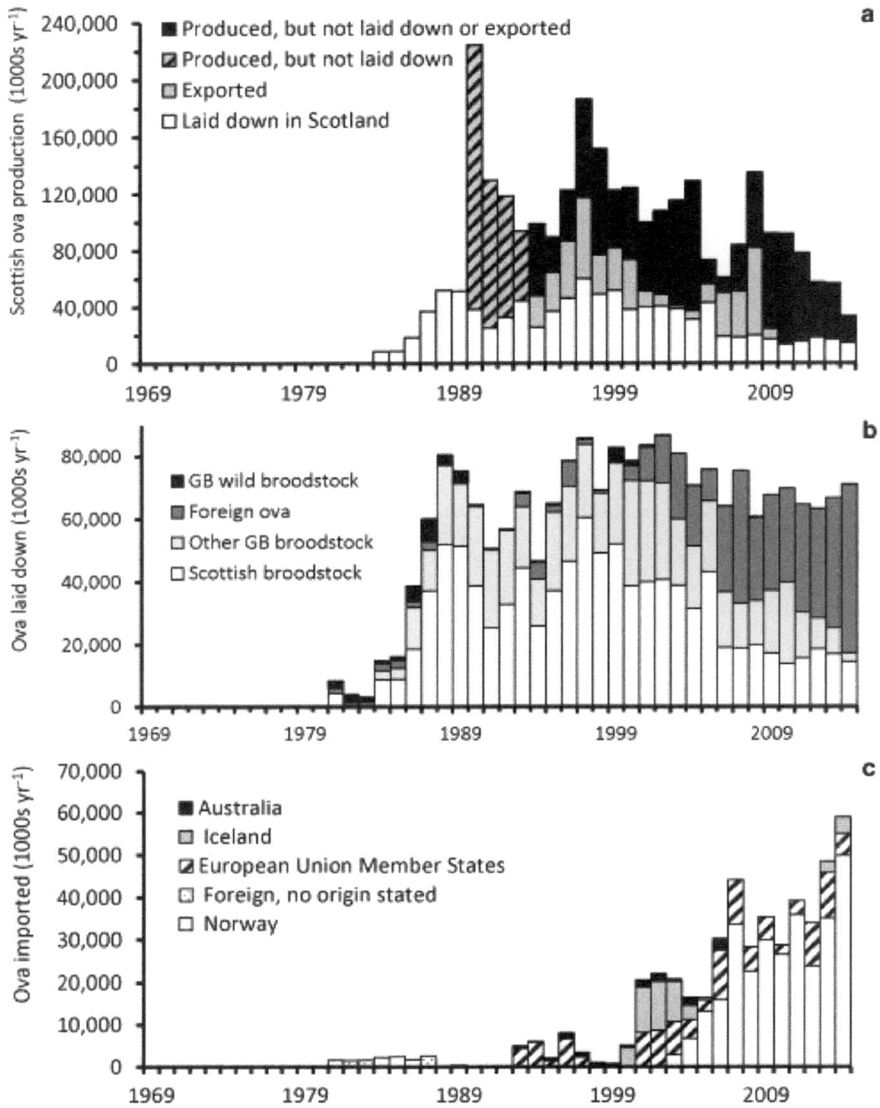

Figure. 2. Data on salmon ova in Scotland. a: Numbers of ova produced in Scotland, subdivided into laid down in Scotland, exported and not laid or exported. Data available 1984/1994/1995–2014. b: Origins of ova laid down to hatch within Scottish salmon industry. Data available 1981–2014. c: Origins of foreign (imported) ova. Data available 1981–2014.

Typically only 31% of ova produced in Scotland are laid down to hatch in Scotland (Fig. 2a). Substantial proportions of stripped ova (mean 49%; 50% in 2014) are unaccounted for (i.e. not reported as laid down or exported) and this proportion has increased over time ($r_s = 0.506$, $n = 21$, $p < 0.05$). Unaccounted ova are likely to represent disposals due to inferior quality, disease status or as surplus (SOAFD, 1991, SOAFD, 1992, Munro and Gauld, 1996 and Stagg and Gauld, 1998).

The total number of ova laid down to hatch in Scotland (i.e. originating from Scottish farmed broodstock and other sources) increased to a peak of 86.7 million in 2002, and has since declined (Fig. 2b). DAFS (1990) noted that the industry took a collective management decision in 1988 to limit production growth by limiting the number of ova laid down. The sources of ova laid down are: in-house broodstock (i.e. Scottish farmed broodstock); out-sourced GB broodstock (farmed broodstock held in England or Wales); GB wild broodstock; foreign ova (including from Northern Ireland, Munro et al., 2014). Key temporal trends for ova laid down are:

- the contribution from wild broodstock has decreased, being replaced by ova from farmed broodstocks;
- the contribution of imported ova, i.e. from foreign farmed broodstocks, has changed over time.

The industry was still dependent upon wild Scottish broodstock for ova in the early 1980s.Munro and Wadell (1981) noted a shortage of ova due to erratic supplies of wild eggs which was exacerbated by loss of a major source where infectious pancreatic necrosis (IPN) virus had been found. They stated that the industry needed to switch from wild eggs to farmed broodstocks, which did occur (DAFS, 1988). Although the salmon farming industry has continued to report wild ova laid down since the 1990s, these statistics are misleading as this has been on behalf of wild fisheries for stock enhancement schemes (Stagg and Allan, 2000, Stagg and Allan, 2001, Walker, 2009, Munro et al., 2014 and Munro and Wallace, 2015).

Imported (mainly Norwegian) eggs were used in the early and mid-1980s (DAFS, 1985 and DAFS, 1987), but their use decreased over time up to the late 1990s (Munro and Gauld, 1996 and Stagg and Allan, 2000); domestic ova from Scottish and GB farmed broodstocks then supplied the bulk of ova, and were considered satisfactory and sufficient to supply the Scottish industry (Munro and Gauld, 1994 and Munro and Gauld, 1996). Munro and Gauld (1997) refer to import of ova from the southern hemisphere (Australia) to support production of out of season smolt, although this was typically minor and ceased in 2006 (Fig. 2c). However, since 2000 there has been a marked increase in the use of foreign ova (Stagg and Allan, 2002, Stagg and Smith, 2003 and Hastings and Smith, 2005). In 2014, the majority (76%) of ova laid down were imported (Fig. 2b), from Norway, Northern Ireland and Iceland (85%, 9% and 6% of imports respectively; Fig. 2c). The current dominance of ova imports (over domestic production) is thought to reflect salmon farming companies centralising broodstock and selective breeding operations elsewhere. Temporal changes in ova imports also reflect the introduction of legislation for disease control over this period: Fish Health Regulations, introduced in 1993 to EU Member States, established conditions for trade in live ova and changes in 2003 enabled import of salmon ova from Norway (Walker et al., 2012 and Munro et al., 2014).

Smolt

Data are available on numbers (and origin and age) of smolt produced in Scotland, imported and exported (but including fry), and put to sea in Scotland. The number of smolt produced in Scottish freshwater farms increased from 1979 to peak at 47.5 million in 2001 (Fig. 3a). The data on Scottish origin smolt put to sea (supplied by seawater farms) are in close agreement with those on smolt production (supplied by Scottish freshwater farms) although the latter are typically 3% higher (Fig. 3a; Munro and Gauld, 1995). Smolt are exported, but the available data (combined with export of fry) indicate that this does not account for the difference. The difference has been noted twice in the annual reports:

- Munro and Gauld (1997) attributed it to variation in the accuracy of the counters and counting methods. However, a consistent difference and valence seems unlikely to be due to counting methods (Aunsmo et al., 2013), although it may be associated with accounting practices (e.g. consistent rounding up v rounding down).
- Hastings and Smith (2005) briefly explained it as smolt not being put to sea in Scotland.

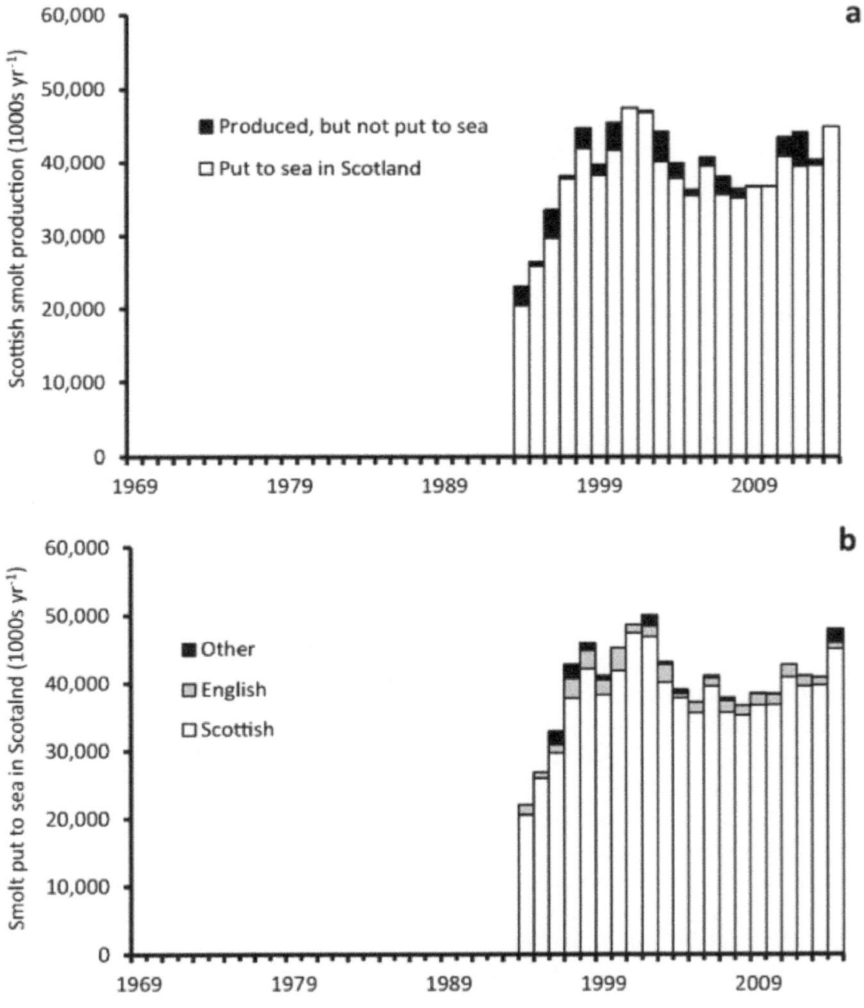

Figure. 3. a: Numbers of smolt produced in Scottish freshwater farms and reported as put to sea in Scotland. b: Origin and numbers of smolt put to sea in Scotland. Data available 1994–2014.

The difference is therefore likely to be explained by either smolt that die (during transport from freshwater to seawater sites, or soon after transfer) not being counted, or surplus/poor quality smolt being culled.

The total number of smolt put to sea increased up to 2002, peaking at 50 million and has since decreased, plateaued and risen again to 48 million in 2014 (Fig. 3b). The bulk of smolt put to sea are produced in Scotland (median 95%), with the balance derived from England and EU member states (Fig. 3b). EU imports of smolt from Ireland first occurred in 1996 (Munro and Gauld, 1997). Nevertheless, Scotland can be considered to be largely self-sufficient with respect to smolt production.

Although smolt size is not reported (Munro and Gauld, 1994), Munro and Gauld (1996)noted that historically S1 smolt were 30–40 g, but size had increased and S1 smolt > 80 g were now commonly produced. [Bergheim et al. (2009) report a similar increase in the size of Norwegian farmed smolt, from 30–50 g in 1985 to 70–120 g in 2000]. This increase in size reflected use of early spawned ova, increased growth associated with temperature control and improved feeding, and early placement of parr into ambient freshwater systems maximising the period of greatest natural growth during spring–autumn (Munro and Gauld, 1996). Larger smolt are favoured by seawater farmers as it enables earlier harvest, or harvest of larger fish (Munro and Gauld, 1994). However, keeping fish for longer in freshwater incurs additional costs which may outweigh the advantages of size and robustness (Munro and Gauld, 1997). Freshwater farmers therefore now cull potential S2 smolt (Munro and Gauld, 1996).

- Time series data for age of smolt is available (Fig. 4), and two long-term trends are apparent:
- The introduction of out-of-season smolt, i.e. S½ and S1½;
- A reduction in age of smolt, i.e. a move from S2 to S1, and S1 to S½.

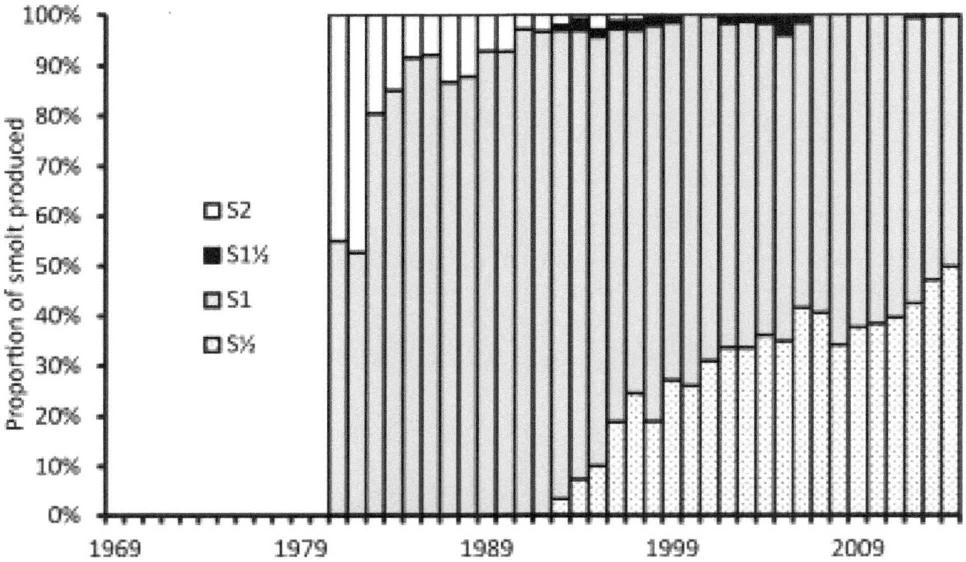

Figure. 4. Proportions of different aged smolt (S½, S1, S1½, S2) produced in Scotland. Data available 1981 to 2014.

In 1981, only S1 and S2 were produced, with the proportion of S2 varying (10%–83%) between farms (DAFS, 1982). The production of out of season S½ and S1½ reflects the introduction of photoperiod control. Out of season smolt were first produced in 1993 (Munro and Gauld, 1994), and there has been a trend towards increased production of S½ (Stagg and Gauld, 1998). Early out of season smolt are produced by extracting the top sizes of normal growing populations (Munro and Gauld, 1995). Production of S½, in conjunction with S1, enables smolt to be put to sea throughout the year and hence more flexible production scheduling (Munro and Gauld, 1995 and Stagg and Gauld, 1998). However, there have been some concerns about out of season smolt. Munro and Gauld, 1995 and Munro and Gauld, 1996 noted that S1 and S2 smolt were larger and more robust than S½ and S1½ smolt, with the latter experiencing more variable and poorer survival in seawater. Although production period and feed costs may be reduced in photoperiod adapted smolt, there are additional costs for lighting and heating (Munro and Gauld, 1995). Munro and Gauld (1997) noted that farmers tended to favour out of season smolt derived from north hemisphere ova, to those from southern hemisphere ova.

Seawater Harvest

The seawater production tonnages reported are the wet weight of fish at harvest (Munro et al., 2014). Over the period for which data are available (1970–2014; NB: 1970–1978 data from http://data.fao.org accessed 5 March 2015) the annual harvest of Scottish farmed salmon has shown a 734 fold increase (from 244 to 179,022 t) (Fig. 5). Harvest initially peaked in 2003 at 169,736 t, then fell, but has recently recovered to a new peak of 179,022 t in 2014. This long term trend has been subject to short term fluctuations, varying between + 70 and − 88% year^{-1}, being greatest during the early development of the industry. Some explanations for the fluctuations in annual growth are provided in the annual reports:

- During the early development of the industry, growth in seawater harvest was limited by a shortage of smolt (Munro et al., 1980 and Munro and Wadell, 1981);

- The reduced annual growth in the early 1990s has been attributed to management decisions within the industry due to loss of confidence in the market for farmed salmon and concern about inability to control disease (DAFS, 1990, SOAFD, 1991,SOAFD, 1992 and SOAFD, 1993);
- The drop in harvest in 1992 was attributed to mortalities from the bacterial disease furunculosis caused by *Aeromonas salmonicida* (Munro and Gauld, 1996). Improved disease control and the introduction of effective vaccines, enabled subsequent growth in harvest (Munro and Gauld, 1994, Munro and Gauld, 1995 and Munro and Gauld, 1996);
- An outbreak of infectious salmon anaemia (ISA) occurred for first time in 1997, which reduced harvest in the late 1990s (Stagg and Gauld, 1998).

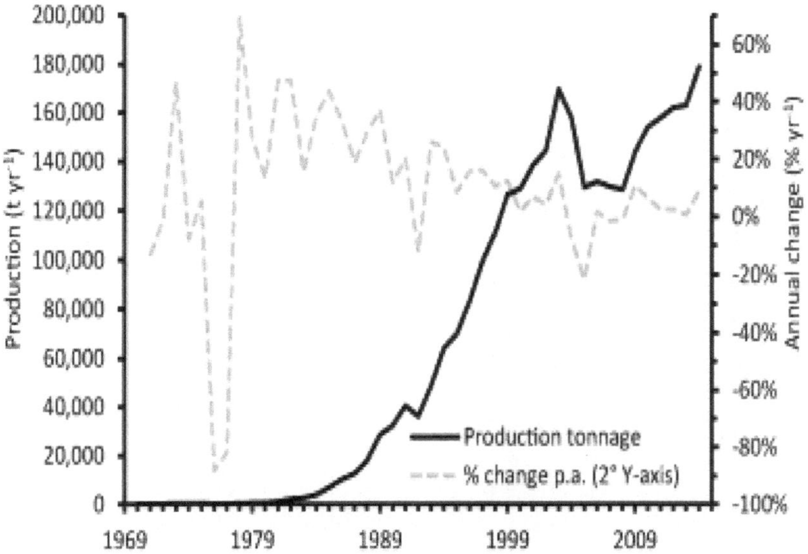

Figure. 5. Harvest (t year^{-1} wet weight whole fish) of Scottish farmed salmon and % change from previous year. Data available 1979–2014 from annual reports and earlier data (1970–1978) from http://data.fao.org (accessed 19/01/15).

The numbers of fish of different ages at harvest are also reported (Fig. 6). Although proportions have changed over time, there is no clear long-term trend. Age will be related to size at harvest, and therefore (partially) reflects market demand for different sized fish — either for whole fresh fish, fresh fillets, or for smoking (Munro and Gauld, 1996). In the early 1990s, marketable fish were harvested early due to a need for cash flow and to reduce losses due to disease (SOAFD, 1991). Input year fish were first harvested in 1993 (Munro and Gauld, 1994); their harvest allows farmers to meet specific market demand whilst simultaneously facilitating control of stocking density (Munro and Gauld, 1997). An increase in harvest of input year fish in 1998 was attributed to fish being put to sea earlier and faster growth (Stagg and Allan, 1999). Increased harvest of year 1 fish has also resulted from compulsory slaughter programmes for the control of infectious salmon anaemia (ISA) (Stagg and Allan, 1999).

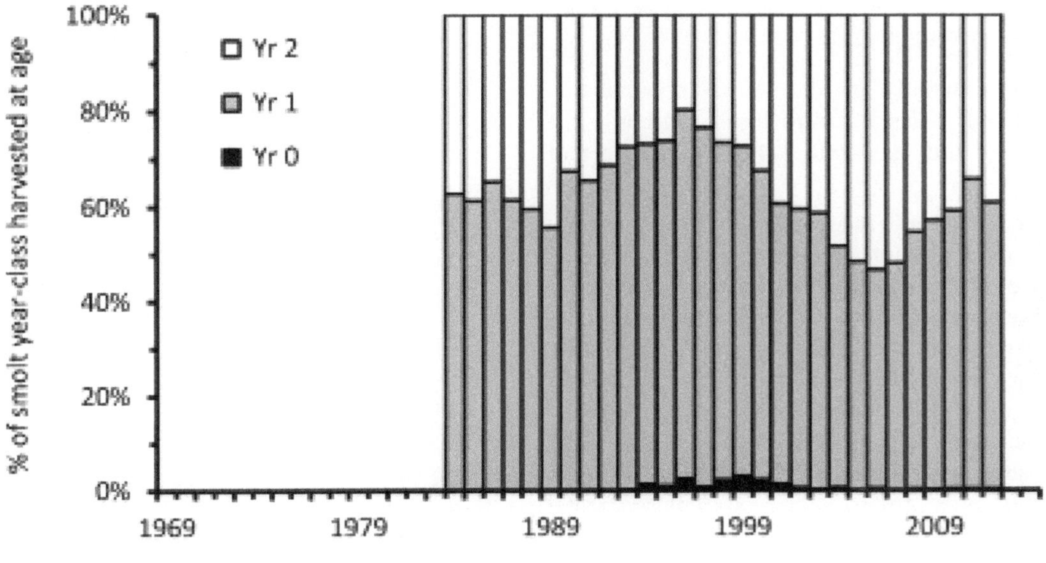

Figure. 6. Proportions of fish harvested from seawater either as year 0 (input year), year 1, or year 2 fish, relative to their year of transfer to seawater. Data available 1984–2012.

PRODUCTION SITE AND SYSTEM METRICS

Numbers and Types of Active Sites

There was a rapid expansion in numbers of active sites (farms) in both freshwater and seawater until the late 1980s–early 1990s; however, numbers have decreased since 1999 (Fig. 7a). Munro and Gauld (1997) noted a trend for companies to concentrate production at individual sites. A contraction in numbers of sites has been suggested to indicate potential for expansion (Munro and Gauld, 1995), although it is likely that vacated (inactive) sites proved unsuitable for successful rearing of salmon for various reasons (Munro and Gauld, 1997).

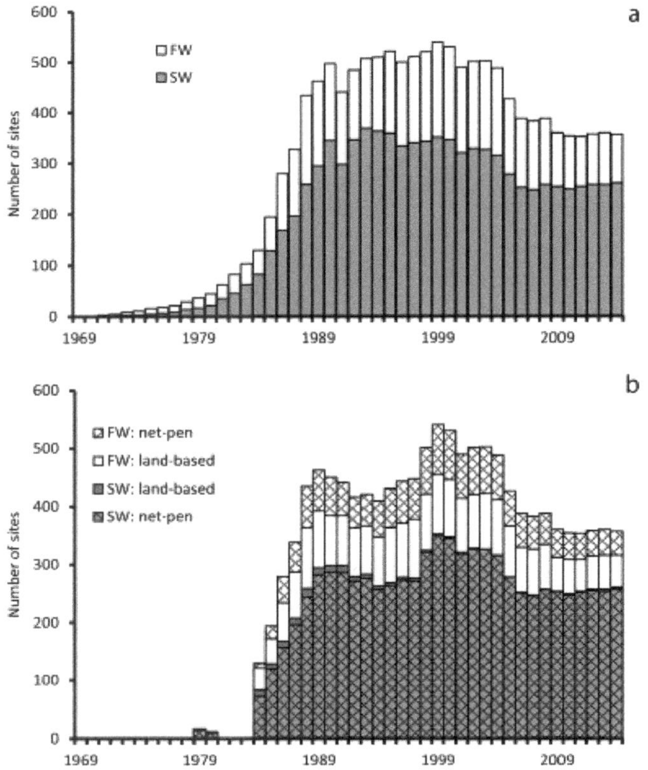

Figure. 7. a: Numbers of freshwater (FW) and seawater (SW) salmon farm sites in Scotland. Data available 1969–2014. b: Numbers of freshwater and seawater sites, subdivided into land-based and net-pen systems. Data available 1979/1984–2014.

Freshwater sites initially (in the 1970s) outnumbered seawater sites, but this balance has reversed. Currently the number of freshwater sites is around 37% of the number of seawater sites.

Sites are categorised as either land-based (tank/trough/raceway systems sited on land) or floating net-pens sited in freshwater lochs or the sea (Fig. 7b). Net-pen systems are considered less expensive to install and run, and are simpler to operate (Munro and Gauld, 1995 and Stagg and Gauld, 1998). Land-based systems are more capital intensive, may depend upon pumps and external energy sources, are more labour intensive, and have higher running costs. However, land-based systems have the advantages that direct observation of stocks is possible (facilitating prompt remedial action), environmental conditions can be controlled or modified (e.g. for production of out of season smolt), and husbandry operations (e.g. handling, grading) are easier (Munro and Gauld, 1995, Munro and Gauld, 1996 and Munro and Gauld, 1997). Stocking density in land-based systems is typically higher than in net-pens to offset higher production costs (Munro and Gauld, 1996).

Seawater farming is dominated by net-pens, with the proportion of land-based seawater sites declining over time (Fig. 7b). New pump-ashore seawater tank sites were developed in Scotland in 1980 (Munro and Wadell, 1981). However, such systems incur high energy costs, are less economically viable than net-pen systems and have largely been redeployed as broodstock sites, or for alternative species, e.g. halibut (*Hippoglossus hippoglossus*) (Munro and Gauld, 1995, Stagg and Gauld, 1998, Stagg and Allan, 1999 and Stagg and Allan, 2002).

In freshwater, hatchery sites require land-based ova incubators and troughs for alevins and fry, whereas nursery sites can use either land-based systems (tanks, raceways) or net-pens (Munro and Gauld, 1994, Munro and Gauld, 1995 and Munro and Gauld, 1996). Freshwater net-pen systems are commonly used in Scotland (Fig. 7b) which contrasts to Norway where their use is rare (Bergheim and Brinker, 2003). Bergheim et al. (2009)provide details on Scottish freshwater net-pen sizes and construction. Munro et al. (1980) noted that land-based freshwater sites were limited by abstraction volumes, and that expansion would require new sites or the introduction of technology. In the 1990s, there were three significant developments in the freshwater sector (Munro and Gauld, 1995, Munro and Gauld, 1996 and Munro and Gauld, 1997):

- Polytunnels were introduced so traditional outdoor (open air) tank systems were under cover, allowing more control of the rearing environment and improving staff working conditions;
- High-tech recirculation aquaculture systems (RAS) were introduced. The introduction of such water reuse systems was considered especially important in areas (such as Shetland) where freshwater supplies are scarce. RAS make efficient use of the available water supply, provide control of the rearing environment, and concentrate particulate wastes within the unit for bulk disposal, reducing discharge to the environment. RAS enabled sites to produce more smolt through increased stocking density and increased growth (enabling the transfer of earlier and/or larger parr to other on-growing net-pen or tank systems) and have been attributed with the increased tank production of smolt in the 1990s (Stagg and Gauld, 1998). Munro and Gauld (1997) noted that: care was needed to ensure efficient operation of biofilters and prevent introduction of pathogens (as treatments can affect biofilters); that increased production was needed to offset higher installation and running costs.
- There was increasing use of remote sensors and computerised control of operations.

Farm Size and Capacity

The average size of net-pen farm sites in both seawater and freshwater has increased over time (r_s = 0.994 and 0.947, n = 27, $p < 0.001$), while no such trends are apparent for land-based sites (Fig. 8a). Several net-pens are typically operated at a single farm site. In seawater, larger circular net-pens of plastic construction (e.g. 100 m circumference × 22 m depth) were introduced in the mid-1990s (Munro and Gauld, 1997 and Stagg and Gauld, 1998). The increases in average net-pen farm sizes over time have contributed to a continual increase in total net-pen capacity in both seawater and freshwater (r_s = 0.963 and r_s = 0.699 respectively, n = 27, p < 0.001; Fig. 8b), despite reductions in farm numbers since 1999.

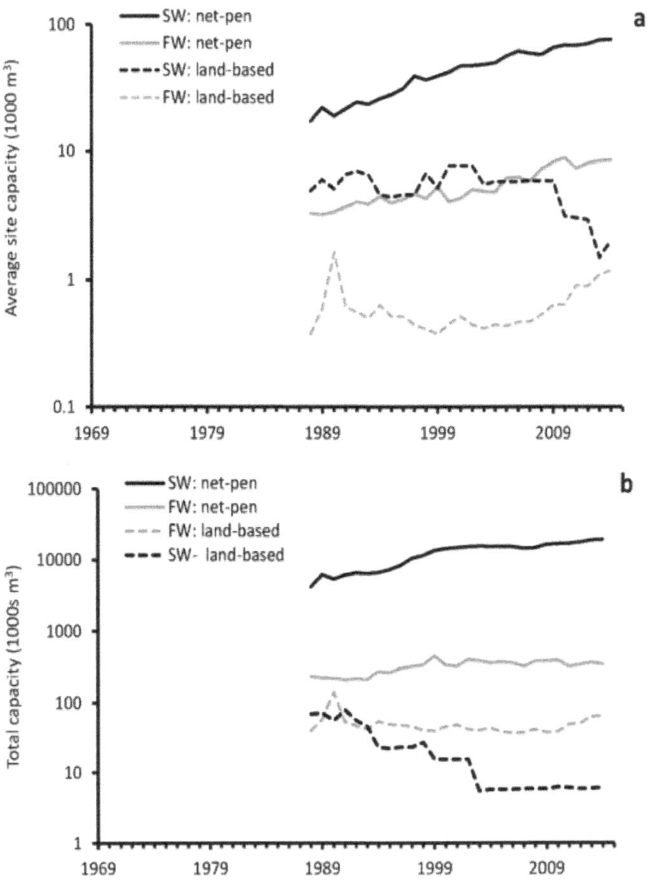

Figure. 8. Average size (a) and total capacity (b) of net-pen and land-based salmon farms in seawater (SW) and freshwater (FW) in Scotland. Please note log scales. Data available 1988–2014.

Against a backdrop of decreasing numbers of seawater and freshwater farms, the proportion of farms with higher production, in both seawater (tonnage) and freshwater (smolt output), has increased (Fig. 9). This illustrates consolidation within the industry, with production being concentrated at fewer, but larger, freshwater and seawater sites (Munro and Gauld, 1996, Stagg and Gauld, 1998, Stagg and Allan, 2001, Stagg and Allan, 2002, Walker, 2010 and Munro and Wallace, 2013). Although concentrating production onto fewer sites provides economic advantages, it does incur the risk that an infectious disease outbreak could result in major financial loss (Munro and Gauld, 1997 and Stagg and Gauld, 1998).

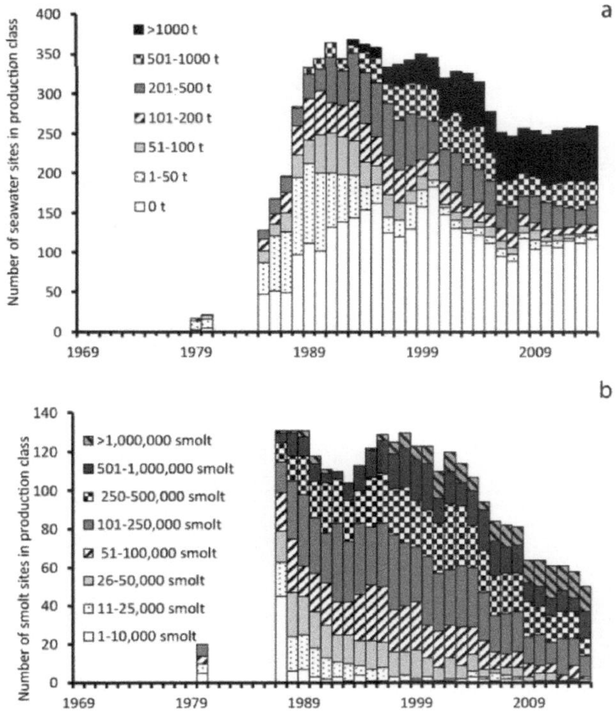

Figure. 9. Numbers of sites, grouped according to production volumes. a: Seawater sites where production is grouped by harvest tonnage. [NB: 0 t category production relates to active farms growing stock, but with no harvest in year.] b: Freshwater sites producing smolt (i.e. excluding hatchery sites) grouped according to smolt production. Data available for 1979/1980, 1985/1987–2014.

Scottish net-pen farms, in both freshwater and seawater, are regulated by the Scottish Environment Protection Agency on the basis of the total biomass of fish (Munro and Gauld, 1997 and Stagg and Gauld, 1998). Freshwater net-pen farms tend to be stocked at or near the maximum permitted biomass (Munro and Gauld, 1997), and an increase in the numbers of smolt produced was accompanied by a reduction in smolt size (Stagg and Gauld, 1998). In the 1990s, the seawater sector introduced a practice so farms operated close to permitted biomass level for as a large a proportion of the production cycle as possible: input smolt were on-grown for 6–9 months before splitting for transfer to other farms (Stagg and Gauld, 1998).

System Productivity

Productivity (reported as output m^{-3} $year^{-1}$) is greater in land-based systems than net-pen systems, in both seawater and freshwater (Fig. 10; Friedman's tests blocked by year: χ^2_r = 27.0, df = 1, p < 0.001 for freshwater; χ^2_r = 8.9, df = 1, p < 0.005 for seawater). Although seawater net-pen productivity has not changed significantly over the entire time series available (r_s = 0.245, n = 27, p > 0.2), this obscures a significant 2-fold increase between 1988 and 1996 (r_s = 0.946, n = 9, p > 0.001) and slight subsequent decrease (r_s = − 0.492, n = 18, p < 0.05). Productivity has increased over time in freshwater land-based systems (r_s = 0.570; n = 27, p < 0.005), although the plot indicates a 2-fold increase between 1988 and 2006, followed by a decrease (Fig. 10). There is no evidence of changes in system productivity over time for freshwater net-pens (r_s = 0.111, n = 27, p > 0.5) or seawater land-based systems (r_s = 0.027, n = 22, p > 0.5).

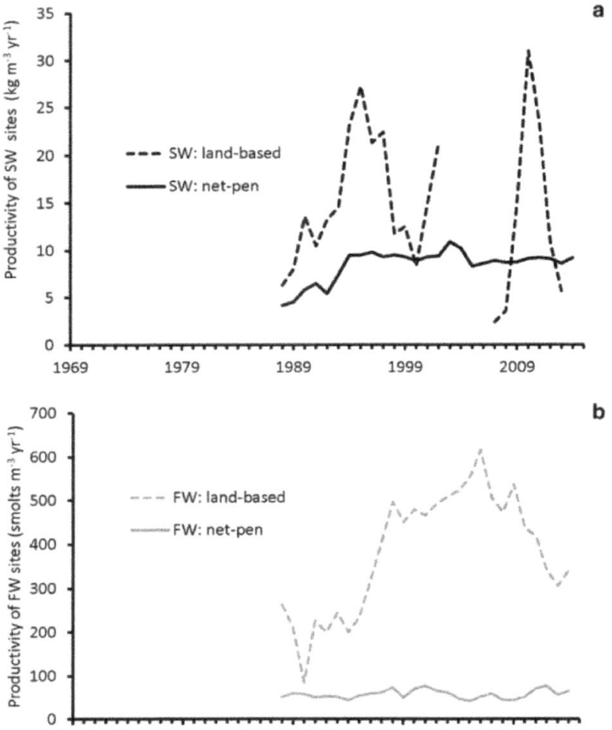

Figure. 10. System productivity in Scottish salmon farming. a: Seawater (SW) net-pens and land-based sites as kg m^{-3} $year^{-1}$. b: Freshwater (FW) net-pens and land-based sites as smolt m^{-3} $year^{-1}$. Data available 1988–2014.

Broodstock Sites

Farm sites holding salmon broodstock (both freshwater and seawater, Stagg and Allan, 2000) are reported separately to production sites. The number of sites holding broodstock has decreased by 80% over time ($r_s = -0.850$, $n = 30$, $p < 0.001$; Fig. 11). This is likely to be due to consolidation of the industry, but may also be related to only maintaining broodstock at sites free from diseases such as IPN virus (SOAFD, 1993).

Figure. 11. Number of freshwater and seawater farm sites in Scotland holding broodstock. Data available 1985–2014.

BIOLOGICAL METRICS

Size at Harvest

Size at harvest has increased over time for grilse ($r_s = 0.968$, $n = 22$, $p < 0.001$), pre-salmon ($r_s = 0.858$, $n = 22$, $p < 0.001$) and salmon ($r_s = 0.925$, $n = 29$, $p < 0.001$), but not for input year fish ($r_s = 0.265$, $n = 20$, $p > 0.2$) (Fig. 12). Munro and Gauld, 1994 and Munro and Gauld, 1996 suggested that various factors contributed to increased harvest weight in the 1990s:

- Increased smolt size and earlier transfer to seawater;
- Enhanced growth rates due to: improved feed and feeding systems (e.g. automatic feeders); reduced inappetance due to improved control of furunculosis; reduced stress due to the better health management through introduction of effective vaccines and management schemes to avoid introduction of infections (e.g. fallowing, group area agreements on single age group stocking over extended areas, stocking of smolt of common health status); reduced stress due to introduction of new husbandry technologies (e.g. air lift removal of mortalities, swim through at net changes).

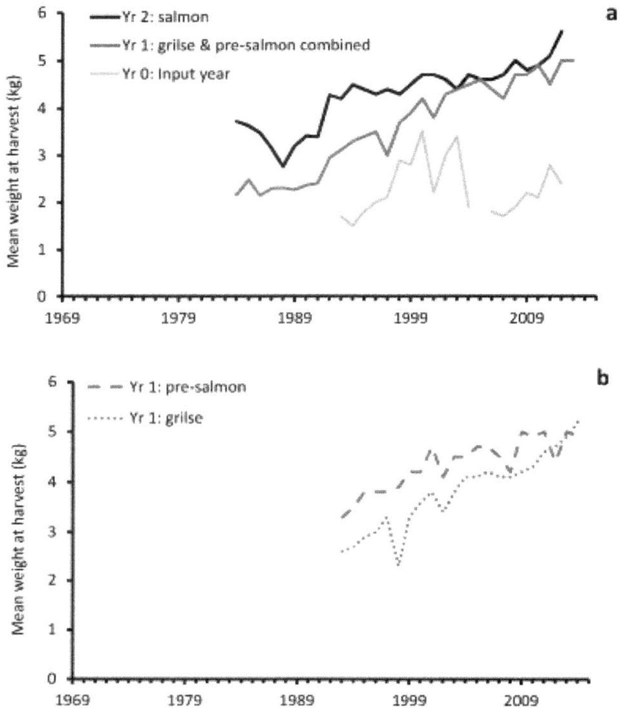

Figure. 12. Mean weight at harvest of different aged seawater salmon. a: Data refer to smolt year-classes and available 1984–2012/2013. b: data refer to year of harvest and available 1993–2014.

Further improvements in these factors, and selective breeding, are likely to account for continuation of the trend of increasing harvest size since the 1990s.

Explanations have been provided for short-term fluctuations in harvest weight, outside the long-term trend. The low harvest weights of salmon in the late 1980s were due to a combination of marketing decisions (based on the need for cash flow) and the early removal of marketable fish to limit losses to disease (DAFS, 1990 and SOAFD, 1991). The subsequent increase in harvest weight was attributed to greater confidence in achieving seawater survival and the market price of larger fish (SOAFD, 1992).

The increase in size (and numbers) of input year fish harvested in 1998 and 1999 was attributed to use of photoperiod "early smolt", rapid growth rates achieved with high energy feeds, and modern husbandry methods (Stagg and Allan, 1999 and Stagg and Allan, 2000).

Survival

Survival data are published for the seawater stage (i.e. % of smolt put to sea in a calendar year that are harvested over years 0, 1 and 2) and the freshwater stage (i.e. ratio of ova laid down: smolt produced within a calendar year). The values provided therefore reflect recovery of fish after mortality, culls and escapes. Although the freshwater stage metric is derived from different year-classes of ova and smolt for reasons of practicality, it does enable assessment of long-term trends.

There is a marked temporal trend for increased survival in freshwater ($r_s = 0.812$, $n = 28$, $p < 0.001$) which has improved threefold from 22% in 1987 to 63% in 2013 (Fig. 13). Increased freshwater survival has been attributed, in part, to reduced mortality between the ova and smolt stages (Munro and Gauld, 1995 and Munro and Gauld, 1996). Mortality at hatching and first feeding stages was estimated at 5–10% in 1996 (Munro and Gauld, 1997). Vaccination of parr against enteric redmouth disease (ERM, caused by the bacterium *Yesinia ruckeri*) was introduced in the 1990s which had previously caused mortalities in young fish in freshwater (Munro and Gauld, 1996). However, the bulk of losses in the 1990s was attributed to inadequate growth in the first summer, resulting in potential S2s (rather than S1s) which were culled (Munro and Gauld, 1995 and Munro and Gauld, 1996). Munro and Gauld (1997) noted that culling and production efficiency could be improved by increasing average growth rate. The continued trend from the 1990s to the present day indicates that culling has indeed been reduced.

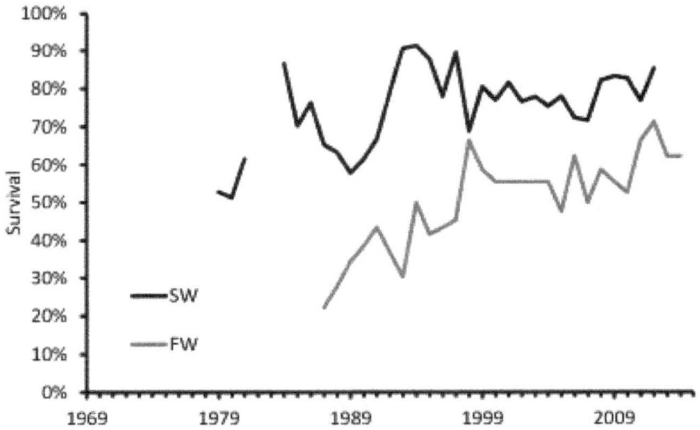

Figure. 13. Survival of Scottish farmed salmon in seawater (SW, i.e. number of year-class harvested: number of smolt put to sea; data available 1979–1981, 1984–2012) and freshwater (FW, i.e. number of smolt produced: number of eggs laid down in calendar year; data available 1987–2014).

Seawater survival was not regularly reported before 1984, although data were provided for the 1979, 1980 and 1981 smolt inputs (DAFS, 1983 and DAFS, 1984). Seawater survival has increased between 1979 and 2014 ($r_s = 0.462$, $n = 32$, $p < 0.01$) to around 80% (Fig. 13). Causes of loss of seawater fish are infectious diseases (furunculosis, pancreas disease, vibriosis), sea-lice, escape in storm damage incidents, a poor ability to tolerate seawater salinity, predation (by seals and birds), jellyfish and plankton blooms, poor husbandry and accidents such as losses during sea lice treatments (Munro and Wadell, 1981, DAFS, 1982, Munro and Gauld, 1996, Stagg and Gauld, 1998, Stagg and Allan, 1999 and Stagg and Allan, 2001). It has been noted that most of the losses occur in first summer, or first 6 months, after seawater transfer (Munro and Wadell, 1981, DAFS, 1982 and Munro and Gauld, 1996).

The initial increase in seawater survival in the early 1980s was attributed to improvements in farm management (DAFS, 1984). The notable decrease in survival that followed in the mid to late 1980s was associated with disease (DAFS, 1990). SOAFD, 1991, SOAFD, 1992 and SOAFD, 1993 noted that furunculosis and sea-lice infestation were the major causes of loss in seawater in the early 1990s. The subsequent increase in survival was attributed to improved disease control through:

- the introduction of effective commercial vaccines for furunculosis (Munro and Gauld, 1995 and Munro and Gauld, 1996);
- the introduction of strategic health management agreements involving collective fallowing, group agreements on single year-classes of smolt, and stocking with vaccinated smolt of tested health status (SOAFD, 1992 and Munro and Gauld, 1994).

These measures were suggested to reduce outbreaks of both furunculosis and sea-lice and reduce resistance to chemotherapeutants (lice to dichlorvos; furunculosis bacterium to antimicrobial medicines).

Munro and Gauld, 1996 and Munro and Gauld, 1997 indicate that large numbers of smolt were lost due to "fading smolt syndrome" (i.e. mortality due to inability to adapt to the marine environment or loss of ability to withstand high salinities) and that the severity of grading may affect smolt survival (although this latter impact was not explained). Disease was also the primary cause of low survival of 1998 smolt: outbreaks of ISA and IPN losses resulted in mortalities and culling (Stagg and Allan, 2001).

Yield per smolt

Yield (in kg) per smolt is reported, and represents the total weight of seawater harvest (as input year, grilse, pre-salmon and salmon) relative to the number of smolt put to sea in a calendar year. It therefore encapsulates seawater survival of smolt and size at harvest. Yield per smolt shows a marked increase over time ($r_s = 0.902$, $n = 29$, $p < 0.001$; Fig. 14), which is attributed to increased survival and weight at harvest (Munro and Gauld, 1994).Munro and Gauld (1996) suggested that yield per smolt was unlikely to increase after 1995, but the data indicate that this performance measure has continued to improve.

Figure. 14. Yield per smolt, i.e. total harvest of year-class relative to smolt input. Data available 1984–2012.

Vaccination

Vaccination was introduced in the mid-1980s — against the bacterial disease furunculosis by intra-peritoneal (ip) injection of freshwater parr (Munro and Gauld, 1996). The industry experienced serious losses of seawater fish due to this disease in the late 1980s and early 1990s (Munro and Wallace, 2013). By 1992 furunculosis vaccines were used at most freshwater sites which reflected how seriously companies viewed the disease and their greater faith in the efficacy of vaccination (SOAFD, 1993). By 1996, 95% of smolt put to sea were vaccinated against furunculosis (Munro and Gauld, 1997). The introduction of vaccines was considered important to boosting industry confidence in controlling disease (SOAFD, 1991). The reports also indicate the introduction of other vaccines:

- in 1995 parr started to be vaccinated against ERM by bath immersion (Stagg and Gauld, 1998);
- In 1996/7, while monovalent vaccine (specifically against furunculosis) use continued, polyvalent ip vaccines were introduced which protected against furunculosis and cold water *Vibrio*/hitra (Munro and Gauld, 1997 and Stagg and Gauld, 1998);
- In 2000, vaccination against infectious pancreatic necrosis (IPN) was undertaken under animal test certificates authorised by the Veterinary Medicines Directorate (Stagg and Allan, 2001).

Currently freshwater sites vaccinate against a range of bacterial (furunculosis, ERM, vibriosis) and viral diseases (IPN, salmonid alphavirus (SAV) the causative agent of pancreas disease, PD) (Munro and Wallace, 2013). The majority of fish are vaccinated against furunculosis and IPN, but smaller numbers receive ERM, *Vibrio* and PD vaccines (Munro et al., 2014).

Data are reported on the number of freshwater sites using vaccines, and number of fish vaccinated (Fig. 15). As these statistics do not indicate the proportion of the industry, the latter has been re-expressed in relation to the number of smolt produced in Scotland (Fig. 15). The number of sites illustrates the introduction and uptake of vaccines from the mid-1980s. Although the number of sites using vaccines has decreased since 1998, this is likely to reflect the decreased number of freshwater farms. Vaccination appears to continue to be universal, with the ratio of numbers of fish vaccinated to smolt produced typically > 1. It is noteworthy that in 1997, several independent, mobile, specialist vaccination teams were established (Stagg and Gauld, 1998).

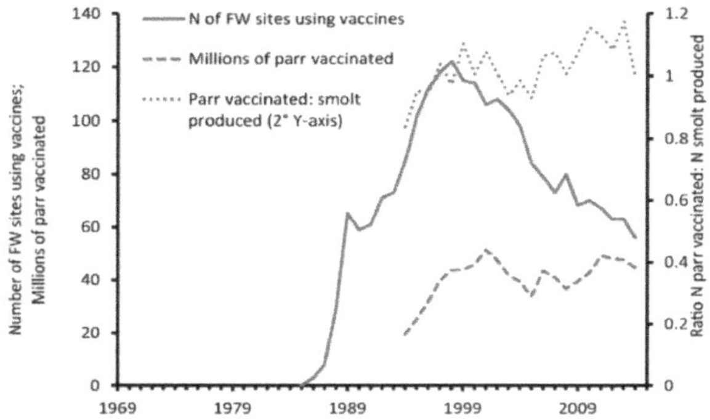

Figure. 15. Vaccination of freshwater salmon parr. Number of freshwater salmon farms using vaccines and number of parr vaccinated year^{-1} (in millions) on 1° Y-axis; ratio of numbers of parr vaccinated: smolt produced on 2° Y-axis. Data available 1985/1994–2014.

Broodstock

Data are reported on the number of female broodstock stripped for ova and the average ova yield fish^{-1} (Fig. 16). Over the 17 years for which data are available, the number of fish stripped year^{-1} has decreased 9-fold ($r_s = -0.917$; n = 17, $p < 0.001$), while the ova yield per female has doubled ($r_s = 0.794$, n = 17, $p < 0.001$). No explanation is provided for the latter trend, although it can be assumed that it is related to larger sized females.Munro and Gauld (1996) noted that low spawning success in 1995 may be related to high water temperatures affecting broodstock.

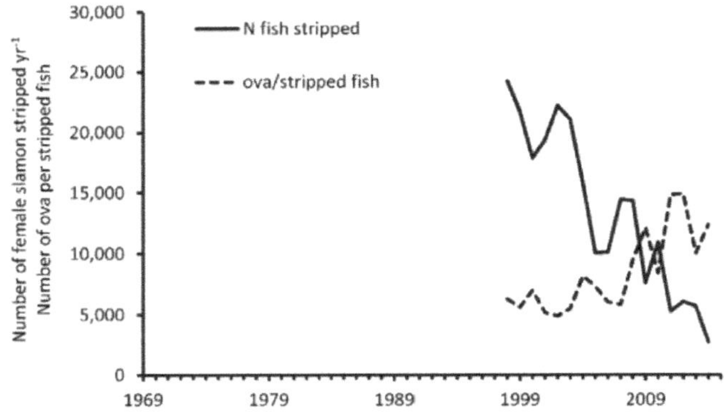

Figure. 16. Annual numbers of female salmon stripped for ova, and numbers of ova obtained per female in Scottish salmon farming industry. Data available 1998–2014.

SOCIO-ECONOMIC METRICS

Number of Companies

Data on the number of companies involved in the seawater and freshwater sub-sectors are available from the mid-1980s (Fig. 17). The data therefore capture only the tail of the initial increase in the 1970s and 1980s, but do illustrate clear subsequent decreases: 15 fold for seawater ($r_s = -0.993$, n = 26, $p < 0.001$) and 3.5-fold for freshwater ($r_s = -0.928$, n = 28, $p < 0.001$). Some additional considerations when interpreting company data are:

- Companies may be double-counted if operating in both freshwater and seawater;
- Large (umbrella) companies may operate under a number of different names for business reasons, retaining the original names of smaller company after buy-outs (Munro and Gauld, 1996 and Munro and Gauld, 1997);
- In 1998, although there were 95 seawater companies, production was dominated by just 7 companies, which accounted for 60% of harvest (Stagg and Allan, 1999). In 2014, the number of seawater companies had reduced to 18, but production was still dominated by just 6 companies, which accounted for 99% of harvest (Munro and Wallace, 2015).

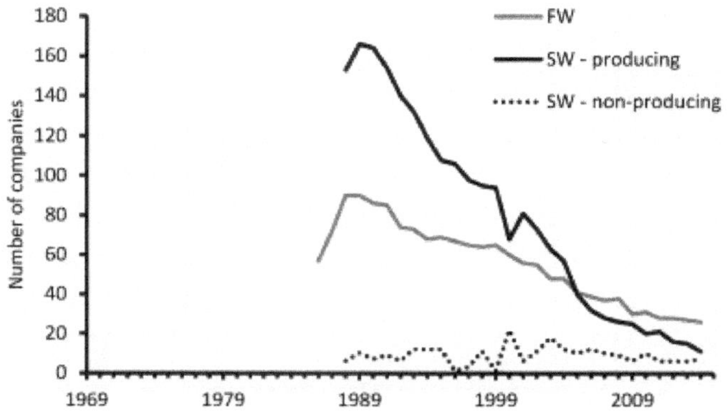

Figure. 17. Number of companies involved in freshwater (FW) or seawater (SW) salmon farming in Scotland. Data available 1986/1988–2014.

There has therefore been a clear trend since 1989 for the numbers of producing companies in both freshwater and seawater to decrease, with production being concentrated within fewer, but larger specialist companies (Munro and Gauld, 1995,Munro and Gauld, 1996, Munro and Gauld, 1997 and Stagg and Allan, 2001). The data also indicate that industry consolidation has been more pronounced in the seawater phase. Currently (2014 data) fewer companies operate in seawater than in freshwater (18 vs 26 respectively) which contrasts with the numbers of active sites (260 vs 96 respectively) (Munro and Wallace, 2015).

The reductions in numbers of companies appear to have occurred through company buy-outs and companies leaving the industry at times of perceived poor trading prospects, market uncertainty, and concern over disease (e.g. ISA) (Munro and Gauld, 1994 and Munro and Gauld, 1995; Stagg and Allan, 1999). Munro and Gauld (1997)reported "*a notable increase in the number of Scottish companies coming under foreign ownership*".

Numbers of Employees

The salmon farming industry is an important source of employment to the communities of the Scottish west coast and Western Isles, and the Orkney and Shetland Islands (Stagg and Gauld, 1998 and Stagg and Allan, 2000). Data on numbers of direct employees have been collated since the first report in 1979, although there was no differentiation between sectors and full- and part-time employees until 1985 (Fig. 18). Employee numbers include site, veterinary, harvesting, maintenance and administrative employees of the farming companies, but not processing or marketing staff (Munro and Gauld, 1995 and Munro et al., 2014). Companies are asked to use their own discretion to categorise full- and part-time staff (Munro and Wallace, 2013). Part-time employees may therefore refer to seasonal and reduced daily hours staff.

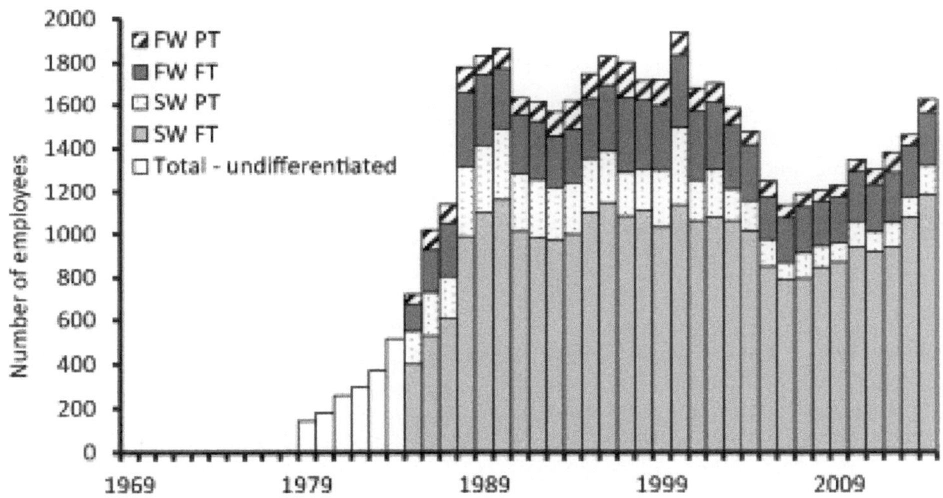

Figure. 18. Numbers of employees (FT = full time; PT = part-time) in the seawater (SW) and freshwater (FW) sectors of the Scottish salmon farming industry. Undifferentiated data available 1979–1984, differentiated data available 1985–2014.

Employment increased from 1979 to around 1700 employees in the period 1990–2000, decreased to 2006, and has since increased (Fig. 18). Employment is greater within the seawater sector than the freshwater sector, and full-time posts outnumber part-time posts. Changes in staffing levels remain largely unexplained by the annual reports, apart from:

- A sharp decrease in part-time freshwater employees in 1998 associated with the industry switching to contract vaccination service instead of direct employment of temporary staff (Stagg and Allan, 1999);
- A sudden increase in employment in 2000 when fallowed seawater sites were re-opened after the 1998 ISA outbreak (Stagg and Allan, 2001).

It has also been suggested that smaller producers employ proportionally more part-time staff and that the introduction of technology may give rise to specialist jobs (Munro and Gauld, 1997).

Employee Productivity

Employee productivity is reported for both sub-sectors, i.e. production (total numbers of smolt from freshwater; total tonnage from seawater) in relation to the total number of employees (sum of full time and part time, without adjustment for hours worked) (Fig. 19). Since 1985, employee productivity has increased 6 fold for freshwater ($r_s = 0.956$, $n = 30$, $p < 0.001$) and 11-fold for seawater ($r_s = 0.946$, $n = 30$, $p < 0.001$).

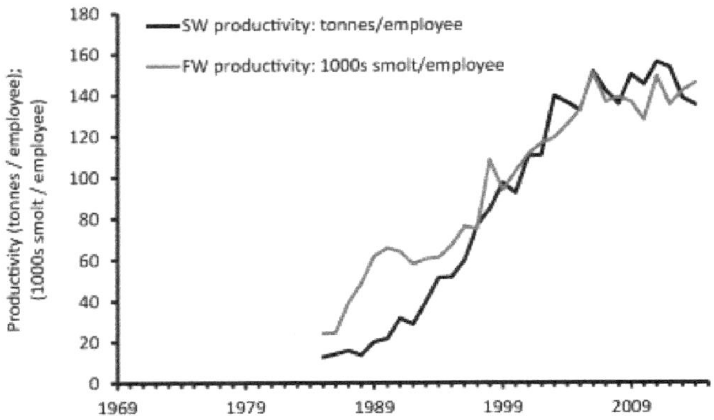

Figure. 19. Productivity of employees of the Scottish salmon industry for seawater (SW) and freshwater (FW) sectors. Data available 1985–2014.

The sudden increase in freshwater productivity in 1998 was associated with a reduction in temporary staff due to the switch to contract vaccination services (Stagg and Allan, 1999). However, the dramatic increases in staff productivity over the longer time period are attributed to:

- Economies of scale associated with larger sites (Munro and Gauld, 1996);
- Economies of scale associated with larger companies (Munro and Gauld, 1994,Munro and Gauld, 1997, Stagg and Gauld, 1998 and Walker, 2010);
- Introduction of technology, automation and mechanisation (Munro and Gauld, 1994 and Stagg and Gauld, 1998);
- Increases in other efficiencies in husbandry (Stagg and Gauld, 1998), assumed to refer to improved survival, growth, and faster throughput of batches of fish.

ENVIRONMENTAL PRESSURES METRICS

Escapes

Munro and Gauld, 1996 and Munro and Gauld, 1997 suggested that escapes from seawater net-pens due to storm damage decreased in the mid-1990s as the industry and insurers gained experience. Statistics on escapes from seawater and freshwater farms have only been published since 1999 (Fig. 20). These data are based upon self-reporting by the industry, and reflect incidents when escapes occurred and, more recently, also incidents when no loss of fish was confirmed. Incidents occur when rearing units fail or are damaged (e.g. due to weather, predators, accidents) or errors are made during fish transfer. These statistics may therefore exclude any additional "trickle" losses of which farm staff are unaware. The number of fish escaping from seawater farms is greater than from freshwater farms (Fig. 20a; Friedman's test blocked by year: $\chi^2_r = 8.07$, df = 1, $p < 0.005$). For the seawater sector, the number of reported escape events has decreased ($r_s = -0.691$, n = 16, p = 0.005) although the total numbers of fish escaping has not reduced ($r_s = -0.409$, n = 16, $p > 0.1$). For the freshwater sector there is no evidence of change in either the number of escape events ($r_s = 0.243$, n = 15, $p > 0.2$) or numbers of escapees ($r_s = -0.157$, n = 15, $p > 0.5$).

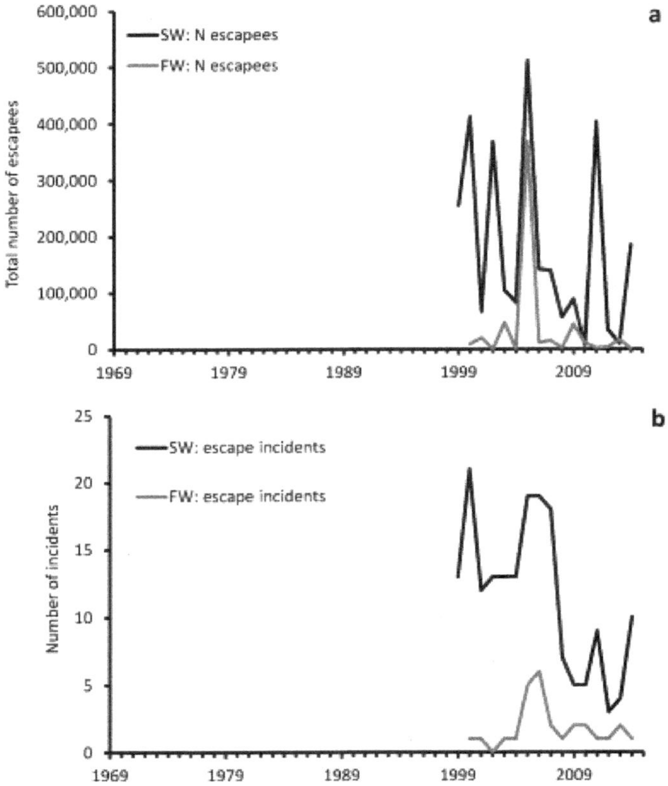

Figure. 20. Reported escapes from Scottish salmon farms in seawater (SW) and freshwater (FW). a: total number of escapees year^{-1} from freshwater (FW) and seawater (SW) farms. Data available 1999/2000–2014. b: Number of escape events (when fish escaped) year^{-1}. Data available 1999/2000–2014.

Fallowing

Fallowing involves the removal of all fish and nets from net-pen sites and was introduced as an industry practice in the late 1980s for disease control purposes (DAFS, 1989 and Munro and Gauld, 1996). A minimum routine fallow period of 6 weeks at the end of the seawater production phase is recommended by the fish health authority, to break any disease and parasite life cycles between successive salmon cohorts (Stagg and Allan, 2001). In 1998, the fish health authority imposed longer mandatory fallow periods to control outbreaks of a notifiable disease: the duration (6 weeks, 3 months or 6 months) depended upon classification of risk and confirmed or suspect disease presence, and simultaneous fallowing was coordinated within high risk zones (Stagg and Allan, 1999).

Data on fallowing of seawater net-pen sites have been reported since 1988 (Fig. 21). In 1989, the number of seawater cage sites using a fallow period was considered to be especially low at 7.4% (DAFS, 1990). However, the proportion of sites with no fallow period has decreased, and the frequency of longer fallow periods has increased. There is possible confusion in the early data due to differences in recording between years, i.e. whether sites are classed as "fallow for 52 weeks" or "not in production" (see Munro and Gauld, 1994).

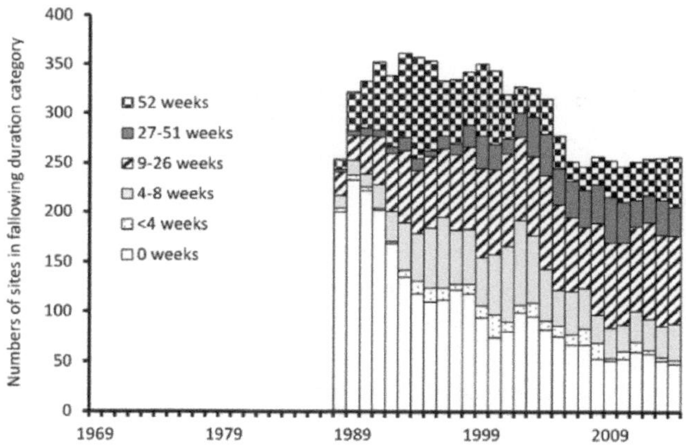

Figure. 21. Fallowing of seawater net-pen sites in Scottish salmon farming industry. Sites are categorised depending upon length of fallow period, and data is presented as number of sites in category. Data available 1988–2014.

Fallowing data were also reported for freshwater net-pen sites for 5 years (1988–1992), but the short time series is not presented here.

Accreditation Schemes

Various accreditation schemes are available to the salmon farming industry that address control of environmental pressures, product quality, fish health, animal welfare, etc. The questionnaire asks salmon farming companies about membership of accreditation schemes, but data are only reported for organic production and since 2010. Organic production in Scotland is minor, representing 2% of seawater harvest and 3% of active seawater net-pen sites in 2014 (Munro and Wallace, 2015).

DISCUSSION

Trends during the Development of Scottish Salmon Farming

This review brings together time series data and explanatory notes from annual reports to illustrate and explain temporal trends in a wide variety of metrics for Scottish salmon farming. Key trends demonstrated during the evolution of this national aquaculture industry are likely to be relevant to other developing aquaculture sectors, i.e.:

- Improved control of production and scheduling illustrated by the switch from wild to farmed ova and the introduction of out-of-season smolt.
- Improved biological performance illustrated by increased survival, size at age (of smolt and harvested seawater fish) and yield per smolt. The reports highlight the role of improvements in husbandry, feed and disease control that have occurred through time. In the second annual report for 1980, Munro and Wadell (1981) recognised that "*salmon farming is growing rapidly and as might be expected with a species one stage removed from the wild, and with an evolving technology, the industry is experiencing some technical problems. However, it appears to be developing solutions and gaining confidence in the process*".
- Reduced duration of production cycle illustrated by the reduction in use of S2 smolt and harvest of input year fish.
- Introduction and uptake of vaccines. The reports indicate vaccines have been a key contributor to the development of salmon farming in Scotland. There are few other case studies that document vaccine usage within aquaculture (Bravo and Midtlyng, 2007).
- Introduction of technology and mechanisation. Although several reports do mention increased automation and use of technology, examples provided are restricted to remote sensors and computerised control of operations, water recirculation systems, automatic feeding systems, air lifts for mortality removal, swim through net changes, and cameras for monitoring stocks (Munro and Gauld, 1994, Munro and Gauld, 1996 and Munro and Gauld, 1997).
- Improved supply chains illustrated by the industry overcoming the smolt supply shortage that restricted seawater production in the late 1970s, and international trade in ova.
- Consolidation. The initial increases in numbers of companies and production sites were followed by reductions. Consolidation of ownership into a few large firms has also occurred in Norway (Bergheim, 2012), and in Scotland is viewed as beneficial having professionalised the industry by improving standards, disease control and financial stability, thereby enabling growth (Marine Scotland, 2014). Naylor and Burke (2005) note vertical integration of salmon companies along a supply chain involving feed manufacture, hatchery and smolt production, seawater grow-out and processing.
- Increased production per farm — due to increased farm size and system productivity.
- Increasing productivity of manpower — attributed to increases in farm size and company size (providing economies of scale), mechanisation and system productivity.

Similar trends have been noted (albeit piecemeal) in the Norwegian salmon farming industry where expansion of production has been attributed to control of the biological production process, improved fish performance and culture systems, and economies of scale that have increased efficiency and reduced production costs (Bergheim and Brinker, 2003, Bergheim et al., 2009, Forster, 2010, Bergheim, 2012, Kristensen et al., 2012, Asche and Roll, 2013, Asche et al., 2013a, Asche et al., 2013b and Shepherd and Little, 2014).

This review of the annual reports on Scottish salmon farming highlights other issues that are likely to be relevant to most aquaculture sectors:

- the importance of disease. Diseases have affected the Scottish salmon farming industry by reducing harvest through mortality of fish, lost growth (due to stress and inappetance), and early harvest. Lack of confidence in controlling disease has also affected investment, and the costs of controlling disease (manpower for application of chemical sea-lice treatments) can prove significant (Munro and Gauld, 1997). Disease has also been noted to affect ova supplies and employment. The importance of diseases to the salmon farming industries in Norway and Chile has been highlighted (Tilseth et al., 1991, Asche et al., 2009, Forster, 2010, Ashe and Bjørndal, 2011 and Asche and Roll, 2013) and FAO (2012) note the common importance of disease to aquaculture. Effective disease management – via reduced stress in stocks, vaccination, fallowing, area management agreements, biosecurity etc. – is recognised to have been vital to the success of Scottish salmon farming.

- the international nature of aquaculture, as illustrated by imports and exports of ova and ownership by foreign companies. The Scottish salmon farming industry is now concentrated within four large companies, predominantly Norwegian-owned, which operate UK-registered subsidiaries within a group structure (Marine Scotland, 2014).
- the role of legislation/regulation, as illustrated by restrictions on net-pen site biomass, and increased use of foreign ova following regulatory changes.
- the occurrence of episodic events: in 1993, for example, contamination of farmed salmon from a major oil spill occurred with 3,549 t of salmon slaughtered and lost to the human food chain, being used as food for mink (Munro and Gauld, 1994).

Development of Scottish Salmon Farming in Relation to Sustainable Intensification

The future challenge for both terrestrial and aquatic farming is to achieve sustainable intensification, i.e. increase production whilst increasing the efficiency of resource use and safeguarding the environment, the nutritional value of products, the welfare of farmed stocks and rural economies (Godfray et al., 2010 and Garnett et al., 2013). The Scottish salmon farming industry has increased harvest production over time (Fig. 5) and shown trends in factors commonly associated with intensification, i.e. increases in system size (i.e. net-pens, Bergheim, 2012), farm size, individual farm output, and employee productivity (Fig. 8, Fig. 9 and Fig. 19).

Increasing Productivity

System productivity (i.e. production m^{-3} $year^{-1}$) has increased for freshwater land-based systems and seawater net-pens (Fig. 10). System productivity is a function of fish stocking density, throughput, and the rate of biomass increase. Freshwater land-based system productivity has increased through increased stocking density and extension of the growing season enabling production of more than one batch of fish each year (Stagg and Allan, 1999). Increases in throughput and stocking density do not appear to account for the increase in seawater net-pen productivity between 1988 and 1996: fallow (i.e. non-productive) periods increased (Fig. 21) and stocking density was decreased in the early 1990s to aid disease management (SOAFD, 1992 and FAWC, 1996). Stocking density in seawater net-pens appears to have remained unchanged since with the introduction of a widely adopted fish welfare accreditation scheme (FAWC, 2014). Increased seawater net-pen productivity can therefore be assumed to be due to improved biological performance (illustrated by size at age, survival and yield relative to input; Fig. 12, Fig. 13 and Fig. 14). Salmon farming is still a young industry (in comparison to agricultural activities) and such improvements can be expected with operational experience (DAFS, 1982 and Marine Scotland, 2014).

Selecting Appropriate Genotypes

A key means of improving biological performance is selecting genotypes best suited to farm conditions (Godfray et al., 2010). Farmed salmon have now been selectively bred over more than eight generations for a number of traits (growth rate, resistance to disease, age at sexual maturity, feed conversion efficiency, fillet yield, flesh quality and colour) and represent one of the most genetically improved stocks within aquaculture (Ashe and Bjørndal, 2011 and Telechea and Fontaine, 2014). However, little information has been published on selective breeding and strains used within the Scottish industry.Munro and Gauld (1997) mention selection of broodstock for low grilsing rates (i.e. reaching sexual maturity after > 1 winter at sea), and spawning early or late within the reproductive season. The wild Scottish salmon strains initially used matured early (i.e. showed a "high grilsing rate") in culture, which was undesirable as it reduced growth and market value. The industry has therefore switched to later maturing Norwegian and hybrid strains (www.fao.org/fishery/culturedspecies/Salmo_salar/en accessed 5 March 2015).

Reducing Waste

The Scottish salmon farming industry has reduced waste associated with mortalities (Fig. 13); continued improvement in disease control and containment would further reduce such waste. Additional reductions in unused ova and smolt produced but not put to sea (Figs. 2a, 3a) may be possible. Data from Norway indicate reductions in feed waste: feed conversion ratio (FCR = mass of feed provided: net increase in fish biomass, within a specific period) has improved for freshwater (around 2 in 1985 to 1 in 2000) and seawater salmon farming (from 3 in 1980 to around 1.1) attributed to improved nutritional formulation, pellet quality, feeding systems and fish survival (Asche et al., 1999, Bergheim and Brinker, 2003, Forster, 2010 and Shepherd and Little, 2014). Such temporal data are lacking for Scotland, but similar reductions are likely to have occurred. However, the seawater FCR in Scotland (1.3) has been suggested to be higher than in Norway (Pelletier et al., 2009).

Reducing Resource Inputs and Environmental Pressures

Environmental pressures from salmon farming are recognised as key sustainability issues for its future in Scotland and elsewhere (Jones et al., 2015). Although the different environmental pressures have each attracted considerable research attention and discussion, consensus is often lacking as to whether they do result in environmental impacts (Forster, 2010). The issues are listed below, and supporting time series data are cited where available (notably from Norway):

- Resource use and emissions: Bergheim and Brinker (2003) indicated a 5 fold improvement in the efficiency of freshwater use in Norwegian salmon smolt farms between 1985 and 2000, due to introduction of water treatment technology. Recent (snapshot) Life Cycle Assessments indicate that the global environmental pressures of farmed salmon products – via resource use (energy) and emissions (greenhouse gases, nitrogen phosphorous) – are similar to those from fishery production, and lower than terrestrial meat farming (Ellingsen and Aanondsen, 2006, Pelletier and Tyedmers, 2007, Hall et al., 2011 and Ytrestøyl et al., 2011). Nevertheless, salmon production in Scotland is judged to have higher impacts than in other countries due to differences in feed ingredients (Pelletier et al., 2009).

- Eutrophication and organic enrichment: Bergheim and Brinker (2003) provided data illustrating substantial reductions in the discharge of suspended solids, nitrogen and phosphorous from Norwegian salmon smolt farms between 1985 and 2000. Comparable data for seawater farms are lacking. Temporal changes in feed formulation, feeding methodology and food conversion, improved site selection, management and regulation, and an increase in fallowing (largely for pathogen management) can be expected to have reduced discharge pressures (Shepherd and Little, 2014).
- Discharge of chemicals. Since the 1980s, there have been substantial reductions in the use and discharge of i) chemicals used to control sea-lice and algal fouling and ii) antibiotics in Norwegian seawater farms (Asche et al., 1999, Tveteras, 2002, Forster, 2010 and Ashe and Bjørndal, 2011); the latter reduction is attributed to the introduction of vaccines and improved disease control (Shepherd and Little, 2014).
- Transfer of disease to wild stocks: Despite improved disease control within farmed stocks, disease pressures on wild stocks, notably from sea-lice, remain a topic of concern (Torrissen et al., 2013 and Shepherd and Little, 2014). Pathogen pressures on wild stocks will depend upon pathogen prevalence in farmed stocks, the size of farmed stocks, and the overlap between farmed and wild fish in space and time (McVicar, 1997). Farmed stock sizes have increased over time within Scotland (Fig. 3b). The Scottish salmon industry is required to maintain sea-lice below threshold infection levels, and its trade body has recently started publication of sea-lice counts (e.g. SSPO, 2015), so data may become available to assess trends in pathogen pressure.
- Escapee impacts on wild populations: There is evidence that the numbers of seawater escape events in Scotland has decreased since 1999, although there are no significant trends for the numbers of freshwater escape events and freshwater and seawater escapees (Fig. 20). Reducing the number of escapees should reduce pressures (competition, introgression) on wild stocks. Concern has been focussed on genetic introgression leading to loss of local adaptations and reduced fitness (McGinnity et al., 2003). Although data are lacking for Scotland, evidence for genetic introgression in wild salmon populations has been reported from Norway, Ireland and North America (Glover et al., 2013). Production of female triploid salmon ova in Scotland was reported between 1989 and 1992 (DAFS, 1990, SOAFD, 1991,SOAFD, 1992 and SOAFD, 1993) and, due to sterility, is being reconsidered to eliminate genetic introgression pressures, although concerns related to performance within aquaculture remain (Maxime, 2008). In contrast, the Scottish rainbow trout (*Oncorhynchus mykiss*) farming industry has embraced the use of triploid (and all female) ova (Munro and Wallace, 2015).
- Use of reduction fishery products in salmon feed. Although reduction fishery products (fish meal and oil) are major salmon feed components, there is evidence for trends of decreasing use (Shepherd, 2012 and Shepherd and Jackson, 2013). Such components of salmon diets are increasingly sourced from certified sustainable fisheries, processing trimmings and by-products, or substituted by other vegetable and land animal sources (Shepherd, 2012, Shepherd and Jackson, 2013 and Shepherd et al., 2015). An index of efficiency of conversion of the fish dietary ingredients to harvested salmonid (i.e. fish in: fish out ratio) has shown a trend for improvement over the period 2000–2010 (Shepherd and Jackson, 2013 and Shepherd and Little, 2014).

Safeguarding Nutritional Value

Salmon, as an oily fish, is a source of high quality animal protein, essential fatty acids and micronutrients to human consumers (Beveridge et al., 2013). Health risks associated with consumption of farmed salmon have been raised (due to concentration of environmental contaminants within the food chain) but it is now accepted that the health benefits of eating farmed salmon far outweigh the risks (Shepherd, 2012). There is an apparent lack of data to assess whether the nutritional value (both health benefits and risks) of Scottish farmed salmon has changed over time with changes in feed composition and production. The need for the Scottish industry to maintain the nutritional value of its products has recently been highlighted (Shepherd et al., 2015).

Safeguarding Animal Welfare

Animal welfare relates to suffering. If the potential for suffering to occur during farming is considered in terms of animal-weeks (animal numbers × production cycle duration), then salmon farming does merit scrutiny of animal welfare:

- the numbers of animals involved in salmon farming is high: the 2014 year-class started with 70.8 million individual ova laid down, and 34.3 million seawater fish were slaughtered for harvest in 2014 (Table 1). Lymbery (2002) recognised the high number of individual salmon farmed, suggesting it was only exceeded by broiler (meat chicken) farming in the UK.
- The production cycle is of a long duration: 0.5–2 years in freshwater, followed by 0.5–2 years in seawater. This compares to 6 weeks from hatch to slaughter for broiler chickens (FAWC, 1992).

The data indicate that performance-based measures of fish welfare, i.e. growth and survival (Fig. 12 and Fig. 13) have improved over time within freshwater and seawater Scottish salmon farms. These improvements (associated with increased productivity; Fig. 10 and Fig. 14) occurred against a backdrop of increasing rearing unit size, farm size and output (Figs. 8a, 9) and reducing staff input relative to production (Fig. 19). This finding counters the common assumption that intensification compromises measures of animal welfare and supports the increasing preference for performance outcome measures over resource inputs for assessing farmed animal welfare (Main et al., 2012). Although it cannot be assumed that such performance measures do address all welfare concerns (Kristensen et al., 2012) increases in survival and growth strongly indicate improvements in health and meeting of dietary and environmental needs (Ellis et al., 2012). Kristensen et al. (2012) also concluded that intensification does not necessarily adversely affect performance measures of salmon welfare on commercial farms in Norway.

Although there is evidence that the welfare of farmed salmon in Scotland has improved over time, there may be potential for further improvement. Recent values for freshwater survival in Scotland (60–70%; Fig. 13) are apparently lower than in Norway where a value of 80% is quoted (Asche and Bjørndal, 2011). Recent values for seawater survival in Scotland (77–85%; Fig. 13) are also below the 90% figure cited by Asche and Bjørndal (2011), although similar to the 80% value suggested by Shepherd and Little (2014).

Safeguarding Rural Economies

No data are available to examine trends in the contribution of salmon farming to rural communities in Scotland. Nevertheless, a recent assessment (Marine Scotland, 2014) indicates that salmon farming in Scotland has become an "anchor industry" sustaining fragile rural communities that are at risk of population decline due to a lack of alternative economic options by providing:

- direct employment for farm-based workers at a range of skill levels. Work is suitable for school leavers and people of child bearing age, and the skills learnt are transferable;
- an income stream to local suppliers and service providers, e.g. divers, hauliers, welders, electricians, engineers and local mechanics servicing vehicles and boats, trainers in boat handling;
- support for infrastructure with shops, schools, housing, transport (roads, ferries, harbours, piers, slipways) and services (haulage, power, broadband) being sustained by the circulation of income and a strengthened local population.

Salmon farming in Scotland is viewed as a natural resource-based means of wealth creation, with socio-economic benefits extending beyond local communities (Marine Scotland, 2014). The industry also has ancillary workers (e.g. logistics and management; marketing; vaccination, veterinary and consultant services; well boat operators) and the supply chain extends both upstream (e.g. feed suppliers, equipment manufacturing, boat and feed barge suppliers, hauliers) and downstream (e.g. harvest stations, processors, retailers). The salmon farming industry in Scotland is therefore considered important for its socio-economic benefits, concurring with a similar recent assessment in Norway (Robertsen et al., 2012).

Maximising Conversion Efficiency

In addition to the strategies associated with sustainable intensification of existing farms, human diets need to change towards products from trophic levels and farming systems that maximise conversion efficiency (Godfray et al., 2010). Although concern has been expressed about the dominance of fed species such as salmon in global aquaculture:

- Food conversion efficiency in farmed salmon has improved over time (see 9.2.4);
- Farmed fish are well recognised as more efficient converters of food to edible product than terrestrial livestock as they have a low energy protein metabolism pathway, do not expend energy on maintaining body temperature (being poikilothermic) or a large bony skeleton (being supported by the water), and the latter attribute also provides a higher yield of edible flesh (Hall et al., 2011 and Shepherd and Little, 2014).
- Farmed salmonids are more efficient converters of food than wild carnivorous fish as they use less energy (Shepherd and Jackson, 2013 and Shepherd and Little, 2014).

The Value of Collecting Data on Aquaculture

The development of aquaculture, be it new or existing sectors, requires examination and planning by governments, regulators, and the industry itself. This review has demonstrated the merit in collecting and publishing consistent time series data on industry wide metrics enabling assessment of trends, identifying areas of underperformance, and assessing resource use, environmental pressures and sustainability. Monitoring of aquaculture sectors is therefore an important contributor to enabling sustainable development. Further examples of the use of performance indicators include:

- Strategic planning: The Scottish salmon industry is becoming increasing reliant on imports of foreign ova (Fig. 2). The annual reports of the Scottish finfish farming industry (e.g. Smith, 2006, Smith, 2007 and Smith, 2008) also include production data for species proposed to diversify the seawater net-pen sector, i.e. halibut and Atlantic cod (*Gadus morhua*); such species have not yet shown substantial expansion.
- Benchmarking: Munro and Gauld (1994) suggested that i) survival higher than 90% was often recorded in individual net-pen populations of seawater salmon, and this should be a standard achievable by all sites, and ii) some companies fell below industry norms for employee productivity and could improve.
- Intra-national comparisons: SOAFD, 1991 and SOAFD, 1993 noted that the Shetland region produced the largest grilse, pre-salmon and salmon, and suggested that the more northern waters favoured salmon growth.
- International comparisons: In early reports, DAFS, 1982 and DAFS, 1983 judged that Scottish salmon farming lagged behind the Norwegian industry, both in terms of production numbers and performance. For example, survival of 1979 and 1980 smolt to harvest in Scotland (53% and 51%) compared poorly to that in Norway of 70% (DAFS, 1983).

Salmon production in Scotland increased until 2003, but then fell and only in 2014 did production exceed that previous peak by 5% (Fig. 5). In contrast, annual production in Norway and Chile has grown by 142% and 38% respectively over a similar period (Fig. 1). Asche and Bjørndal (2011) suggest access to new farm sites is limiting expansion of salmon farming in Scotland, and increases in harvest need to come from further increases in system productivity.

Gaps in the published statistics for Scotland relate to some biological metrics (e.g. smolt size, use of cleaner fish for biological control of sea-lice), system types (e.g. RAS or flow through land-based systems; inshore or offshore net-pens) and environmental and socio-economic performance indicators. Extending publication to include environmental pressures would allow open scrutiny of evidence and trends to be assessed. Data and trends provide evidence to both highlight problems within, and respond to criticisms of, aquaculture (e.g. Naylor and Burke, 2005, Shepherd and Jackson, 2013 and Shepherd and Little, 2014). Economic data on salmon farming in Norway has been collected and published for a number of years enabling analysis (Forster, 2010, Ashe and Bjørndal, 2011, Asche et al., 2013a, Oglend, 2013 and Shepherd and Little, 2014) but similar information is not readily available for Scotland. Within the annual reports, information is limited to a single statement for 1996, referring to a decrease in market price of salmon and profitability (Munro and Gauld, 1997). Gathering valid data on environmental and financial indicators may, however, prove difficult due to practical and confidentiality issues.

ACKNOWLEDGEMENTS

This review represents an output from two projects: SARF082 Scottish aquaculture's utilisation of environmental resources, funded by the Scottish Aquaculture Research Forum; AW1209 Study to identify ways of reducing farmed fish fin damage in hatcheries, funded by the Department of Environment, Food and Rural Affairs. Thanks to: Jackie Brown (MSS library) for providing pdfs of the historical DAFS/SOAFD/FRS annual production reports, and Sandy Murray (MSS) for commenting on a draft of the manuscript. The authors have no conflicts of interest to declare.

REFERENCES

1. Amberg, S.M., Hall, T.E., 2008. Communicating risks and benefits of aquaculture: a content analysis of US newsprint representations of farmed salmon. J. World Aquacult. Soc. 39, 143–157.
2. Asche, F., Roll, K.H., 2013. Determinants of inefficiency in Norwegian salmon aquaculture. Aquac. Econ. Manag. 17, 300–321.
3. Asche, F., Guttormsen, A.G., Tveterås, R., 1999. Environmental problems, productivity and innovations in Norwegian salmon aquaculture. Aquac. Econ. Manag. 3, 19–29.
4. Asche, F., Hansen, H., Tveteras, R., Tveteras, S., 2009. The salmon disease crisis in Chile. Mar. Resour. Econ. 24, 405–411.
5. Asche, F., Guttormsen, A.G., Nielsen, R., 2013a. Future challenges for the maturing Norwe gian salmon aquaculture industry: an analysis of total factor productivity changes from 1996 to 2008. Aquaculture 396–399.
6. Asche, F., Roll, K.H., Sandvold, H.N., Sorvig, A., Zhang, D., 2013b. Salmon aquaculture: larg er companies and increased production. Aquac. Econ. Manag. 17, 322–339.
7. Asche, F., Bjørndal, T., 2011. The Economics of Salmon Aquaculture. second ed. Wiley-Blackwell, Chichester, UK.
8. Aunsmo, A., Skjerve, E., Midtlyng, P., 2013. Accuracy and precision of harvest stock esti mation in Atlantic salmon farming. Aquaculture 396–399, 113–118.
9. Bergheim, A., 2012. Recent growth trends and challenges in the Norwegian aquaculture industry. Lat. Am. J. Aquat. Resour. 40, 800–807.
10. Bergheim, A., Brinker, A., 2003. Effluent treatment for flow through systems and European environmental regulations. Aquac. Eng. 27, 61–77.
11. Bergheim, A., Drengstig, A., Ulgenes, Y., Fivelstad, S., 2009. Production of Atlantic salmon smolt in Europe — current characteristics and future trends. Aquac. Eng. 41, 46–52.
12. Beveridge, M.C.M., Thilsted, S.H., Phillips, M.J., Metian, M., Troell, M., Hall, S.J., 2013. Meet ing the food and nutrition needs of the poor: the role of fish and the opportunities and challenges emerging from the rise of aquaculture. J. Fish Biol. 83, 1067–1084.
13. Bravo, S., Midtlyng, P.J., 2007. The use of fish vaccines in the Chilean salmon industry 1999–2003. Aquaculture 270, 36–42.
14. DAFS, 1982. Salmon and Trout Farming in Scotland — Report of Results of DAFS Annual Survey 1981. DAFS (2 pp.).
15. DAFS, 1983. Salmon and Trout Farming in Scotland — Report of Results of DAFS Annual Survey 1982. DAFS (7 pp.).
16. DAFS, 1984. Salmon and Trout Farming in Scotland — Report of Results of DAFS Annual Survey 1983. DAFS (11 pp.).
17. DAFS, 1985. Salmon and Trout Farming in Scotland — Report of Results of DAFS Annual Survey 1984. DAFS (6 pp.).
18. DAFS, 1986. Salmon and Trout Farming in Scotland — Report of Results of DAFS Annual Survey 1985. DAFS (8 pp.).
19. DAFS, 1987. Salmon and Trout Farming in Scotland — Report of Results of DAFS Annual Survey 1986. Department of Agriculture and Fisheries for Scotland (8 pp.).
20. DAFS, 1988. Salmon and Trout Farming in Scotland — Report of Results of DAFS Annual Survey 1987. Department of Agriculture and Fisheries for Scotland (11 pp.)
21. DAFS, 1989. Report of the DAFS Annual Survey of Fish Farms for 1988. DAFS (11 pp.). DAFS, 1990. Report of the DAFS Annual Survey of Fish Farms for 1989. DAFS (14 pp.). Ellingsen, H., Aanondsen, S.A., 2006.

Environmental impacts of wild caught cod and farmed salmon — a comparison with chicken. Int. J. Life Cycle Assess. 1, 60–65. Ellis, T., Berrill, I., Lines, J., Turnbull, J., Knowles, T.G., 2012. Mortality and fish welfare. Fish Physiol. Biochem. 38, 189–199.

22. FAO, 2007. Aquaculture — the only way to fill the coming "fish gap": outcomes of the high level special event on aquaculture. FAO Aquacult. Newsl. 39, 4–5 (http://www.fao. org/docrep/014/i0148e/i0148e00.htm accessed 5 March 2015).
23. FAO, 2012. The State of World Fisheries and Aquaculture 2012 (Rome. 209 pp.). FAO, 2014. The State of World Fisheries and Aquaculture 2014 (Rome. 223 pp.).
24. FAWC, 1992. Report on the Welfare of Broiler Chickens. Farm Animal Welfare Council, London (37 pp.).
25. FAWC, 1996. Report on the Welfare of Farmed Fish. Farm Animal Welfare Council, London (52 pp.).
26. FAWC, 2014. Opinion on the Welfare of Farmed Fish. Farm Animal Welfare Committee, London (40 pp.).
27. Forster, J., 2010. What can U.S. open ocean aquaculture learn from salmon farming? Mar. Technol. Soc. J. 44, 68–79.
28. Garnett, T., Appleby, M.C., Balmford, A., et al., 2013. Sustainable intensification in agricul ture: premises and policies. Science 341, 33–34.
29. Glover, K.A., Pertoldi, C., Besnier, F., Wennevik, V., Kent, M., Skaala, Ø., 2013. Atlantic salm on populations invaded by farmed escapees: quantifying genetic introgression with a Bayesian approach and SNPs. BMC Genet. 14, 74.
30. Godfray, H.C.J., Beddington, J.R., Crute, I.R., et al., 2010. Food security: the challenge of feeding 9 billion people. Science 327, 812–818.
31. Hall, S.J., Delaporte, A., Phillips, M.J., Beveridge, M., O›Keefe, M., 2011. Blue Frontiers: Man aging the Environmental Costs of Aquaculture. The WorldFish Center, Penang, Malaysia (103 pp.).
32. Hastings, T.S., Smith, R.J., 2005. Scottish Fish Farms Annual Production Survey, 2003. Fish eries Research Services (53 pp.).
33. Hawkins, S.J., Firth, L.B., McHugh, M., et al., 2013. Data rescue and re-use: recycling old in formation to address new policy concerns. Mar. Policy 42, 91–98.
34. Jones, A.C., Mead, A., Kaiser, M.J., et al., 2015. Prioritization of knowledge needs for sus tainable aquaculture: a national and global perspective. Fish Fish. 16, 668–683.
35. Kristensen, T., Haugen, T.O., Rosten, T., Fjellheim, A., Atland, A., Rosseland, B.O., 2012. Ef fects of production intensity and production strategies in commercial Atlantic salmon smolt (Salmo salar L.) production on subsequent performance in the early sea stage. Fish Physiol. Biochem. 38, 273–282.
36. Lymbery, P., 2002. In Too Deep — the Welfare of Intensively Farmed Fish. Compassion in World Farming Trust, Hampshire (56 pp.).
37. Main, D.C.J., Mullan, S., Atkinson, C., Bond, A., Cooper, M., Fraser, A., Browne, W.J., 2012. Welfare outcomes assessment in laying hen farm assurance schemes. Anim. Welf. 21, 389–396.
38. Marine Scotland, 2014. An Assessment of the Benefits to Scotland of Aquaculture. Marine Scotland (164 pp. http://www.gov.scot/Topics/marine/Publications/ publicationslatest/ farmedfish/AqBenefits accessed 5 March 2015).
39. Maxime, V., 2008. The physiology of triploid fish: current knowledge and comparison with diploid fish. Fish Fish. 9, 67–78.
40. McGinnity, P., Prodöhl, P., Ferguson, A., et al., 2003. Fitness reduction and potential extinc tion of wild populations of Atlantic salmon, Salmo salar, as a result of interactions with escaped farm salmon. Proc. R. Soc. Lond. B 270, 2443–2450.
41. McVicar, A.H., 1997. Disease and parasite implications of the coexistence of wild and cul tured Atlantic salmon populations. ICES J. Mar. Sci. 54, 1093–1103.
42. Munro, A.L.S., Gauld, J.A., 1994. Scottish Fish Farms Annual Production Survey, 1993. SOAFD (32 pp.).
43. Munro, A.L.S., Gauld, J.A., 1995. Scottish Fish Farms Annual Production Survey 1994. Ma rine Laboratory Aberdeen (35 pp.).
44. Munro, A.L.S., Gauld, J.A., 1996. Scottish Fish Farms Annual Production Survey 1995. Ma rine Laboratory Aberdeen (36 pp.).
45. Munro, A.L.S., Gauld, J.A., 1997. Scottish Fish Farms Annual Production Survey 1996. Ma rine Laboratory Aberdeen (35 pp.).
46. Munro, A.L.S., Wadell, I.F., 1981. Annual Report for 1980 on the Production of the Scottish Salmon and Trout Farming Industry. DAFS (4 pp.).
47. Munro, L.A., Wallace, I.S., 2013. Scottish Fish Farm Production Survey 2012. Marine Scot land Science (54 pp.).
48. Munro, L.A., Wallace, I.S., 2015. Scottish Fish Farm Production Survey 2014. Marine Scot land Science (53 pp.).
49. Munro, A.L.S., Wadell, I.F., Elso, K.G.R., 1980. Report on the Growth of Scottish Salmonid Fish Farms and their Production and Manpower in 1979. DAFS Marine Lab (8 pp.).
50. Munro, L.A., Warwick, A.L., Wallace, I.S., 2014. Scottish Fish Farm Production Survey 2013. Marine Scotland Science (48 pp.).
51. NASCO, 2014. Report of the thirty-first annual meeting of the council, Saint-Malo, France, 3–6 June 2014. (CNL(14)58, 291 pp. http://www.nasco.int/reports_ annual.html accessed 5 March 2015).
52. Naylor, R., Burke, M., 2005. Aquaculture and ocean resources: raising tigers of the sea. Annu. Rev. Environ. Resour. 30, 185–218
53. Oglend, A., 2013. Recent trends in salmon price volatility. Aquac. Econ. Manag. 17, 281–299.
54. Pelletier, N., Tyedmers, P., 2007. Feeding farmed salmon: is organic better? Aquaculture 272, 399–416.

55. Pelletier, N., Tyedmers, P., Sonesson, U., et al., 2009. Not all salmon are created equal: life cycle assessment (LCA) of global salmon farming systems. Environ. Sci. Technol. 43, 8730–8736.

56. Robertsen, R., Andreassen, O., Iversen, A., 2012. Havbruksnæringens ringvirkninger i Troms. Rapport/Report 28/2012. Nofima (52 pp. [In Norwegian]. http://www. nofima.no/filearchive/Rapport%2028–2012.pdf accessed 5 March 2015).

57. Scottish Executive, 2002. Review and Synthesis of the Environmental Impacts of Aquacul ture. Scottish Executive Central Research Unit (62 pp. http://www. gov.scot/Resource/ Doc/46951/0030621.pdf accessed 5 March 2015).

58. Seafish, 2011. Market based questions (as at April 2011). What is the most popular fish in the UK? (http://www. seafish.org/media/486918/faqs%20web%20version%20 010711. pdf accessed 5 March 2015).

59. Seafish, 2012. Responsible Sourcing Guide: Farmed Atlantic Salmon. http://www.seafish.org.uk/media/Publications/SeafishResponsibleSourcingGuide_ Atlanticsalmon_ 201208.pdf (accessed 5 March 2015).

60. Shepherd, J., 2012. Aquaculture: are the criticisms justified? Feeding fish to fish. World Agric. 3 (2), 11–18.

61. Shepherd, C.J., Jackson, A.J., 2013. Global fishmeal and fish-oil supply: inputs, outputs and markets. J. Fish Biol. 83, 1046–1066.

62. Shepherd, C.J., Little, D.C., 2014. Aquaculture: are the criticisms justified? II — aquaculture›s environmental impact and use of resources, with specific reference to farming Atlantic salmon. World Agric. 4 (2), 37–52.

63. Shepherd, J., Monroig, O., Tocher, D.R., 2015. Production of high quality, healthy farmed salmon from a changing raw material base, with special reference to a sustainable Scottish industry. Scottish Aquaculture Research Forum (159 pp. http://www.sarf. org.uk/cms-assets/documents/216181-554802.sarfsp007.pdf accessed 8 February 2016).

64. Smith, R.J., 2006. Scottish Fish Farms Annual Production Survey, 2005. Fisheries Research Services (53 pp.).

65. Smith, R.J., 2007. Scottish Fish Farms Annual Production Survey, 2006. Fisheries Research Services (53 pp.).

66. Smith, R.J., 2008. Scottish Fish Farms Annual Production Survey, 2007. Fisheries Research Services (53 pp.).

67. Smith, R.J., Bland, M.D., Hastings, T.S., 2005. Scottish Fish Farms Annual Production Survey, 2004. Fisheries Research Services (53 pp.).

68. SOAFD, 1991. Annual Survey of Fish Farms — 1990 (Revised). DAFS (16 pp.).

69. SOAFD, 1992. Report of the SOAFD Annual Survey of Fish Farms for 1991. SOAFD (14 pp.). SOAFD, 1993. Report of the SOAFD Annual Survey of Fish Farms for 1992. SOAFD (15 pp.). SSPO, 2015. Fish health management report October to December 2014. http://scottishsalmon.co.uk/wp-content/uploads/2015/02/Fish-Health-Managementreport-Q4-2014.pdf (accessed 5 March 2015).

70. Stagg, R.M., Allan, C.E.T., 1999. Scottish Fish Farms Annual Production Survey 1998. Fish eries Research Services (36 pp.).

71. Stagg, R.M., Allan, C.E.T., 2000. Scottish Fish Farms Annual Production Survey 1999. Fish eries Research Services (36 pp.).

72. Stagg, R.M., Allan, C.E.T., 2001. Scottish Fish Farms Annual Production Survey 2000. Fish eries Research Services (36 pp.).

73. Stagg, R.M., Allan, C.E.T., 2002. Scottish Fish Farms Annual Production Survey, 2001. Fish eries Research Services (53 pp.).

74. Stagg, R., Gauld, J.A., 1998. Scottish Fish Farms Annual Production Survey 1997. Fisheries Research Services (40 pp.).

75. Stagg, R.M., Smith, R.J., 2003. Scottish Fish Farms Annual Production Survey, 2002. Fisher ies Research Services (53 pp.).

76. Telechea, F., Fontaine, P., 2014. Levels of domestication in fish: implications for the sus tainable future of aquaculture. Fish Fish. 15, 181–195.

77. Tilseth, S., Hansen, T., Møller, D., 1991. Historical development of salmon culture. Aqua culture 98, 1–9.

78. Torrissen, O., Jones, S., Asche, F., et al., 2013. Salmon lice — impact on wild salmonids and salmon aquaculture. J. Fish Dis. 36, 171–194.

79. Tveteras, R., 2002. Norwegian salmon aquaculture and sustainability: the relationship be tween environmental quality and industry growth. Mar. Resour. Econ. 17, 121–132.

80. Walker, A.J., 2009. Scottish Fish Farms Annual Production Survey 2008. Marine Scotland (53 pp.).

81. Walker, A.J., 2010. Scottish Fish Farm Production Survey 2009 Report. Marine Scotland Science (58 pp.).

82. Walker, A.J., McAlister, J.M., 2011. Scottish Fish Farm Production Survey 2010 Report. Ma rine Scotland Science (56 pp.).

83. Walker, A.J., Wallace, I.S., Munro, L.A., 2012. Scottish Fish Farm Production Survey 2011. Marine Scotland Science (58 pp.).

84. Ytrestøyl, T., Aas, T.S., Berge, G.M., Hatlen, B., Sørensen, M., Ruyter, B., Thomassen, M., Hognes, E.S., Ziegler, F., Sund, V., Åsgård, T., 2011. Resource utilisation and eco efficiency of Norwegian salmon farming in 2010. Nofima Report 53/2011 (65 pp.).

SECTION II - SUSTAINABLE FISH NUTRITION WITH FISH-INGREDIENT-FREE SOURCES.

CHAPTER 4 SUSTAINABLE AQUAFEEDS

Pallab K. Sarker , Anne R. Kapuscinski, Alison J. Lanois, Erin D. Livesey, Katie P. Bernhard, Mariah L. Coley

INTRODUCTION

Aquaculture has been rapidly expanding all over the world in the last few decades, at an average rate of 8–10% per year [1]. Between 1980 and 2010, aquaculture contribution to global fishery output for human consumption rose from nine percent to 47 percent and its use of artificial feeds rose from 50 to 66 percent of production [2,3]. Analysts expect aquaculture to provide an additional 23 million tonnes of aquatic foods by 2030 to maintain the current per-capita level of consumption of aquatic foods for a growing global population [3]. Responsible expansion of aquafeeds, *inter alia*, requires finding sustainable alternatives to fishmeal and fish oil ingredients, of which aquaculture is the largest user. Aquaculture feeds currently use over 80% of the world's fishmeal and fish oil [3], which are extracted from small ocean-caught fish. To date, fish oils (FO) are still regarded as good lipid sources of aquafeed formulations [4]. However, due to the short-fall from wild-harvested marine fish stocks for the production of fish oil [3], alternative lipid sources for the production of fish feed are essential to meet the demand of the growing aquaculture industry [5].

This research project focuses on reformulation of diets for Nile tilapia (*Oreochromis niloticus*), the second most farmed fish worldwide [6] and the most widely farmed in the United States [7]. This species is pre-adapted to thrive on fish-ingredient-free aquafeed because of its low trophic status. Commercial diets for tilapia have already greatly reduced inclusion levels of fish oil by replacing them with terrestrial vegetable oils, which have relatively low prices and large production volumes. Scientists have reported considerable success in partially or totally replacing fish oil with vegetable oil in many fish species [8, 9]. Several studies have shown that replacement of high-quality fish oil with vegetable oil sources negatively affects growth performance, fatty acid profile or health parameters of fish, such as gilthead seabream (*Sparus aurata*) [10] and European sea bass (*Dicentrarchus labrax* L.) [11], the magnitude of which depends on the source and level of replacement. Moreover, unbalanced essential amino acids, low levels of n3 PUFAs, lack of DHA and EPA, and high levels of anti-nutritional factors [12] limit the inclusion rates of terrestrial plant ingredients, even in diets for omnivorous species like tilapia [13,14].

Recent commercial-scale production of microalgae for biofuels and human nutritional supplements has also stimulated interest in microalgae for animal feeds [15]-[16]. Increasing attention has focused on marine microalgae for aquaculture feeds because of their elevated fatty acid profiles. Additionally, microalgae are at the base of the aquatic food

Citation: Sarker PK, Kapuscinski AR, Lanois AJ, Livesey ED, Bernhard KP, Coley ML (2016) Sustainable Aquafeeds. doi:10.1371/journal.pone.0156684

chain that produce the food resources that tilapia are adapted to consume. In contrast to terrestrial plant sources, microalgae are relatively high in essential long chain n-3 polyunsaturated fatty acids (n3 LC PUFA) such as DHA (C22:6n3) and EPA (C20:5n3), which are important both for maintaining fish health and imparting neurological, cardiovascular and anticancer benefits to humans [17,18]. Our recent evaluation of a marine microalga, *Schizochytrium* sp. (Sc), in tilapia found significantly higher digestibility of lipid and all unsaturated fatty acid fractions from Sc compared to freshwater microalgae [19]. This was the first study to report that dried whole cells of Sc are a highly digestible lipid source for Nile tilapia. However, the response of fish growth performance, feed utilization efficiency, and fatty acid deposition in muscle to dietary inclusion of Sc is unknown. Building on our previous findings on digestibility of Sc in tilapia [19], the study reported below determined whether whole-cell Sc is a quality fish oil substitute in Nile tilapia feed. We conducted a nutritional feeding trial with diets containing dried whole cells of this microalga designed to identify the optimum level of fish oil substitution (partial or complete) to support high growth and feed efficiency, and to measure whether inclusion of dried whole-cell Sc can raise n3 LC PUFAs deposition in juvenile tilapia fillets.

MATERIALS AND METHODS

Experimental Design, Fish Rearing and Feeding

We obtained Nile tilapia (*O. niloticus*) juveniles from Americulture Inc. (Animas, New Mexico, 88020, USA). We conducted this experiment in a temperature controlled wet lab at the Life Sciences Center, Dartmouth College (Hanover, NH, USA) using fifteen indoor, static-water 114-L cylindro-conical tanks. We filled each tank with charcoal filtered de-chlorinated tap water and provided aeration through an air stone diffuser via a low-pressure electrical blower. Each tank contained bio-ball and sponge biological filters. Prior to the start of the experiment, we randomly assigned 40 tilapia to each tank, with an initial mean weight of 1.52 ± 0.2 g/fish and accustomed the fish to a photoperiod cycle of 10 h light and 14 h dark. We acclimated fish to the experimental conditions for two weeks before starting the experiment, during which we fed them the control diet (see section, feed formulation and pellet preparation). We randomly allocated the five experimental diets to 15 tanks and fed each diet to three replicate tanks ($n = 3$) in a completely randomized design. We hand-fed fish three times daily at 0930, 1300 and 1700h for 84 days. At the start of the trial, we administered feed at a rate of 10% of body weight. We recorded the daily satiation ration and used this to guide a gradual reduction of the feeding rate to 4% of body weight at the end of the trial, following the approach of Hussein et al. 2012 and Karapanagiotidis et al. 2007 for tilapia [20,21], and NRC 2011 [4]. The experimental design and fish use protocol were approved by the Institutional Animal Care and Use Committee (IACUC) of Dartmouth College.

All water quality parameters, monitored during the course of the study, confirmed that we maintained excellent conditions for tilapia. We exchanged 10–15% of the tank water each week. We kept water temperature throughout the experiment within the range 26.4–28.2°C with a thermostat-regulated, immersion heater in each tank (Hagen Marina Submersible Pre-Set Aquarium Heater, 150W, model UPC 015561112345). The range of values for other variables were pH 7.17 to 7.60, dissolved oxygen 6.18 to 7.13 mg L^{-1}, nitrite 0.10 to 0.20 mg L^{-1}, and total ammonia nitrogen 0.23 to 0.53 mg L^{-1}.

Feed Formulation and Pellet Preparation

We formulated five iso-nitrogenous (38% crude protein), iso-energetic (14 kJ/g) and iso-lipidic (10% lipid) experimental diets, following the requirements for optimum growth of juvenile Nile tilapia [4]. We prepared a control diet for juvenile tilapia, adapted from a proven high quality formulation [22]. The diets differed from each other in their relative amounts of fish oil (menhaden-derived) and dried whole cells of *Schizochytrium* sp. (Sc). We used wheat flour (only 11% protein) as filler [20] and all experimental diets were iso-nitrogenous, iso-energetic, iso-lipidic composition, and also balance protein/lipid ratio differences between fish oil and its replacer (Sc) in the diets. We designated control feed containing fish oil (Sc0), and substituted 25% fish oil (Sc25), 50% fish oil (Sc50), 75% fish oil (Sc75), and 100% fish oil (Sc100) with dried whole cells of *Schizochytrium* sp. (Table 1). We obtained dried *Schizochytrium* sp. from Algamac™, Aquafauna Bio-marine, Inc., Hawthorne, CA, USA. We obtained menhaden fish oil from Double Liquid Feed Service, Inc., Danville, IL, USA. We produced the diets in accordance with our previous work [19], by weighing and mixing oil and dry ingredients in a stand mixer (Hobart Corporation, Tory, OH, USA) for 15 min; then blending water (330 ml kg^{-1} diet) into the mixture to attain a texture appropriate for pelleting; and running each diet through a meat grinder (Panasonic, MK-G20NR) to create 2 mm-diameter pellets. After pelleting, we dried the diets to a moisture content of 80–100 g kg^{-1} under a chemical fume hood at room temperature for 12 h and then stored the finished diets at -20°C. Table 2 reports the proximate composition, gross energy, and amino acid profiles of dried Sc and Table 3 reports the fatty acid profile of dried Sc and menhaden fish oil. Table 4 reports the fatty acid profiles of the five experimental diets.

Table 1. Formulation (g/100g diet) and proximate composition (%) and essential amino acids (% in the weight of diet as is) of five experimental diets for juvenile tilapia

	Diet				
Ingredient	**Control (Sc0)**	**Sc25**	**Sc50**	**Sc75**	**Sc100**
Fish meal*	20	20	20	20	20
Corn gluten meal	20	20	20	20	20
Soybean meal	20	20	20	20	20
Wheat flour	26.25	24.5	22.75	20.5	19.15
CaH_2PO_4	0.75	0.75	0.75	0.75	0.75
Vitamin mix[1]	1	1	1	1	1
Mineral mix[2]	1	1	1	1	1
Fish oil	9	6.75	4.5	2.25	0
Schizochytrium sp.	0	4	8	12.5	16.1
Choline chloride	2	2	2	2	2
Proximate composition (%)					
Dry matter	88.2	89	90.2	89.4	91.2
Crude protein	38.2	38.6	39.3	39.1	39.3
Lipid	11.1	11.1	10.6	10.7	10.2
Ash	7	7.4	7.2	8	8.6
NFE	31	31.8	32.2	31.4	32
Gross energy (kJ g-1)	14	14.1	14.2	14	14.1
Amino acids (% in the weight of diet as is)					
Arginine	2	2	2.1	2	2.2
Lysine	2	2	2.1	1.8	2.1
Isoleucine	1.4	1.4	1.6	1.4	1.5
Leucine	3.7	3.9	4.1	3.7	4.2
Histidine	0.8	0.8	0.8	0.8	0.8
Methionine	0.8	0.8	0.9	0.9	0.8
Cystine	0.5	0.5	0.5	0.5	0.5
Phenyle alanine	1.8	1.9	2	1.8	2.1
Threonine	1.3	1.3	1.5	1.2	1.5
Tryptophan	0.2	0.2	0.2	0.1	0.1
Valine	1.6	1.6	1.8	1.6	1.8

[1]Vitamin premix (mg kg˙1 dry diet unless otherwise stated):vitamin A (as acetate), 7500 IU kg˙1 dry diet; vitamin D3 (as cholecalcipherol), 6000 IU kg˙1 dry diet; vitamin E (as DL-a-tocopherylacetate), 150 IU kg˙1 dry diet; vitamin K (as menadione Na-bisulphate), 3; vitamin B12 (as cyanocobalamin), 0.06; ascorbic acid (as ascorbyl polyphosphate), 150; D-biotin, 42; choline (as chloride), 3000; folic acid, 3; niacin (as nicotinic acid), 30; pantothenic acid, 60; pyridoxine, 15; riboflavin, 18; thiamin, 3

[2]Mineral premix (mg kg˙1 dry diet unless otherwise stated):ferrous sulphate, 0.13; NaCl, 6.15; copper sulphate, 0.06; manganese sulphate, 0.18; potassium iodide, 0.02; zinc sulphate, 0.3; carrier (wheat middling or starch).

*Omega Protein, Inc. Houston, Texas 77042, as manufacturer specification, the guaranteed gross composition analysis: crude protein, 60%; crude fat, 6%; fiber, 2%.

Table 2. Proximate chemical composition, gross energy, essential, and non-essential amino acids of whole dried cell of *Schizochytrium* sp

	Ingredient
Composition	***Schizochytrium* sp.**
Dry matter	96.5
Crude protein	11.9
Lipid	54.1
Ash	8.7
Crude fibre	2.4
Energy	17.7
Essential amino acids (% in the weight of ingredient as is)	
Arginine	0.8
Lysine	0.5
Isoleucine	0.4
Leucine	0.7
Histidine	0.3
Methionine	1.2
Phenylealanine	0.4
Threonine	0.4
Tryptophan	0.2
Valine	0.6
Non-essential amino acids ((% in the weight of ingredient as is)	
Aspartic acid	1.2
Serine	0.4
Glutamic acid	1.9
Glycine	0.5
Tyrosine	0.3
Alanine	0.8
Hydroxyproline	0.0
Proline	0.5

Table 3. Fatty acid content (% of total fatty acids) of lipid sources (menhaden fish oil and whole cell dried *Schizochytrium* **sp)** used in the experimental diets

	Lipid source	
Fatty acids (% of TFA)	**Fish oil**	***Schizochytrium* sp**
14:00	8.1	9.3
15:00	0.6	0.5
16:00	17.9	24.4
17:00	0.6	ND
18:00	3.1	0.5
20:00	0.2	0.1
22:00	0.1	0.1
24:00	ND	ND
Total SFA	6.6	9.2
16:1n9	0.2	ND
16:1n7	13.9	0.2
18:1n9	5.2	0.1
18:1n7	3.3	ND
20:1n9	0.6	ND
20:1n7	0.2	ND
22:1n11	ND	ND
22:1n9	0.1	ND
24:1n9	0.4	1.4
Total MUFA	23.9	1.7
18:2n6	1.5	ND
18:3n6	0.3	0.2
20:2n6	0.2	ND
20:3n6	0.2	0.3
20:4n6 ARA	1.3	1.4
22:4n6	0.3	0.1
22:5n6	0.5	15.8
Total n6 PUFA	4.3	17.8
18:3n3 ALA	1.5	ND
18:4n3	2.7	0.6
20:3n3	0.2	0.1
20:4n3	1.4	0.8
20:5n3 EPA	14.9	0.8
22:5n3	2.6	0.4
22:6n3 DHA	13	43.2
Total n3 PUFA	36.3	45.9
Total PUFA	40.6	63.7
Total n6 LCPUFA	2.5	17.6
Total n3 LCPUFA	32.1	45.3
n3:n6 PUFA ratio	8.4	2.6

SFA, saturated fatty acids (sum of all fatty acids without double bonds); MUFA, monounsaturated fatty acids (sum of all fatty acids with a single bond); PUFA, polyunsaturated fatty acids (sum of all fatty acids with ≥2 double bonds); n6 PUFA, omega 6 polyunsaturated fatty acids (18:2, 18:3, 20:2, 20:3, 20:4, 22:4, 22:5); n6 LCPUFA, omega 6 long chain polyunsaturated fatty acids (20:2, 20:3, 20:4, 22:4, 22:5), n3 PUFA, omega 3 polyunsaturated fatty acids (18:3, 18:4, 20:3, 20:4, 20:5, 22:5, 22:6); n3 LCPUFA, omega 3 long chain polyunsaturated fatty acids (20:3, 20:4, 20:5, 22:5, 22:6); EPA, eicosapentaenoic acid; DHA, docosahexaenoic acid; ND, not detectable (<0.1% of total fatty acids).

Table 4. Fatty acid content (% of total fatty acids) of experimental diets

			Diet		
Fatty acids (% of TFA)	**Control (Sc0)**	**Sc25**	**Sc50**	**Sc75**	**Sc100**
14:00	7.6	8	8.2	8	8
15:00	0.5	0.6	0.5	0.5	0.5
16:00	18.4	19.6	20.7	21.7	22.8
17:00	0.6	0.5	0.4	0.3	0.1
18:00	3.2	3.0	2.4	1.8	1.3
20:00	0.2	0.2	0.2	0.1	0.2
22:00	0.2	0.1	0.1	0.1	0.2
24:00	0.1	0.1	0.1	0.1	0.2
Total SFA	30.8	32.1	32.6	32.6	33.3
16:1n9	0.2	0.2	0.2	ND	ND
16:1n7	11	9.1	6.8	4.1	2.0
18:1n9	7.8	6.7	5.6	5.0	4.2
18:1n7	2.9	2.4	1.8	1.2	0.7
20:1n9	0.5	0.4	0.3	0.2	0.1
20:1n7	0.2	0.1	0.1	0.3	ND
22:1n11	ND	ND	ND	ND	ND
22:1n9	ND	ND	ND	ND	ND
24:1n9	0.3	0.5	0.6	0.9	1.0
Total MUFA	23	19.4	15.4	11.7	8.0
18:2n6	10.1	9.4	9.1	9.2	9.6
18:3n6	0.3	0.2	0.2	0.2	0.1
20:2n6	0.2	0.1	0.1	ND	ND

20:3n6	0.2	0.2	0.2	ND	0.3
20:4n6 ARA	1.2	1.3	1.2	1.2	1.3
22:4n6	0.2	0.2	0.1	0.1	0.1
22:5n6	0.8	3	5.7	8.5	10.8
Total n6 PUFA	13	14.4	16.6	19.2	22.2
18:3n3 ALA	1.6	1.4	1.1	0.9	0.7
18:4n3	1.9	1.6	1.2	0.8	0.4
20:3n3	0.2	0.1	0.1	0.1	ND
20:4n3	1.0	1.0	0.8	0.8	0.6
20:5n3 EPA	11.1	9.22	7.0	4.6	2.5
22:5n3 DPA	2.1	1.8	1.3	1.0	0.6
22:6n3 DHA	10.4	15	20.7	26.2	30.8
Total n3 PUFA	28.3	30.12	32.2	34.4	35.6
Total PUFA	41.3	44.52	48.8	53.6	57.8
Total n6 LCPUFA	2.6	4.8	7.3	9.8	12.5
Total n3 LCPUFA	24.8	27.12	29.9	32.7	34.5
n3:n6 PUFA ratio	2.2	2.1	1.9	1.8	1.6

SFA, saturated fatty acids (sum of all fatty acids without double bonds); MUFA, monounsaturated fatty acids (sum of all fatty acids with a single bond); PUFA, polyunsaturated fatty acids (sum of all fatty acids with $\geq$2 double bonds); n6 PUFA, omega 6 polyunsaturated fatty acids (18:2, 18:3, 20:2, 20:3, 20:4, 22:4, 22:5); n6 LCPUFA, omega 6 long chain polyunsaturated fatty acids (20:2, 20:3, 20:4, 22:4, 22:5), n3 PUFA, omega 3 polyunsaturated fatty acids (18:3, 18:4, 20:3, 20:4, 20:5, 22:5, 22:6); n3 LCPUFA, omega 3 long chain polyunsaturated fatty acids (20:3, 20:4, 20:5, 22:5, 22:6); EPA, eicosapentaenoic acid; DHA, docosahexaenoic acid; ND, not detectable (<1% of total fatty acids).

Biological Sampling Procedures, Fillet Preparations and Growth Measurements

We bulk weighed the fish at the beginning of the experiment, and then every 3 weeks until the end of the experiment (84 days). We stopped feeding the fish for 24 h prior to each bulk weight-sampling event. We sampled 5 fish per tank at day 42 (middle) and day 84 (final) for fillet fatty acid composition analysis. We immediately euthanized the fish by single cranial pithing, filleted the fish from a standardized dorso-anterior landmark, packaged in sterile polythene bags (Whirl-pak, Naso, Fort Atkinson, Wisconsin) and stored frozen (-20°C) until fatty acid analysis. At the end of the feeding experiment, we sampled 10 fish per tank, pooled, ground into a homogeneous slurry, freeze-dried, reground and stored at—20°C until analyzed for the whole-body proximate analysis.

We determined the dietary effects on growth by evaluating final weight, weight gain, feed conversion ratio (FCR), specific growth rate (SGR), protein efficiency ratio (PER) and percent survival. We calculated the indices as follows: Weight gain (g) = final weight–initial weight; FCR, feed conversion ratio = feed intake (as fed basis)/ weight gain; SGR, specific growth rate (%/day) = 100 x (ln final wet weight (g)–ln initial wet weight (g)) / Time (days); PER, protein efficiency ratio = weight gain (g)/protein fed (g); and % survival rate = (Final number of fish / Initial number of fish) x 100.

Biochemical Analysis

We sent the five types of samples (pure microalgae, fish oil, diets, wholebody and fillet) to New Jersey Feed Laboratory, Inc. (Ewing, NJ, USA) for the following types of analysis [23]: moisture (AOAC, 930.15), crude protein (AOAC 990.03), lipid (AOAC 920.39), ash (AOAC 942.05), crude fiber (AOAC 1978.10), The nitrogen-free extract (NFE; i.e. carbohydrate) was determined by difference: NFE = 100 − (% protein + % lipid + % fiber + %moisture + % ash), energy (automated oxygen bomb calorimeter), amino acids (high-performance liquid chromatography, HPLC analysis, via AOAC methods 994.12, 985.28, 988.15, and 994.12) and fatty acid (fatty acids methyl esters, FAME analysis, via AOAC method 963.22.). Table 3reports the fatty acid content of the two lipid sources used, menhaden fish oil and Sc whole dried cells; and Table 4 reports the fatty acid content of the five experimental diets.

Statistical Analysis

We conducted one-way analysis of variance (ANOVA) of growth performance and feed utilization parameters, whole body proximate composition, fillet fatty acid deposition and, when significant differences were found, compared the treatment means using Tukey's test of multiple comparisons with 95% level of significance. Data were expressed as the mean ± SE of three replicates. We conducted a repeated measure analysis within the general linear model (GLM) framework for 42 and 84 days of fillet fatty acids (%TFA) data to determine whether there were differences among dietary treatments, sampling time, or main effect interactions (diet x time). We carried out all statistical analyses using the IBM Statistical Package for the Social Sciences (SPSS) program for Windows (v. 21.0, Armonk, NY, USA). We computed the Pearson correlation coefficient between Sc inclusion level in a diet and tissue deposition of DHA, weight gain and FCR data.

RESULTS

We conducted a 84-day growth experiment with dried whole-cells of Sc and found (Table 5): significantly higher weight gain (g) and protein efficiency ratio (PER), and lower (improved) feed conversion ratio (FCR) and feed intake when fish oil was fully replaced by Sc (Sc100 diet) compared to control diet containing fish oil (Sc0 diet). Tilapia appeared healthy at the end of the experiment, and showed no difference in SGR and survival rate among all diets. Weight gain was in the range of 23.8–27.3 g. Feed intake ranged from 23.6 to 27.0 g/fish. Feed conversion ratios (FCR) were within the range 0.9–1.1 and protein efficiency ratios (PER) were within the range 2.4–3.1 among all dietary treatments. We found a strong proportional linear relationship between the Sc content of the diet and weight gain ($y = 0.0193x + 23.895$; $r = 0.970$; $P < 0.01$). The FCR decreased (improved) as the dietary Sc content increased, showing an inverse relationship ($y = -0.013x + 1.101$; $r = 0.970$; $P<0.05$).

Table 5. Initial weight, final weight, weight gain, percentage weight gain, feed conversion ratio (FCR), specific growth rate (SGR), protein efficiency ratio (PER), feed intake, and survival rate of tilapia fed experimental diets.

	Diet[1]					ANOVA	
	Sc0	Sc25	Sc50	Sc75	Sc100	F value	P value
Initial weight (g)	1.4 ± 0.1	1.42 ± 0.1	1.7 ± 0.2	1.5 ± 0.1	1.6 ± 0.1	1.5	0.27
Final weight (g)	25.3 ± 0.3bc	26.4 ± 0.4bc	27.2 ± 0.7ab	27.4 ± 0.3ab	28.8 ± 0.2a	9.9	<0.01
Weight gain (g)[2]	23.8 ± 0.4bc	24.9 ± 0.3bc	25.5 ± 0.7ab	25.8 ± 0.2ab	27.3 ± 0.2a	9.3	<0.01
FCR[3]	1.1 ± 0.0bc	1.0 ± 0.0bc	1.0 ± 0.1ab	0.9 ± 0.0ab	0.9 ± 0.0a	10.2	<0.01
SGR[4]	3.4 ± 0.1	3.5 ± 0.1	3.3 ± 0.0	3.4 ± 0.1	3.5 ± 0.1	0.8	0.52
PER[5]	2.4 ± 0.1bc	2.6 ± 0.1bc	2.7 ± 0.2ab	2.8 ± 0.1ab	3.1 ± 0.1a	9.6	<0.01
Feed intake (g/fish)	27.0 ± 0.3a	25.9 ± 0.4ab	25.1 ± 0.6bc	24.9 ± 0.2bc	23.6 ± 0.2c	10.4	<0.01
%Survival rate[6]	92.7 ± 0.37	95.2 ± 1.5	95.8 ± 0.8	92.3 ± 0.2	96.3 ± 2.6	0.9	0.57

Values are means of ±SE of three replicate groups (n = 3)
[1]Mean values not sharing a superscript letter in the same row differ significantly (P < 0.05).
[2]Weight gain (g) = final wet weight–initial wet weight
[3]FCR, feed conversion ratio = feed intake/ weight gain
[4]Specific growth rate SGR (%/day) = 100 x (ln final wet weight (g)–ln initial wet weight (g)) / Time (days);
[5]PER, protein efficiency ratio = weight gain (g)/protein fed (g)
[6]%Survival = (Final number of fish / Initial number of fish) x 100.

The whole body proximate composition of Nile tilapia fillets did not differ among dietary treatments (data not shown). This included moisture (ranged from 70.5 to 71.0%), crude protein (ranged from 16.1 to 16.6.0%), ash (ranged from 4.5 to 4.8%) and total lipid (ranged from 6.2 to 6.7%).

The fillet fatty acids composition (% of total fatty acids) was significantly influenced either by the dietary treatment or the length of the experiment or both factors (Table 6). With the exception of 18:0, all SFA fractions showed significant time effects and were higher at the middle (42 days) than at the end of the experiment (84 days). The composition of four SFA fractions (15:0, 18:0, 20:0, and 22:0) did not differ across dietary treatments. One SFA, palmitic acid (16:0), had the highest final concentration in the fillet irrespective of dietary treatment, as well as significantly higher amounts deposited in the flesh of tilapia fed the Sc100 diet compared to fish fed the Sc0 diet (P<0.01). It also showed a significant diet and time interaction (P = 0.03). Concentrations of 14:0 and total saturated fatty acid (SFA) were significantly higher in the Sc100-fed fish than in Sc0-fed fish at 42 days, as well as at 84 days (P<0.01). These results were probably due to the higher supply of these two components in the Sc100 diet compared to the Sc0 diet (Table 4).

Table 6. Fatty acid (% of total fatty acids) content of fillets from Nile tilapia fed experimental diets for 84 days and 42 days sampling events; average ± SE for 15 fish per diet (5 fish/replicate with 3 replicates/diet)§

	Control (Sc0)		Sc25		Sc50		Sc75		Sc100		P value‡		
Fatty acid (% TFA)	84 days	42 days	84 days	42 days	84 days	42 days	84 days	42 days	84 days	42 days	Diet	Time	Diet x Time interaction
14:00	5.6 ± 0.1^{b}	7.6 ± 0.7^{b}	6.1 ± 0.9^{b}	8.8 ± 0.2^{ab}	6.1 ± 0.3^{ab}	9.6 ± 0.4^{a}	6.9 ± 0.6^{b}	10.5 ± 0.2^{a}	7.5 ± 0.3^{b}	9.6 ± 0.1^{a}	0.01	<0.01	0.12
15:00	0.6 ± 0.0	0.7 ± 0.0	0.5 ± 0.0	0.8 ± 0.0	0.5 ± 0.0	0.8 ± 0.0	0.5 ± 0.0	0.6 ± 0.0	0.4 ± 0.0	0.6 ± 0.0	0.06	<0.01	0.47
16:00	19.9 ± 0.1^{c}	28.6 ± 1.1^{b}	24 ± 2.0^{bc}	32.2 ± 0.9^{b}	24.0 ± 0.6^{bc}	38.8 ± 0.4^{a}	25.6 ± 2.4^{bc}	41.2 ± 0.5^{a}	30.0 ± 1.6^{b}	42.3 ± 0.6^{a}	<0.01	<0.01	0.03
17:00	0.7 ± 0.0^{a}	0.8 ± 0.0^{a}	0.6 ± 0.0^{b}	0.8 ± 0.0^{a}	0.5 ± 0.0^{ab}	0.7 ± 0.0^{a}	0.4 ± 0.0^{c}	0.5 ± 0.0^{bc}	0.3 ± 0.0^{c}	0.4 ± 0.0^{c}	<0.01	<0.01	0.21
18:00	6.9 ± 0.2	7.9 ± 0.2	6.1 ± 0.5	8.0 ± 0.5	5.7 ± 0.2	8.3 ± 0.3	5.5 ± 0.4	8.2 ± 0.2	6.1 ± 0.3	7.9 ± 0.3	0.99	<0.01	0.65
20:00	0.3 ± 0.0	0.3 ± 0.0	0.3 ± 0.0	0.4 ± 0.0	0.3 ± 0.0	0.3 ± 0.0	0.2 ± 0.0	0.3 ± 0.0	0.3 ± 0.0	0.4 ± 0.0	0.41	0.08	0.2
22:00	0.1 ± 0.0	0.1 ± 0.0	0.1 ± 0.0	0.1 ± 0.0	0.1 ± 0.0	0.1 ± 0.0	ND	0.1 ± 0.0	0.1 ± 0.0	ND			
24:00:00	0.1 ± 0.0	ND	0.1 ± 0.0	0.1 ± 0.0	0.1 ± 0.0	ND	0.1 ± 0.0	0.1 ± 0.0	0.1 ± 0.0	0.1 ± 0.0	0.99	<0.01	0.65
Total SFA	34.2 ± 0.2^{c}	46.0 ± 2.3^{b}	37.8 ± 3.6^{c}	51.2 ± 1.8^{b}	37.3 ± 0.6^{c}	58.6 ± 0.6^{a}	39.2 ± 3.5^{c}	61.5 ± 0.5^{a}	44.8 ± 2.3^{b}	61.3 ± 0.9^{a}	<0.01	<0.01	0.07
16:1n9	0.4 ± 0.2^{a}	0.4 ± 0.0	0.3 ± 0.0^{ab}	0.4 ± 0.0	0.3 ± 0.0^{ab}	0.3 ± 0.0	0.3 ± 0.0^{ab}	0.3 ± 0.0	0.2 ± 0.0^{b}	0.2 ± 0.1	0.13	0.72	0.96
16:1n7	8.7 ± 0.1^{b}	11.1 ± 0.3^{a}	6.7 ± 1.1^{c}	9.5 ± 0.1^{a}	5.8 ± 0.2^{c}	7.6 ± 0.3^{b}	5.3 ± 0.6^{c}	5.8 ± 0.3^{c}	2.9 ± 0.0^{d}	2.9 ± 0.2^{d}	<0.01	<0.01	0.09
18:1n9	9.8 ± 0.2^{b}	14.4 ± 0.0^{a}	9.9 ± 0.6^{b}	12.8 ± 0.4^{a}	8.3 ± 0.5^{b}	12.1 ± 0.6^{a}	9.8 ± 1.0^{b}	12.2 ± 0.6^{a}	8.3 ± 0.4^{b}	9.1 ± 0.6^{b}	0.01	<0.01	0.05
18:1n7	4.1 ± 0.1^{b}	5.1 ± 0.0^{a}	3.5 ± 0.3^{bc}	4.5 ± 0.1^{a}	3.1 ± 0.1^{bc}	4.0 ± 0.1^{b}	2.8 ± 0.2^{c}	3.3 ± 0.1^{bc}	2.1 ± 0.0^{c}	2.5 ± 0.0^{c}	<0.01	<0.01	0.15
20:1n9	0.6 ± 0.0^{b}	0.8 ± 0.0^{a}	ND	0.8 ± 0.0^{a}	0.5 ± 0.0^{b}	0.6 ± 0.0^{b}	0.5 ± 0.0^{b}	0.7 ± 0.0^{a}	0.4 ± 0.0^{b}	0.4 ± 0.0^{b}	<0.01	<0.01	0.08
20:1n7	0.2 ± 0.0^{a}	0.1 ± 0.0^{b}	0.2 ± 0.0^{a}	0.1 ± 0.0^{b}	0.1 ± 0.0^{b}	ND	0.1 ± 0.0^{b}	0.1 ± 0.0^{b}	ND	ND	<0.01	<0.01	<0.01
24:1n9	0.3 ± 0.0^{a}	0.1 ± 0.0^{b}	0.2 ± 0.0^{ab}	0.1 ± 0.0^{b}	0.1 ± 0.0^{b}	ND	0.1 ± 0.0^{b}	0.1 ± 0.0^{b}	ND	ND	0.03	<0.01	<0.01
Total MUFA	24.1 ± 0.3^{b}	32.0 ± 0.7^{a}	21.2 ± 2.2^{b}	28.2 ± 0.4^{a}	18.2 ± 0.7^{bc}	24.6 ± 0.9^{b}	18.9 ± 1.9^{bc}	22.5 ± 1.1^{b}	14.0 ± 0.6^{c}	15.1 ± 0.9^{c}	<0.01	<0.01	0.13
18:2n6	8.6 ± 0.0^{a}	5.0 ± 0.8^{b}	6.9 ± 0.7^{b}	4.2 ± 0.4^{bc}	7.2 ± 0.2^{a}	2.8 ± 0.2^{c}	6.6 ± 0.2^{b}	2.3 ± 0.1^{c}	5.8 ± 0.3^{b}	4.0 ± 0.6^{c}	<0.01	<0.01	0.18
18:3n6	0.3 ± 0.0^{a}	ND	0.2 ± 0.0^{ab}	0.1 ± 0.0^{b}	0.2 ± 0.0^{ab}	ND	0.2 ± 0.0^{ab}	ND	0.1 ± 0.0^{b}	ND	0.02	<0.01	0.03
20:2n6	0.4 ± 0.0^{a}	0.2 ± 0.0^{ab}	0.3 ± 0.0^{a}	0.2 ± 0.0^{ab}	0.3 ± 0.0^{a}	ND	0.3 ± 0.0^{a}	0.1 ± 0.0^{b}	0.2 ± 0.0^{ab}	0.1 ± 0.0^{b}	<0.01	<0.01	<0.01
20:3n6	0.4 ± 0.0^{a}	0.2 ± 0.0^{b}	0.4 ± 0.0^{a}	0.2 ± 0.0^{b}	0.3 ± 0.0^{ab}	ND	0.4 ± 0.0^{a}	0.1 ± 0.0^{b}	0.3 ± 0.0^{ab}	ND	<0.01	<0.01	0.13
20:4n6 ARA	1.7 ± 0.0	0.9 ± 0.0	1.6 ± 0.1	0.9 ± 0.0	1.7 ± 0.1	0.8 ± 0.1	1.5 ± 0.1	0.8 ± 0.0	1.6 ± 0.0	1.1 ± 0.0	0.6	<0.01	0.07
22:4n6	0.5 ± 0.0^{a}	0.2 ± 0.0^{b}	0.4 ± 0.0^{a}	ND	0.4 ± 0.0^{a}	ND	0.4 ± 0.0^{a}	ND	0.3 ± 0.0^{ab}	ND	0.04	<0.01	0.22
22:5n6	0.8 ± 0.0^{d}	0.4 ± 0.0^{d}	4.2 ± 1.0^{b}	1.5 ± 0.1^{c}	5.6 ± 0.2^{ab}	2.0 ± 0.2^{c}	6.2 ± 1.0^{a}	2.3 ± 0.2^{c}	7.8 ± 0.5^{a}	3.9 ± 0.1^{ab}	<0.01	<0.01	<0.01
Total n6 PUFA	12.6 ± 0.1	6.9 ± 1.0	13.9 ± 1.8	7.1 ± 0.5	15.6 ± 0.5	5.6 ± 0.4	15.5 ± 1.4	5.6 ± 0.5	16.1 ± 0.9	9.1 ± 0.6	0.43	<0.01	<0.01
18:3n3 ALA	1.0 ± 0.0^{a}	0.4 ± 0.0^{b}	0.6 ± 0.1^{ab}	0.3 ± 0.0^{b}	0.7 ± 0.0^{ab}	0.1 ± 0.0^{c}	0.5 ± 0.0^{b}	0.1 ± 0.0^{c}	0.3 ± 0.0^{b}	0.2 ± 0.1^{c}	<0.01	<0.01	<0.01
18:4n3	0.7 ± 0.0^{a}	0.4 ± 0.0^{b}	0.5 ± 0.0^{a}	0.2 ± 0.0^{bc}	0.5 ± 0.0^{a}	0.1 ± 0.0^{c}	0.3 ± 0.0^{bc}	0.1 ± 0.0^{c}	0.1 ± 0.0^{c}	ND	0.03	<0.01	0.1
20:3n3	0.2 ± 0.0^{ab}	ND	0.1 ± 0.0^{b}	ND	0.2 ± 0.0^{ab}	ND	0.1 ± 0.0^{b}	ND	0.1 ± 0.0^{b}	ND	0.02	<0.01	0.02
20:4n3	0.9 ± 0.0^{a}	0.3 ± 0.0^{c}	0.6 ± 0.0^{b}	0.2 ± 0.0^{c}	0.6 ± 0.0^{b}	ND	0.5 ± 0.0^{a}	0.1 ± 0.0^{c}	0.3 ± 0.0^{c}	ND	<0.01	<0.01	0.03
20:5n3 EPA	3.9 ± 0.1^{a}	1.8 ± 0.3^{b}	2.2 ± 0.3^{b}	1.3 ± 0.1^{c}	1.9 ± 0.1^{b}	0.7 ± 0.1^{d}	1.3 ± 0.2^{c}	0.5 ± 0.0^{d}	0.6 ± 0.0^{d}	0.4 ± 0.1^{d}	<0.01	<0.01	<0.01
22:5n3 DPA	6.8 ± 0.1^{a}	2.7 ± 0.5^{c}	4.6 ± 0.7^{b}	1.9 ± 0.2^{c}	4.2 ± 0.1^{b}	1.0 ± 0.1^{cd}	3.0 ± 0.5^{bc}	0.7 ± 0.0^{d}	1.8 ± 0.1^{cd}	0.6 ± 0.0^{d}	<0.01	<0.01	<0.01
22:6n3 DHA	13.0 ± 0.3^{b}	7.7 ± 1.0^{c}	16.7 ± 2.7^{ab}	8.4 ± 0.6^{c}	18.9 ± 0.5^{a}	8.1 ± 0.8^{b}	19.0 ± 2.1^{a}	8.0 ± 0.7^{c}	20.5 ± 1.4^{a}	12.4 ± 0.7^{b}	<0.01	<0.01	<0.01
Total n3 PUFA	26.5 ± 0.3	13.3 ± 0.0	26.9 ± 3.6	12.3 ± 0.0	27.0 ± 0.7	10.0 ± 0.0	24.7 ± 3.7	10.4 ± 0.0	26.1 ± 1.7	13.6 ± 0.0	0.6	<0.01	0.15
Total PUFA	39.1 ± 0.4	20.2 ± 3.0	39.2 ± 5.6	19.5 ± 1.7	42.6 ± 1.2	15.6 ± 1.4	40.2 ± 5.2	16.0 ± 1.1	39.8 ± 2.7	22.7 ± 0.8	0.46	<0.01	0.14
Total n6 LCPUFA	3.8 ± 0.1^{c}	1.9 ± 0.1^{c}	6.8 ± 1.2^{b}	2.9 ± 0.1^{c}	8.3 ± 0.4^{ab}	2.8 ± 0.3^{c}	8.7 ± 1.3^{ab}	3.3 ± 0.2^{c}	10.2 ± 0.6^{a}	5.1 ± 0.1^{b}	<0.01	0	<0.01
Total n3 LCPUFA	24.8 ± 0.3	12.5 ± 1.8	24.2 ± 3.5	11.8 ± 1.0	25.8 ± 0.6	9.8 ± 0.9	23.9 ± 3.7	9.3 ± 0.8	23.3 ± 1.7	13.4 ± 0.8	0.67	0	0.06
n3:n6 PUFA ratio¶	2.1 ± 0.0^{a}	1.9 ± 0.0^{a}	1.8 ± 0.0^{b}	1.7 ± 0.0^{b}	1.7 ± 0.0^{b}	1.7 ± 0.0^{b}	1.5 ± 0.1^{bc}	1.8 ± 0.0^{b}	1.4 ± 0.0^{c}	1.4 ± 0.1^{c}	<0.01	0.07	<0.01
DHA:EPA ratio	3.3 ± 0.2^{c}	4.2 ± 0.3^{c}	7.6 ± 0.9^{bc}	6.5 ± 0.4^{bc}	9.9 ± 0.6^{b}	11.6 ± 2.1^{b}	14.6 ± 2.4^{b}	16 ± 0.7^{b}	34.2 ± 1.7^{a}	31 ± 4.9^{a}	<0.01	0.43	0.84

SFA, saturated fatty acids (sum of all fatty acids without double bonds); MUFA, monounsaturated fatty acids (sum of all fatty acids with a single bond); PUFA, polyunsaturated fatty acids (sum of all fatty acids with ≥2 double bonds); n6 PUFA, omega 6 polyunsaturated fatty acids (18:2, 18:3, 20:2, 20:3, 20:4, 22:4, 22:5); n6 LCPUFA, omega 6 long chain polyunsaturated fatty acids (20:2, 20:3, 20:4, 22:4, 22:5), n3 PUFA, omega 3 polyunsaturated fatty acids (18:3, 18:4, 20:3, 20:4, 20:5, 22:5, 22:6); n3 LCPUFA, omega 3 long chain polyunsaturated fatty acids (20:3, 20:4, 20:5, 22:5, 22:6); EPA, eicosapentaenoic acid; DHA, docosahexaenoic acid, and n3:n6 ratio calculated for total n3 PUFA: total n6 PUFA. Mean values across the row not sharing a common superscript were significantly different as determined by Tukey's HSD test, P<0.05. ND, not detectable (<0.1% of total fatty acids).

§ in many cases ± 0.0 (S.E) values are rounding error.

¶We first computed the ratio for each replicate and then computed an average and SEM for the diet treatment.

‡ Significance probability associated with F-statistics.

With respect to MUFAs in the fillet, total MUFA content and all fractions, except for 16:ln9, had significant time effects and generally showed a higher content at the middle (42 days) of the experiment compared to the end of the experiment (84 days). The concentrations of 16:1n7, 18:1n7, 20:1n9, 20:1n7, and total MUFAs were significantly affected by dietary treatments and time. Fish fed the Sc0 diet displayed the highest amount of MUFA, which was directly related to the MUFA content in the experimental diets. Irrespective of the diet, oleic acid (18:1n9) was the most abundant monounsaturated fatty acid (MUFA) in the fillet.

Most of the individual polyunsaturated fatty acids (PUFAs) varied greatly among five dietary treatments and time. Regarding n6 fatty acids, all n6 fractions including total n6 PUFA content had significant time effects and were higher at the end of the experiment (84 days) than at the middle (42 days). The total n6 PUFA content was not affected by the diet ($P = 0.43$) but showed a significant time effect and interaction between diet and time ($P<0.01$). From 42 days to 84 days, total n6 PUFA content increased by 12.6% in fish fed the Sc0 diet, and by 16.1% in fish fed the SC100 diet. The Sc0 fed fish contained the highest amounts of 18:2n6, 18:3n6, and 20:3n6. In contrast, fish fed the Sc0 diet had significantly decreased amounts of 22:5n6 compared to fish fed the Sc25, Sc50, Sc75, and Sc100 diet. At the end of the experiment, concentrations of 20:4n6 in the fillet did not differ among five dietary treatments ($P = 0.6$).

With respect to the n3 fatty acids, most of the n3 PUFA including 22:6n3 DHA and 20:5n3 EPA were significantly influenced by the dietary treatment or time or both. The n3 PUFA content was generally higher at the end of the experiment (84 days) than at the middle (42 days). For example, from 42 days to 84 days, the 22:6n3 DHA content increased by 13.0% in fish fed the Sc0 diet, and by 20.5% in fish fed the SC100 diet. Tilapia fed the Sc100 diet had the highest contents of 22:6n3 DHA in the fillet lipids at the end of the experiment, and reflected the higher 22:6n3 DHA supplied by this diet. Furthermore, increasing the levels of Sc (Sc50, Sc75 and Sc100), which corresponded to reduced levels of fish oil, resulted in significant increases in the fillet 22:6n3 DHA compared to the control diet (Sc0) at the end of the feeding experiment. Tilapia fed the Sc0 diet had significantly increased amounts of 18:3n3 compared to the Sc75 and the Sc100 diet. They also exhibited significantly increased amounts of 20:5n3 EPA and 22:5n3 DPA compared to the four Sc inclusion diets due to a higher concentration of 18:3n3 in the Sc0 diet. However, the amounts of total n3 PUFA, total PUFA, and total n3 LC PUFA were not significantly different ($P>0.01$) in any of the diets. The n3:n6 and DHA: EPA ratios were significantly influenced by the dietary treatment. The n3:n6 PUFA ratio in the fillet was the highest in fish fed the Sc0 diet. Tilapia fed the Sc100 diet had significantly increased DHA level, hence increased DHA: EPA ratio compared to the Sc0, Sc50, and Sc75 diet.

Amounts of 22:6n3 DHA deposited in the fish fillet (mg/100g fillet) significantly increased in fish fed the Sc100 diet compared to the Sc0 diet (Table 7). With increasing Sc inclusion levels in the diet, the deposition of DHA in the fillet increased from 143.5 mg/100g for the inclusion level of 0 g Sc /kg diet (Sc0 diet) to 261.8 mg/100g for the inclusion level of 161 g Sc /kg diet (Sc100 diet). The relationship between Sc inclusion level in the diet and deposition of DHA (mg/100g) in the fillets of tilapia was positively correlated ($y = 7.1423x + 155.8$; $r = 0.9459$; $P<0.01$; Fig 1).

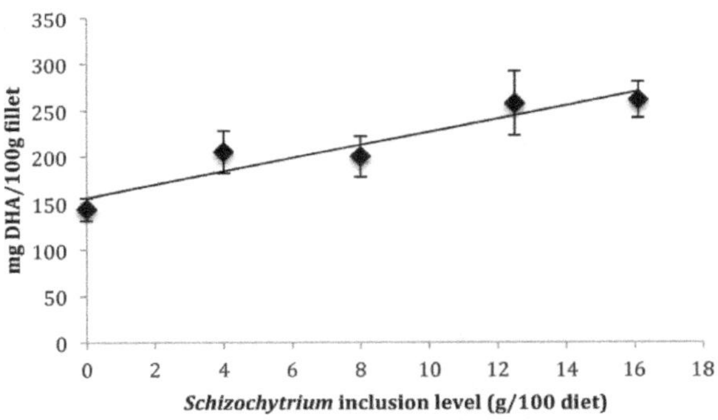

Figure 1. Linear correlation between the Sc inclusion levels in diet and DHA (mg/100g fillet) deposition in fish fillet.

Each point represents the average value (±SE) of three tanks per diet. The amount of DHA is described by the function of $y = 7.1423x + 155.8$. The correlation coefficient (r) was 0.9459 ($p<0.01$).

Table 7. Amounts of lipid and major n3 and n6 PUFA in the fillet (wet weight basis) of Nile tilapia fed experimental diets for 84 days; average ± SE for 15 fish per diet (5 fish/replicate with 3 replicates/diet)

	Fillet					F value	P value
Composition	Control (FO)	SC25	SC50	SC75	SC100		
Lipid (g/100g)	2.2 ± 0.20	2.2 ± 0.74	2.3 ± 0.26	2.3 ± 0.18	2.2 ± 0.06	0.03	0.99
18:2n6 LA (mg/100g)	96.7 ± 13.9	70.7 ± 14.1	77.5 ± 11.2	76.2 ± 11.2	55.1 ± 7.2	1.6	0.24
20:4n6 ARA (mg/100g)	3.8 ± 0.6	3.3 ± 0.4	3.1 ± 0.2	2.9 ± 0.3	2.7 ±0.1	1.2	0.36
18:3n3 ALA (mg/100g)	13.6 ± 2.4	6.4 ± 1.4	6.9 ± 1.2	6.4 ± 0.9	3.2 ± 0.4	6.8	0.006
20:5n3 EPA (mg/100g)	52.0 ± 7.9[a]	22.1 ± 5.0[b]	20.5 ± 3.4[b]	16.1 ± 2.9[b]	5.9 ± 0.5[c]	13.5	<0.01
22:6n3 DHA (mg/100g)	143.5 ± 12.2[c]	205.3 ± 23.0[b]	200.4 ± 22.3[b]	258.0 ± 34.3[a]	261.8 ± 19.3 [a]	4.2	0.02

LA, linoleic Acid; ARA, arachidonic acid; ALA, alpha linolenic acid; EPA, eicosapentaenoic acid; DHA, docosahexaenoic acid. Mean values across the row not sharing a common superscript were significantly different as determined by Tukey's HSD test, P<0.05.

DISCUSSION

This study is the first report of improved feed utilization indices, weight gain, and beneficial fatty acid profiles in Nile tilapia when fish oil is fully replaced in tilapia diets by 16% of dried whole-cells of a marine microalga species, *Schizochytrium* sp. (Sc). Additionally, SGR and survival rates were high and showed no significant difference among all diets at the experiment's end. These results indicate that *Schizochytrium* sp. provides a high-quality substitute for fish oil and a supplement of long-chain polyunsaturated fatty acids (PUFA), especially 22:6n3 DHA, that improves n3LC PUFA deposition in Nile tilapia fillets. These results also point to the possibility of formulating ecologically and socially sustainable aquafeeds, with greatly reduced or no fish oil (from marine fisheries) and without having to switch to vegetable oils (from industrially farmed crops). Commercial realization of this potential will require advances in strategies to reduce non-renewable inputs (e.g., inorganic fertilizers and fossil fuels) and monetary costs of large-scale production of marine microalgae.

Regarding improved growth, fish fed the Sc100 diet exhibited significantly higher final weight and weight gain compared to those fed the Sc0 diet. The specific growth rate SGR (%/day), however, did not differ among all diets, which is likely due to slight differences (not significant) in the initial body weight among treatments, and subsequently the final body weight and thereby the calculated weight gain was higher in fish a higher initial body weight, and the SGR remained almost unaltered by the diets. During this 84-day experiment, mean SGR ranged from 3.3 to 3.5, fitting with literature values for the exponential growth phase [24, 25].

Our results for improved weight gain may seem surprising in light of previous reports that a high dietary level of n3 LC PUFA depressed growth [26] or did not increase growth performance of tilapia [27]. We have identified two possible reasons for this difference between our results and previous research. Firstly, elevated levels of Sc exhibited improved FCR, consistent with our prior observation that the nutrient of Sc is highly digestible to tilapia [19]. Thus, the better feed conversion by Nile tilapia fed the Sc100 dietary nutrient was reflected in their improved weight gain; and this is similar to a recent report that dietary inclusion of Sc stimulated muscle or tissue development of Atlantic salmon [28]. Our observed effect of Sc on the growth of tilapia is also consistent with findings in shrimp and barramundi, which demonstrated an algal derived DHA stimulated growth performance [29, 30]. Our followup studies are examining effects of Sc-inclusion diets in tilapia fed up to market size.

Secondly, dried whole cell Sc may contain high levels of a micronutrient [31, 32], such as the caroteniod astanxanthin and bioactive compounds, which we did not measure but could have contributed to the growth benefit of whole cell Sc.

The relatively high values of FCR and low value of PER in fish fed the control diet, compared to fish fed Sc inclusion diets, could be due to lower nutrient digestibility of the control diet containing fish oil compared to Sc-inclusion diets. In our previous digestibility study with Nile tilapia [19], we confirmed that the Sc diet, with high DHA amounts (302g kg^{-1} of total fatty acids), did not depress the digestibility of protein, amino acids, and unsaturated fatty acids in tilapia compared to a reference diet containing only fish oil. We also found high digestibility of lipid and all unsaturated fatty acid fractions in the Sc diet compared to the reference diet containing fishmeal and fish oil, and suggested this may be because tilapia more easily digest the polar fraction of phospholipids prevalent in marine microalgae than they digest neutral class lipids (triglycerides) prevalent in fish oil (reviewed in [19]).

Depositions of total n3 PUFA and total n3 LC PUFA did not differ among all dietary treatments. Thus, even the diet replacing 100% of the fish oil with Sc (Sc100) did not reduce these totals compared to the control diet (Sc0). Total n3 LC PUFAs in tilapia fillets increased from 13% to 20.5% of total fatty acids when we fed the Sc100 diet that replaced all the fish oil with dried Sc (at 16% of the Sc100 diet) compared to the Sc0 diet containing only fish oil (at 9% of the Sc0 diet), although this difference was not significant.

Length of feeding time significantly influenced the deposition of most of the n3 PUFA, including 22:6n3 DHA and 20:5n3 EPA. The fillet contents for all the n3 and n6 PUFA content was higher at the end of the experiment (84 days) than at the middle (42 days), whereas fillet contents of SFA and MUFA fractions was lower at 84 days. These results fit with the understanding that the fatty acid composition of the fish is largely determined by the digestible fatty acid intake of the animal. In fatty fish, the overall change in the fatty acid profile is determined by triglycerides while in lean fish, like tilapia, the muscle lipid is dominated by phospholipids (75.5%) [33, 34, 35]. Our prior results for digestibility of whole cells of Sc in tilapia showed that 22:6n3 DHA phospholipid is highly digestible (93%) [19]. The high digestibility of 22:6n3 DHA is reflected in this study, given that tilapia fed Sc100 deposited the highest amount of 22:6n3 DHA in muscle at the end of the experiment (84 days).

Our results regarding fillet deposition of two important n3 LC PUFAs, DHA and EPA, can be explained by prior research and the relative abundance of these two fatty acids in Sc. Regardless of the treatment, 22:6n3 DHA was the major n3 LC PUFA found in the fillet lipids, supporting the conclusions of [21] that DHA, rather than EPA, was preferentially incorporated into the fillet of tilapia. This result along with the significantly higher weight gain of tilapia fed the Sc100 diet, compared to the Sc0 diet, leads us to propose the following hypothesis: the Sc100 diet, containing elevated levels of dietary 22:6n3 DHA, reduces energy expenditure that would have been spent on *de novo* DHA biosynthesis and consequently supports the growth performance of tilapia. Thus, future research is needed to elucidate the mechanism by which dietary 22:6n3 DHA affects energy expenditure, fatty acid metabolism, and growth in tilapia.

The Sc0 diet's level of 20:5n3 was 11.1% of total fatty acids but tilapia fed the Sc0 diet deposited a lower level of 20:5n3 in the fillet (3.9% of total fatty acid), suggesting that 20:5n3 EPA will likely be extensively oxidized in Nile tilapia. This is in agreement with the notion that fish, including tilapia, selectively use 20:5n3 EPA as substrate for β-oxidation to produce sufficient energy [21].

Using dried whole-cells of a marine microalga as a rich source of LC n3 PUFA, instead of an extracted oil, offers several benefits regarding the human nutritional value of tilapia fillets. One benefit is reduced lipid oxidation in fillets, given that marine-derived fish oil is highly susceptible to oxidation.

Aquafeeds enriched with fish oil increase the susceptibility of fish fillets to lipid oxidation, yielding undesirable lipid oxidation products that adversely affect fillet quality [36,37]. The benefit of using DHA-rich whole cell *Scizochytrium* sp. biomass in fish feeds is the natural encapsulation provided by the cell wall, which should protect valuable fatty acids from exposure to oxidative agents. Moreover, this alga is a potential source of natural antioxidants such as astaxanthin and carotenoid pigments [31,38], which could be readily deposited in fish tissue and enhance oxidative stability.

Inclusion of dried whole-cell Sc in aquafeeds offers a second human nutritional benefit of the fish product. Recent mammalian studies have revealed higher anti-inflammatory and immuno-enhancer properties of *Schizochytrium* sp., and attributed these findings to the docosapentaenoic acid (DPAn6) and DHA present in high levels in this microalga [38]. Such higher biological activity could also occur in tilapia fed diets with dried whole-cell Sc, and could translate to a human nutritional benefit of eating the fillet.

CONCLUSION

This is the first study to use Sc as a complete fish oil replacement for Nile tilapia. Total replacement of the fish oil used in commercial diets with dried whole-cell Sc resulted in significantly higher weight gain (g) and protein efficiency ratio (PER), and lower (improved) feed conversion ratio (FCR) and feed intake when compared to the control feed containing fish oil (Sc0); and no significant change in specific growth rate and survival rate among all diets. Our results are interesting for tilapia farming by showing how microalgae inclusion can help optimize the composition of tilapia starter and juvenile feed. Diet formulation with dried whole-cell Sc is also an innovative strategy to raise the n3 LC PUFAs level and thus rebalance the n3:n6 and DHA: EPA ratios in farmed tilapia, which will benefit the health of human consumers. In our study, complete substitution of fish oil with Sc led to n3:n6 and DHA:EPA ratios of 1.4:1 and 34.2:1, respectively in tilapia fillets, suggesting the relative ease by which inclusion of whole cell Sc in tilapia diets can manipulate and tailor the n3 LC PUFA composition of tilapia fillets. However, in this study we used juvenile fish over a narrow growth range. Feed manufacturers can explore this approach to develop aquafeeds for aquaculturists aiming to cater to the consumer willing to pay a premium for health enhanced foods. Niche markets may therefore be developed for these nutritionally enhanced tilapia fillets.

In order for such nutritionally enhanced tilapia to succeed in the market, researchers have to find the ways to cut the high production cost of microalgae [39, 40]. Towards this end, we are now focusing our research on reducing production costs by using organic waste streams, such as on-site fish culture effluent, as a partial replacement for expensive, external inputs of inorganic compounds normally used to grow microalgae. Furthermore, ongoing research to lower costs of producing and processing microalgae for biofuel and human nutraceuticals may lead to higher market availability and lower price of this microalga in the near future [41,42,43]. Our findings also open up a pathway to reduce or eliminate ingredients extracted from marine forage fish and from industrially farmed crops in aquafeeds. Our lab is now investigating combinations of different marine microalgae to reduce or eliminate fish oil, as well as fishmeal, in aquafeeds, towards the goal of improving life-cycle ecological and social sustainability of aquafeeds.

ACKNOWLEDGMENTS

We thank the faculty members and staff of the Department of Biology, Life Sciences Center for use of their facilities and for technical assistance. We also thank to Jianjun Hua, Statistical consultant, Dartmouth College, for helping us on the statistical analyses.

AUTHOR CONTRIBUTIONS

Conceived and designed the experiments: PKS ARK. Performed the experiments: PKS AJL EDL KPB MLC. Analyzed the data: PKS. Contributed reagents/materials/analysis tools: AJL EDL KPB MLC. Wrote the paper: PKS ARK. Reviewed manuscript drafts: AJL EDL KPB MLC.

REFERENCES

1. Tacon AGJ, Hasan MR, Metian M (2011) Demand and Supply of Feed Ingredients for Farmed Fish and Crustaceans: Trends and Prospects (Food and Agriculture Organization, Rome).
2. Hall SJ, Delaporte A, Phillips MJ, Beveridge M, O'Keefe M (2011) Blue frontiers: managing the environmental costs of aquaculture. The WorldFish Center, Penang, Malaysia.
3. FAO (Food and Agriculture Organization of the United Nations) (2012) The state of world fisheries and aquaculture (SOFIA). FAO, Rome, Italy.
4. NRC (National Research Council) (2011) Nutrient Requirements of Fish and Shrimp. National Academies Press, Washington, D. C. , USA.
5. Tacon AGJ, Metian M (2008) Global overview on the use of fish meal and fish oil in industrially compounded aquafeeds: Trends and future prospects. Aquaculture 285(1–4):146–58. doi: 10. 1016/j. aquaculture. 2008. 08. 015
6. FAO (Food and Agriculture Organization of the United Nations) (2010) The State of World Fisheries and Aquaculture.
7. Weaver KL, Ivester P, Chilton JA, Wilson MA, Pandey P, Chilton FH (2008) The content of favorable and unfavorable polyunsaturated fatty acids found in commonly eaten fish. Journal of the American Dietetic Association 108: 1178–1185. doi: 10. 1016/j. jada. 2008. 04. 023. pmid:18589026
8. Ng WK (2005) Lipid nutrition of farmed tilapia. Glob. Aquacult. Advocate, October, 60–61.
9. Turchini GM, Torstensen BE Ng WK (2009) Fish oil replacement in finfish nutrition. Reviews in Aquaculture 1: 10–57. doi: 10. 1111/j. 1753-5131. 2008. 01001. x
10. Menoyo D, Izquierdo MS, Robaina L, Gines R, Lopez-Bote CJ, Bautista JM (2004) Adaptation of lipid metabolism, tissue composition and flesh quality in gilthead seabream (*Sparus aurata*) to the replacement of dietary fish oil by linseed and soyabean oils. British Journal of Nutrition 92: 41–52. pmid:15230986 doi: 10. 1079/bjn20041165

11. Mourente G, Good JE, Thompson KD, Bell JG (2007) Effects of partial substitution of dietary fish oil with blends of vegetable oils, on blood leucocyte fatty acid compositions, immune function and histology in European sea bass (*Dicentrarchus labrax* L.). British Journal of Nutrition 98: 770–779. pmid:17466094 doi: 10. 1017/s000711450773461x

12. Sarker PK, Bureau DP, Drew M, Hua K, Forster I, et al. (2013) Sustainability issues related to feeding salmonids: a Canadian perspective. Reviews in Aquaculture 5: 1–21. doi: 10. 1111/raq. 12013

13. Shiau S, Lin S, Yu S, Lin A, Kwok C (1990) Defatted and full-fat soybean meal as partial replacements for fish meal in tilapia (*Oreochromis niloticus* x *O. aureus*) diets at low protein level. Aquaculture 86: 401–407. doi: 10. 1016/0044-8486(90)90328-k

14. Thompson KR, Velasquez A, Patterson JT, Metts LS, Webster CD, et al. (2012) Evaluation of plant and animal protein sources as partial or total replacement of fish meal in diets for Nile tilapia fry and juvenile stages. North American Journal of Aquaculture 74: 365–375. doi: 10. 1080/15222055. 2012. 675999

15. Hemaiswarya S, Raja R, Ravi-Kumar R, Ganesan V, Anbazhagan C (2011) Microalgae: a sustainable feed source for aquaculture. World Journal of Microbiology and Biotechnology 27: 1737–1746. doi: 10. 1007/ s11274-010-0632-z

16. Ryckebosch E, Bruneel C, Muylaert K, Foubert I (2012) Microalgae as an alternative source of omega-3 long chain polyunsaturated fatty acids. Lipid Technology 24: 128–130. doi: 10. 1002/lite. 201200197

17. Peet M, Stokes C (2005) Omega-3 fatty acids in the treatment of psychiatric disorders. Drugs 65(8): 1051–9. pmid:15907142 doi: 10. 2165/00003495-200565080-00002

18. Brasky TM, Till C, White E, Neuhouser ML, Song X, et al. (2011) Serum Phospholipid Fatty Acids and Prostate Cancer Risk: Results From the Prostate Cancer Prevention Trial. American Journal of Epidemiology (doi: 10. 1093/aje/kwr027).

19. Sarker PK, Gamble MM, Kelson S, Kapuscinski AR (2016) Nile tilapia (*Oreochromis niloticus*) show high digestibility of lipid and fatty acids from marine *Schizochytrium sp.* and of protein and essential amino acids from freshwater *Spirulina sp.* feed ingredients. Aquaculture Nutrition 22: 109–119. doi: 10. 1111/anu. 12230

20. Hussein EE, Dabrowski K, El-Saidy DMSD, Lee BJ (2012) Enhancing the growth of Nile tilapia larvae/ juveniles by replacing plant (gluten) protein with algae protein. Aquaculture Research 44:937–949. doi: 10. 1111/j. 1365-2109. 2012. 03100. x

21. Karapanagiotidis IT, Bell MV, Little DC, Yakupitiyage A (2007) Replacement of dietary fish oils by alpha-linolenic acid-rich oils lowers omega-3 content in tilapia flesh. Lipids 42, 547–559. pmid:17473942 doi: 10. 1007/s11745-007-3057-1

22. Trushenski JT, Boesenberg J, Kohler CC (2009) Influence of grow-out feed fatty acid composition on finishing success in Nile tilapia. North American Journal of Aquaculture 71:242–251. doi: 10. 1577/a08-051. 1

23. Association of Official Analytical Chemists, AOAC (1995) Agriculture chemicals; contaminants, drugs, Vol. I. *In* Official Methods of Analysis of AOAC International, 16th edn. AOAC International, Arlington, VA.

24. Vechklang K, Boonanuntanasarn S, Ponchunchoovong S, Pirarat N, Wanapu C (2011) The potential for rice wine residual as an alternative protein source in a practical diet for Nile tilapia (*Oreochromis niloticus*) at the juvenile stage. Aquaculture Nutrition 17: 685–694. doi: 10. 1111/j. 1365-2095. 2011. 00870. x

25. Yones AM, El-Saidy DMSD, Abdel-Hakim NF (2013) Effects of fish oil substitution with vegetable oils in diets of juvenile Nile tilapia, *Oreochromis niloticus* (*L.*) on growth performance, nutrients utilization and muscle fatty acids contents. Merit Research Journal of Food Science and Technology 1 (1): 9–18.

26. Huang CH, Huang MC, Hou PC (1998) Effect of dietary lipids on fatty acid composition and lipid peroxidation in sarcoplasmic reticulum of hybrid tilapia,*Oreochromis niloticus*x *O. aureus*. Comparative Biochemistry and Physiology 120: 331–336. doi: 10. 1016/s0305-0491(98)10022-6

27. Takeuchi T, Satoh S, Watanabe T (1983) Requirement of Tilapia nilotica for essential fatty acids. Bulletin of the Japanese Society of Scientific Fisheries 49: 1127–1134.

28. Kousoulaki K, Mørkøre T, Nengas I, Berge RK, Sweetman J (2016) Microalgae and organic minerals enhance lipid retention efficiency and fillet quality in Atlantic salmon (*Salmo salar* L.). Aquaculture 451: 47–57 doi: 10. 1016/j. aquaculture. 2015. 08. 027

29. Glencross BD, Smith DM (2001) Optimising the essential fatty acids, eicosapentaenoic and docosahexaenoic acid in the diet of the prawn, Penaeus monodon. Aquaculture Nutrition 7, 101–112. doi: 10. 1046/j. 1365-2095. 2001. 00158. x

30. Glencross B. D. & Rutherford N. (2011) A determination of the quantitative requirements for docosahexaenoic acid for barramundi (Lates calcarifer). Aquaculture Nutrition 17, e536–e548. doi: 10. 1111/j. 1365-2095. 2010. 00795. x

31. Yokoyama R, Honda D (2007) Taxonomic rearrangement of the genus Schizochytriumsensulato based on morphology, chemotaxonomic characteristics, and18SrRNA gene phylogeny (Thraustochytriaceae, Labyrinthulomycetes) emendation for Schizochytrium and erection of Aurantiochytrum and Oblongichytrium gen. nov. Mycosciences 48: 199–211. doi: 10. 1007/ s10267-006-0362-0

32. Raghukumar S (2008) Thraustochytrid Marine Protists: Production of PUFAs and Other Emerging Technologies. Marine Biotechnology 10(6): 631–640. doi: 10. 1007/ s10126-008-9135-4. pmid:18712565

33. Sargent J, Bell G, McEvoy L, Tocher D, Estevez A (1999) Recent developments in the essential fatty acid nutrition of fish. Aquaculture 177: 191–199 doi: 10. 1016/s0044-8486(99)00083-6

34. Bureau DP, Hua K, Harris AM (2008) The effect of dietary lipid and long-chain n-3 PUFA levels on growth, energy utilization, carcass quality and immune function of rainbow trout (*Oncorhynchus mykiss*). Journal of the World Aquaculture Society 39: 1–21. doi: 10. 1111/j. 1749-7345. 2007. 00146. x

35. Suloma A, Ogata HY, Garibay ES, Chavez DR el-Haroun ER (2008) Fatty acid composition of Nile Tilapia *Oreochromis niloticus* muscles: a comparative study with commercially important tropical freshwater fish in Philippines. 8th International Symposium on Tilapia in Agriculture, Egypt Ministry of Agriculture, Cairo, Egypt, pp. 921–932.

36. Menoyo D, López-Bote CJ, Obach A, Bautista JM (2005) Effect of dietary fish oil substitution with linseed oil on the performance, tissue fatty acid profile, metabolism, and oxidative stability of Atlantic salmon. Journal of Animal Sciences 83: 2853–286.

37. Ng WK, Bahurmiz OM (2009) The impact of dietary oil source and frozen storage on the physical, chemical and sensorial quality of fillets from market-size red hybrid tilapia, *Oreochromis* sp. Food Chemistry 113: 1041–1048. doi: 10. 1016/j. foodchem. 2008. 08. 060

38. Nauroth JM, Liu YC, Van Elswyk M, Bell R, Hall EB, Chung G, et al. (2010) Docosahexaenoic acid (DHA) and docosapentaenoic acid (DPAn-6) algal oils reduce inflammatory mediators in human peripheral mononuclear cells in vitro and paw edema in vivo. Lipids 45 (5): 375–84. doi: 10. 1007/s11745-010-3406-3. pmid:20364438

39. Walker DA (2009) Biofuels, facts, fantasy, and feasibility. Journal of Applied Phycology 21: 509–517. doi: 10. 1007/s10811-009-9446-5

40. Barrows FT, Silverstein J (2010) From fish meal-dependent to fish meal-free: feeds research is producing the alternative diets of the future for trout. p 15–16 In: Rust MB, Barrows FT, Hardy RW, Lazur A, Naughten K, Silverstein J, editors. The Future of Aqua-feeds, NOAA/ USDA Alternative Feeds Initiative, Maryland.

41. Chauton MS, Reitan KI, Norsker NH, Tveterås R, Kleivdal HT (2015) A techno-economic analysis of industrial production of marine microalgae as a source of EPA and DHA-rich raw material for aquafeed: Research challenges and possibilities. Aquaculture 436: 95–103. doi: 10. 1016/j. aquaculture. 2014. 10. 038

42. Gross M, & Jarboe D, Wen Z (2015) Biofilm-based algal cultivation systems. Applied Microbiology and Biotechnology 99: 5781–5789. doi: 10. 1007/s00253-015-6736-5. pmid:26078112

43. White RL, Ryan RA (2015) Long-Term Cultivation of Algae in Open-Raceway Ponds: Lessons from the Field. Industrial Biotechnology 11 (4): 213–220. doi: 10. 1089/ ind. 2015. 0006

SECTION III - ECOSYSTEM ECONOMY, RESPONSIBLE FISH FARMING AND VULNERABILITIES OF NATURAL RESOURCES.

CHAPTER 5

AQUACULTURE: ENVIRONMENTAL IMPACTS AND TROUBLESHOOTING ALTERNATIVES

Marcel Martinez-Porchas and Luis R. Martinez-Cordova

INTRODUCTION

Aquaculture, the farming of aquatic organisms, has been the agroindustrial activity with the highest growth rate worldwide in the last four decades. From 1970 to 2008 the production of aquaculture organisms grew at a rate of 8.3% per year, compared to less than 2% of fisheries, and 2.9% of livestock [1]. The annual aquaculture production is at present over 60 million tons (including marine plants), with an approximate value of 85 billion dollars [2]. The last FAO report revealed that the world population increased by 6.3% from 2004 to 2009, whereas the production of aquatic organisms by aquaculture increased by 31.5% in the same period (Figure 1) [2].

Despite the undeniable benefits of aquaculture such as the provision of good quality and accessible food for population and the generation of millions of jobs and billion dollars in budget for the developing countries, the activity is one of the most criticized worldwide, mainly because of the environmental impacts that have been and can be caused. Thus, the predominant and unavoidable question is: could aquaculture be a truly sustainable activity?

Understanding sustainability as "the ability to meet the needs of the present without compromising the ability of the future generations to meet their own needs" [3], many researchers, aquaculturists, and governmental instances have considered that a sustainable aquaculture is possible, but it depends on the way that the activity will be managed [4]. Additionally, other authors [5] have argued that "the sustainability of aquaculture not only requires neutral or benign effects on the environment, but also economic feasibility."

The present paper is a review of the world aquaculture and its environmental impacts. It analyzes the situation of aquaculture production up to date and summarizes the main problems faced by the activity, as well as the strategies suggested, evaluated, and proven to contribute to achieve a sustainable activity.

Citation: Marcel Martinez-Porchas and Luis R. Martinez-Cordova, "Aquaculture: Environmental Impacts and Troubleshooting Alternatives doi:10.1100/2012/389623

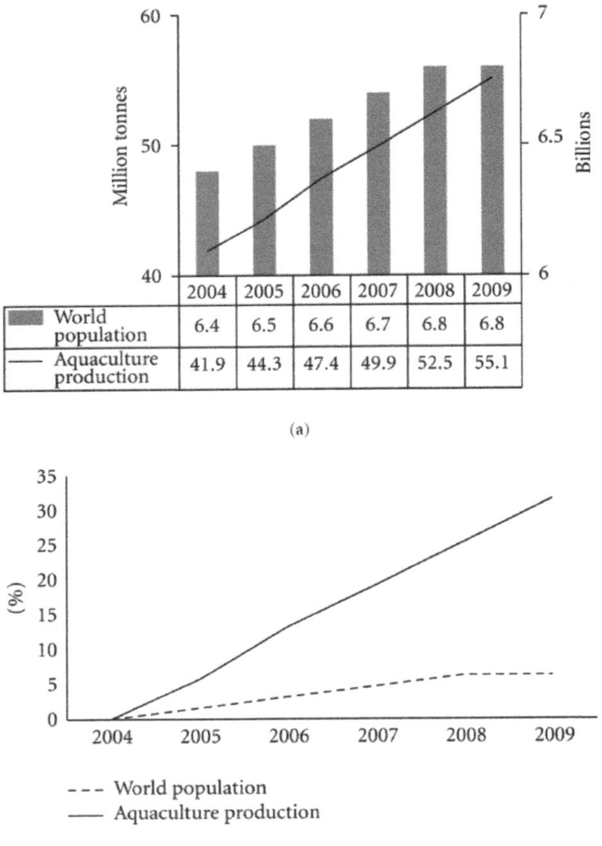

Figure 1: Growth behavior of world population and aquaculture production during the last six years. (a) illustrates the total world population by year (billions) and the total production of aquatic organisms by aquaculture (million tonnes). (b) compares the percentage of annual increase of world population and aquaculture production, considering year 2004 as the basepoint. Data obtained from FAO Report 2010: World Review of Fisheries and Aquaculture.

Aquaculture Is an Essential Activity for World's Welfare

Aquaculture is considered as a double-edged sword, because it has not only tremendous benefits for the humanity, but also great repercussions to the environment. Considering the benefits, seafood produced by fisheries and aquaculture contributes with 15 to 20% of average animal protein consumption to 2.9 billion people worldwide [6] without considering the contribution of freshwater or brackish water species. The nutritional quality of aquatic products has a high standard and represents an important source of macro- and micronutrients for people from developing countries [7]. Additionally, aquaculture and fisheries are recognized as a source of employment; for instance, near to 43.5 million people were employed in 2006, and 520 million people relied on income from seafood production [6, 8].

Aquaculture products have also high trade potential as food commodities in the international market; fish and shellfish exports from developing countries have a greater value than the combination of important products such as coffee, tea, tobacco, meat, cocoa, rubber, and rice [6, 8]. In many cases, the incomes generated by aquaculture exceed those from other agricultural activities, due to the high price market that some products can achieve and due to the most effective bioenergetics of some aquatic species. However, despite all these benefits, aquaculture is actually not considered a sustainable activity in the perception of the scientific community and the average population.

WHY AQUACULTURE IS CONSIDERED A NONSUSTAINABLE ACTIVITY?

With or without valid arguments, aquaculture has been accused to be the cause of many environmental, social, economic, and inclusively esthetic problems. Ecosystems are not always as fragile as could be considered, instead, they have remarkable capacity of resiliency, and as long as basic processes are not irretrievably upset, ecosystems will continue to recycle and distribute energy [9]. However, irreversible damages have been already caused due to inadequate management of the activity. The main negative impacts attributed to the activity are as follows.

(1). Destruction of Natural Ecosystems, In Particular Mangrove Forests to Construct Aquaculture Farms [4, 10, 11]

The mangrove forests are important ecosystems considered as the main source of organic matter to the coastal zone [12, 13]; they are also nursery areas for many aquatic species ecologically and/or economically important, as well as refuge or nesting areas for bird, reptiles, crustaceans, and other taxonomic groups [14]. Mangroves are additionally accumulation sites for sediments, contaminants, nitrogen, carbon and offer protection against coastal erosion [15]. According to environmentalists [16], mangroves support diverse local fisheries and also provide critical nursery habitat and marine productivity which support wider commercial fisheries. These forests also provide valuable ecosystem services that benefit coastal communities, including coastal land stabilization and storm protection. The cover of mangrove forest has decreased worldwide from 19.8 million hectares in 1980 to less than 15millions in 2000. The annual deforestation rate was 1.7% from 1980 to 1990 and 1.0% from 1990 to 2000 [17], and the problem continues up today. Some authors have documented that aquaculture has been responsible for the deforestation of millions hectares of mangrove forest in Thailand, Indonesia, Ecuador, Madagascar, and other countries [18, 19]. From 1975 to 1993, the construction of shrimp farms in Thailand diminished the mangrove cover from 312,700 to 168,683 ha [20]. Philippines has reconverted 205,523 ha of mangrove and wetlands into aquaculture farms, Indonesia 211,000 ha, Vietnam 102,000 ha, Bangladesh 65,000 ha, and Ecuador 21,600 ha [21].

(2) Salinization/Acidification of Soils

Aquaculture farms are sometimes abandoned by multiple problems (operative, economic, sanitary, and etc.), and the soil from those former farms remain hypersaline, acid and eroded [22]. Therefore, those soils cannot be used for agricultural purposes and are unusable for long periods. In addition, the application of lime and other chemicals used in aquaculture to treat the soil can also modify its physicochemical characteristics, which could aggravate the problem [23].

(3) Pollution of Water for Human Consumption

Although few studies have been conducted in relation with such topic, there are some signs indicating that inland aquaculture has been responsible for the deterioration of water bodies used for human consumption [21]. For instance, preliminary calculations revealed that an intensive aquaculture system farming three tons of freshwater fish can be compared, in respect to waste generation, to a community of around 240 inhabitants [24].

Although most of the aquaculture farms produce marine species, there is a growing sector of aquaculture farms producing freshwater species, which is a point of concern considering the above information.

(4) Eutrophication and Nitrification of Effluent Receiving Ecosystems

The eutrophication or organic enrichment of water column is mainly produced by nonconsumed feed (especially due to overfeeding), lixiviation of aquaculture feedstuffs [25, 26], decomposition of died organisms, and overfertilization [27–30]. It is well documented that from the total nitrogen supplemented to the cultured organisms, only 20 to 50% is retained as biomass by the farmed organisms, while the rest is incorporated into the water column or sediment [31, 32], and eventually discharged in the effluents toward the receiving ecosystems, causing diverse impacts such as phytoplankton blooms (sometimes of toxic microalgaes, such as red tides) [33], burring, and death of benthic organisms, as well as undesirable odors and the presence of pathogens in the discharge sites [34]. The impact may be more or less severe depending on some factors such as the intensification of the system (density of organisms), which is directly related to the amount of feed supplied [26, 35]. The feed conversion ratio (FCR) is a well indicator of the effectiveness of feeding and, consequently, of the retention of nitrogen and carbon as biomass of the farmed organisms. For instance, farms culturing the tiger shrimp Penaeus monodon usually report FCRs ranging from 1 to more than 2.5; such huge difference is later reflected in the amount of organic matter, nitrogen, and phosphorous discharged in the effluents, which may range from 500 to 1625 kg, 26 to 117 kg, and 13 to 38 kg, respectively, for each ton of shrimp harvested [28]. The estimated mean FCR worldwide for shrimp aquaculture is 1.8, which means that, for a world annual shrimp production around 5 million tons, 5.5 million tons of organic matter, 360,000 tons of nitrogen, and 125,000 tons of phosphorous are annually discharged to the environment. Unfortunately, these data considers only shrimp production, which represents around 8% of the total aquaculture production; if we assume that the FCRs are similar for the other farmed organisms and the diet formulations have some similitude [36], the total discharge of wastes may be multiplied by 12.5 from a very preliminary perspective. The nutrification is considered as the nutrient (N, P, C) enrichment of water column, mainly due to fertilization, mineralization of organic matter, resuspension of sediments, and excretion of organisms into the ponds. The greatest concern in this aspect is the increasing production of nitrogenous metabolites especially ammonia, which is highly toxic in its unionized form (NH_3) for many aquatic organisms [37].

(5) Ecological Impacts in Natural Ecosystems because of the Introduction of Exotic Species

The negative impacts of the "biological contamination" for the introduction of exotic aquacultural species on the native populations have been well documented [18, 38, 39]. The main reported problems are the displacement of native species, competition for space and food, and pathogens spread. To cite an example, recent reports have revealed a parasite transmission of sea lice from captive to wild salmon [40]. The authors of such study have hypothesized that "if outbreaks continue, then local extinction is certain, and a 99% collapse in pink salmon abundance is expected in four salmon generations."

(6) Ecological Impacts Caused by Inadequate Medication Practices

Farmers usually expose their cultured organisms to medication regimes, for different purposes such as avoiding disease outbreaks and improving growth performance. However, monitoring studies have detected low or high levels of a wide range of pharmaceuticals, including hormones, steroids, antibiotics, and parasiticides, in soils, surface waters, and groundwaters [41]. These chemicals have caused imbalances in the different ecosystems. In particular, the use of hormones in aquaculture and its environmental implications have been scarcely studied.

(7) Changes on Landscape and Hydrological Patterns

The agricultural and aquacultural activities have contributed to the degradation of ecosystems including important modification on landscape [10, 18, 22, 42]. The construction of shrimp farms in the river beds has modified the hydrological patterns in many regions of the world with the consequent impacts on the regional ecosystems and the local weather.

(8) Trapping and Killing of Eggs, Larvae, Juveniles, and Adults of Diverse Organisms

It has been estimated that, for each million of shrimp postlarvae farmed, four to seven millions of other organisms are killed by trapping in the nets of farms inlet [18, 43].

(9) Negative Effect on Fisheries

Although aquaculture has been proclaimed as a solution to avoid overfishing, it has contributed in more or less proportion to the fisheries collapse. Fishermen who work in places near to aquaculture farms argue that the contamination produced by farms has decreased the population of aquatic organisms and in consequence their volume captures. Additionally, another problem of similar magnitude is the extremely high aquaculture's dependence of fishmeal and fish oil, which could be another nonsustainable practice in aquaculture. The proportion of fishmeal supplies used for fish production have increased from 10% in 1988 to more than 30% in the last years, which classifies aquaculture as a potential promoter of the collapse of fisheries stocks worldwide [24].

(10) Some Other Accusations

Some other accusations for aquaculture include the production of fish and shellfish with high concentrations of toxins and/or heavy metals; genetic pollution and infestation of nondesirable phytoplankton and/or zooplankton species [44–47].

(11) In Its Role as Food Producer, Aquaculture Is Far from Complying an Adequate Distribution of Food

Overlaying net exports, governance, and undernourishment suggest that "seafood's contribution as a source of protein and livelihood is precarious" [6]. Moreover, it has been revealed that some countries with undernourishment and weak governance usually play a role as baler of seafood from countries well-nourished and with strong economic capacity [6].

For these above a reason remains a generalized perception that the sustainability of aquaculture is at present being threatened or, in some cases, far from being reached.

WHAT TO DO FOR A SUSTAINABLE AQUACULTURE?

Many strategies have been suggested, evaluated, and/or proven in order to advance in the sustainability of aquaculture. Basically, all of them respond to the criticisms and are possible solutions to the problems attributed to the activity. The main aspects that have to be performed to advance toward such goal are the correct selection of the farming sites and species; the implementation of the most adequate culture system; use of the best feed and feeding practices; the use of bioremediation systems; decreasing the dependence of fishmeal and fish oil; adequate management of the effluents; achieving certification of compliance with sustainability; improving research and legislation related to evaluation and solutions for aquaculture impacts.

(1) In the context of the site selection it is necessary to consider the following.

(a). The vocation of the selected site. It would be absurd to select a site for aquaculture purposes if it is excellent for agriculture or livestock. Unfortunately, this is the case in many regions of the world, where agricultural lands have been reconverted to aquaculture farms. The vocation of a selected site is determined for many aspects (which can change from region to region) such as physical and chemical soil characteristics; water availability, soil fertility, topography, wild vegetal and animal communities, proximity to cities, towns, tourism zones, and so forth; priorities of the region or country (food, fuels, tourism budget, aquaculture budget, and etc.).

(b) The carrying capacity of the water bodies from the sites considered to supply the farms or used as effluent discharge places. It is very important to evaluate how much water can be taken from a particular water body or how much effluents it can receive without important alterations on its ecological equilibrium [48]. The use of advanced technologies such as remote sensing could be an excellent auxiliary in this field [11].

(2) For the selection of species it is crucial to consider the following.

(a) It is always better to select native instead of exotic species. The introduction of exotic species causes many and diverse problems as mentioned in the previous section. Additionally, the obtaining and maintenance of broodstock of exotic species could be difficult and expensive.

(b) It is necessary to have the most possible knowledge about the biology and ecology of the organism that is pretended to be farmed (life cycle, feeding habits and nutritional requirements, tolerance to environmental parameters, and etc.).

(c) It is important to select organisms with a good market and price when farmed for commercial purposes.

(3) Regarding implementation of the best culture system, the main aspects to consider include the following.

(a) The type and size of farming structure [49]. Depending on the species, intensity, land and water availability, and economic investment, it is possible to use different types of farming structures for the culture of the same species or group. Some of them are more adequate and sustainable. For the case of shrimp farming, for instance, it has been suggested that floating or submerged cages could have a lower impact on the environment than earthen ponds. The same suggestion is applicable for culture of fishes or mollusks. Regarding size of production units, small ponds or farming structure is easier to manage in aspects such as feeding, monitoring, cleaning, pond bottom management, and harvesting. Such considerations usually lead to lower environmental impacts.

(b). Intensity. The stocking density and the consequent biomass harvested are absolutely related to the sustainability of aquaculture. The increase of the intensity implies an increase in the supplemental feed and in consequence, in the organic matter, nitrogen, and phosphorous in the effluents. Additionally, intensive or super intensive systems require the use of diverse chemicals (antibiotics, algaecides, parasiticides, and etc.), which also contribute to increasing the pollution [50]. The most adequate intensity depends on the land and water availability, as well as the carrying capacity of the water body or terrestrial ecosystems which will receive the effluents. However, recalculating and zero water exchange systems can eliminate the environmental impact while maintaining extremely high densities of aquatic organisms. Promising results have been achieved in the culture of fish and crustaceans using biofloc systems with zero water exchange [51].

(c). An adequate design of the water inlet and outlet systems, considering the water quality, weather conditions, marine currents and tide patterns (for sea water), and hydrological patterns (for continental waters) [52]. The modifications of oceanic currents patterns may have implications on the sediment transport and consequently on the beaches conformation.

(d). The possibility of farming simultaneously two or more species (polycultures or integrated multitrophic aquaculture (IMTA)). This strategy has proven to be one of the most effective ways to recuperate the carbon, nitrogen, and phosphorous supplied to the system as biomass of the farmed organisms and to diminish the environmental impacts caused by the effluents [53–55]. Polyculture is commonly referred to organisms of the same environment (marine, brackish wáter, or continental waters) and trophic level, while IMTA is mostly referred to organisms from different trophic levels and inclusively different environments. The implementation of such alternative systems improves the nutrient cycling within the culture units. In short, while in a traditional aquaculture system, 25 to 35% of the nitrogen supplied is recuperated as biomass of the farmed organisms, in a polyculture or IMTA, the recuperation could be increased by more than 50%. A pilot project made aware and informed a group of participants about the benefits of IMTA; the authors revealed that 50% of the participants were willing to pay an extra 10% of products labeled as of "IMTA products." Moreover, the authors were optimistic regarding the social impacts caused by the implementation of IMTA as a sustainable practice [5].

(4) Since supplemental feed is considered the main source of contamination of aquaculture systems and effluent receiving ecosystems [56], the improvement of these feed, as well as the feeding, strategies could be considered as an important part of the solution for a sustainable aquaculture [28, 57]. The main aspects in which the feedstuffs must be improved include the following.

(a) Better and more precise formulations for the particular species to be farmed, which consider the best concentration and quality of the nutrients. A common practice of world aquaculture is the use of diets with protein contents higher than those required, thus affecting not only the price of the feed but also increasing the pollution potential, considering that protein catabolism produces ammonium nitrogen as the main metabolite. Regarding nutrient quality, it is important to use ingredients with high digestibility; the low digestibility of ingredients (protein, lipid, carbohydrate) is partially the responsible for a low retention of those nutrients in the farmed organisms and their increase in the water column and sediment, augmenting the polluting potential [58].

(b). Higher hydrostability. One of the most important causes of nutrient losses of aquafeeds is the low hydrostability, which provoke fast disintegration and lixiviation, decreasing the nutrient incorporation efficiency by the farmed organisms and increasing the concentration in the water column. Fishes are faster swimmers and can consume a formulated feed within minutes, but crustaceans are usually less active and can consume the formulated feed within minutes or even hours. The hydrostability of feedstuffs can be improved by incorporation of effective binders and/or for the use of special fabrication processes [59].

(c) Better attractability and palatability. It is necessary to produce feeds which can be consumed as soon as possible to avoid nutrient losses. This is possible with the incorporation of effective attractants and improving the palatability with ingredients such as fish oils and others. Many of these ingredients have been sufficiently proven [60].

(5) Regarding to the feeding strategies some important advances have been achieved but there are yet much more to advance in aspects such as forms to supply the feed, adjustment of the ration, and frequency of feeding.

(a) The use of feeding trays and the increase of feeding frequency have been demonstrated to diminish the pollution potential of the effluents in shrimp farms [37]; however these strategies are suitable only for high-intensity systems (intensive or superintensive), but not economically feasible for extensive, semiextensive of semiintensive systems.

(b) The promotion, management, and rational utilization of natural feed, including microorganisms (biofilm, biofloc), are considered as a promising strategy for the culture of shrimp, fishes, and mollusks. Some authors [61–63] have successfully enhanced the production of zooplankton and benthos in shrimp ponds and demonstrated their great contribution not only in the production response, but also in the nutritional, sanitary, and immune condition of the farmed organisms. Additionally, the use and contribution of microorganisms associated to biofilms and bioflocs for the nutrition of farmed organisms have been also documented [24, 64–66]. Such practice may also decrease the dependence of fishmeal and fish oil; however other strategies such as the use of plant ingredients and the use of bioflocflour have been tested and proposed to substitute at different rates the fishmeal in formulated feeds [51].

(c) The practice of subfeeding or intermittent-feeding regimes is a strategy aimed to achieve average growth performances in aquatic organisms, but supplying significantly lower amounts of formulated feed. Such alternative takes advantage of the compensatory growth process of shrimp and crustaceans [67].

(6) The adequate management of effluents is indubitably one of the central aspects to consider for a sustainable aquaculture. Diverse strategies have been proven or suggested to minimize the environmental impacts of

effluents. The most promising are settling lagoons [34], treatments with septic tanks [68], the implementation of systems with low or zero water exchange [69], the utilization of recirculation systems [70, 71], the use of mangrove forests as sinks for nutrients, organic matter, and contaminants [72], the polyculture or integrated multitrophic aquaculture systems [55, 73], and the bioremediation [54, 74].

(a) It is considered as bioremediation the use of individual or combined organisms (including animal, vegetal, and bacteria) to minimize the contaminating charge of effluents from any activity (including aquaculture). This practice takes advantage of the natural or modified abilities of those organisms to reduce and/or transform waste products [75].

(b) There are different ways to conduct bioremediation: in situ, ex situ, biostimulation, bioaugmentation, and others. Many successfully examples of bioremediation practices can be mentioned: the use of plants (phytoremediation), macroalgae, microalgae, filter feeders, biofilters (polymer spheres with immobilized microorganisms), biofilms, and bioflocs [76, 77]. There are also combined systems which use two or more of these practices. Many studies have been conducted to use individual or combined organisms for bioremediation [78–80]. However, the ideal strategy would be the decreasing or complete halting of effluent discharge and using zero water exchange systems.

(7) Achieve certification of compliance with sustainability.

A combination of analyses has been suggested to evaluate the sustainability of commercial aquaculture farms [81]. For example, the authors of such contribution suggest the calculation of mass balances and undesirable outputs of shrimp farms; calculation of the input distance function approach which provides a complete characterization of the structure of multiinput, multioutput efficient production technology and provides a measure of the distance from each producer to that efficient-sustainable technology; finally a productivity measurement with and without undesirable outputs. However, the analysis of the socioeconomic impacts caused by farms ought to be included in the list.

Additionally, certification processes can be followed to assure the sustainability of aquaculture or to compare the standards established by the different agencies and check if the practices of any farm cope with those standards. The certification of aquaculture is performed by the International Standards Organization (ISO), the WTO Technical Barriers to Trade (TBT), the FAO Guidelines for the Ecolabelling of Fish and Fishery Products from Marine Capture Fisheries, and the Network of Aquaculture Centres in Asia-Pacific (NACA) and others [82].

According to the FAO criterion [82], certification "is a procedure through which written or equivalent assurance states that a product, process, or service conforms to specified requirements. Within the aquaculture sector certification can be applied to a process followed by a production unit (pond, cage, farm, processing plant), a specific product or commodity or to the inputs being applied to the system before or during production."

(8) Finally, there is an unavoidable need to improve research and legislation regarding evaluation and solutions for aquaculture impacts.

(a). One of the reasons of the severe environmental impacts of aquaculture is that scientific research in some developing countries is firstly focused on increasing biomass production (improvement of formulated feeds, production systems, genetically improved organisms, etc.) and later on the environmental impacts; however, it is desirable to evaluate the potential impacts of any farm that is pretended to be installed, rather than monitoring the pollution that is already being caused by any farm constructed without considering its environmental impact.

(b). In addition, there is a great heterogeneity regarding policies and legislation of aquaculture impacts among different countries; while some developed countries have complete and concrete legislation for aquaculture in order to avoid environmental impacts, others have weak policies that do not protect their environment from aquaculture wastes; under such scenario ecological imbalances and disasters have been caused, with some of them being irreversible. Herein, Smith et al. asserted that "some developing countries often lack the institutions necessary to prevent deleterious ecosystem impacts of seafood production and to sustain trade benefits" [6]; they also argued that "the developed countries have a history of these problems as well, but with less-obvious consequences." In the same report, the authors revealed that with base in the World Bank indicators, more than 60% of the countries had inefficient governance regarding the regulation of aquaculture and fisheries activities; a possible cause of such result is related to corruption and regulatory quality. Thus, it is absolutely essential for the future of aquaculture that the governments and the producers be attuned with each other to reach agreements that resolve the problems of this activity.

(c). Finally, contrasting actions have been observed by governmental instances, while some instances try to protect the environment and achieve a sustainable aquaculture, others have directly or indirectly supported the unsustainable aquacultural practices; for instance, according to a recent report [16] "the conversion of mangroves to aquaculture ponds has been fuelled by governmental support, private sector investment and external assistance from multilateral development agencies such as the World Bank and Asian Development Bank" [83, 84].

In conclusion, aquaculture is a possible panacea, but at present is also responsible for diverse problems related with the environmental sanity; however the new strategies proposed during the last decade have proven that it is possible to reach a sustainable aquaculture, but such strategies should be supported and proclaimed by the different federal environmental agencies from all countries. Only under such scenario, aquaculture will be a sustainable practice. The implementation of the different alternatives stated above would depend on particular circumstances of any farm. Fortunately, there are reports of some aquaculture farms along the world on sustainable practices.

REFERENCES

1. L. Luchini and S. Panné-Huidobro, Perspectivas en Acuicultura: Nivel Mundial, Regional y Local, Dirección de Acuicultura. Subsecretaría de Pesca y Acuicultura, Buenos Aíres, Argentina, 2008.
2. FAO, World Review of Fisheries and Aquaculture, 2010.
3. World Commission on the Environment and Development (WCED), Our Common Future, Oxford University Press, New York, NY, USA, 1987.
4. R. R. Stickney and J. P. McVey, Responsible Marine Aquaculture, World Aquaculture Society, New York, NY, USA, 2002.
5. K. Barrington, N. Ridler, T. Chopin, S. Robinson, and B. Robinson, "Social aspects of the sustainability of integrated multi-trophic aquaculture," Aquaculture International, vol. 18, no. 2, pp. 201–211, 2010.
6. M. D. Smith, C. A. Roheim, L. B. Crowder et al., "Sustainability and global seafood," Science, vol. 327, no. 5967, pp. 784–786, 2010.
7. N. Roos, M. A. Wahab, C. Chamnan, and S. H. Thilsted, "The role of fish in food-based strategies to combat vitamin A and mineral deficiencies in developing countries," Journal of Nutrition, vol. 137, no. 4, pp. 1106–1109, 2007.
8. FAO, The State of the World Fisheries and Aquaculture 2008, FAO, Rome, Italy, 2009.
9. A. Frankic and C. Hershner, "Sustainable aquaculture: developing the promise of aquaculture,"Aquaculture International, vol. 11, no. 6, pp. 517–530, 2003.
10. B. R. DeWalt, J. R. Ramirez-Zavala, L. Noriega, and E. González, "Shrimp aquaculture, the people and the environment in coastal Mexico," Tech. Rep., World Bank, NACA, WWF y FAO Consortium Program on Shrimp Farming and the Environment, 2002.
11. K. Rajitha, C. K. Mukherjee, and R. Vinu Chandran, "Applications of remote sensing and GIS for sustainable management of shrimp culture in India," Aquacultural Engineering, vol. 36, no. 1, pp. 1–17, 2007.
12. J. H. Tidwell and G. L. Allan, "Fish as food: aquaculture's contribution. Ecological and economic impacts and contributions of fish farming and capture fisheries," EMBO Reports, vol. 2, no. 11, pp. 958–963, 2001.
13. E. J. Olguín, M. E. Hernández, and G. Sánchez-Galván, "Contaminación de manglares por hidrocarburos y estrategias de biorremediación, fitorremediación y restauración," Revista Internacional de Contaminacion Ambiental, vol. 23, no. 3, pp. 139–154, 2007.
14. F. Páez-Osuna, "Retos y perspectivas de la camaronicultura en la zona costera," Revista Latinoamericana de Recursos Naturales, vol. 1, pp. 21–31, 2005.
15. D. M. Alongi, "Present state and future of the world's mangrove forests," Environmental Conservation, vol. 29, no. 3, pp. 331–349, 2002.
16. B. B. Walters, P. Rönnbäck, J. M. Kovacs et al., "Ethnobiology, socio-economics and management of mangrove forests: a review," Aquatic Botany, vol. 89, no. 2, pp. 220–236, 2008.
17. FAO, El Estado Mundial de la Pesca y Acuicultura 2006, FAO, Rome, Italy, 2007.
18. R. L. Naylor, R. J. Goldburg, J. H. Primavera et al., "Effect of aquaculture on world fish supplies," Nature, vol. 405, no. 6790, pp. 1017–1024, 2000.
19. G. J. Harper, M. K. Steininger, C. J. Tucker, D. Juhn, and F. Hawkins, "Fifty years of deforestation and forest fragmentation in Madagascar," Environmental Conservation, vol. 34, no. 4, pp. 325–333, 2007.
20. E. Barbier and S. Sathirathai, Shrimp Farming and Mangrove Loss in Thailand, Edward Elgar, 2003.
21. F. PÁez-Osuna, "The environmental impact of shrimp aquaculture: causes, effects, and mitigating alternatives," Environmental Management, vol. 28, no. 1, pp. 131–140, 2001.
22. J. A. Rodríguez-Valencia, D. Crespo, and M. López-Camacho, "La camaronicultura y la sustentabilidad del Golfo de California," 2010,
23. L. R. Martínez-Córdova, M. Martínez-Porchas, and S. Pedrín-Avilés, "Selección de sitios, construcción y preparación de estanques," in Camaronicultura Sustentable, L. R. Martínez-Córdova, Ed., chapter I, p. 179, Trillas, D.F., Mexico, 2009.
24. Y. Avnimelech, Biofloc Technology. A Practical Guide Book, The World Aquaculture Society, Baton Rouge, La, USA, 2009.
25. S. Focardi, I. Corsi, and E. Franchi, "Safety issues and sustainable development of European aquaculture: new tools for environmentally sound aquaculture," Aquaculture International, vol. 13, no. 1-2, pp. 3–17, 2005.
26. R. Crab, Y. Avnimelech, T. Defoirdt, P. Bossier, and W. Verstraete, "Nitrogen removal techniques in aquaculture for a sustainable production," Aquaculture, vol. 270, no. 1–4, pp. 1–14, 2007.
27. M. A. Burford and K. C. Williams, "The fate of nitrogenous waste from shrimp feeding," Aquaculture, vol. 198, no. 1-2, pp. 79–93, 2001.
28. A. G. J. Tacon and I. P. Forster, "Aquafeeds and the environment: policy implications," Aquaculture, vol. 226, no. 1–4, pp. 181–189, 2003.
29. Y. Y. Feng, L. C. Hou, N. X. Ping, T. D. Ling, and C. I. Kyo, "Development of mariculture and its impacts in Chinese coastal waters," Reviews in Fish Biology and Fisheries, vol. 14, no. 1, pp. 1–10, 2004.
30. A. Gyllenhammar and L. Håkanson, "Environmental consequence analyses of fish farm emissions related

to different scales and exemplified by data from the Baltic—a review," Marine Environmental Research, vol. 60, no. 2, pp. 211–243, 2005.

31. C. Jackson, N. Preston, P. J. Thompson, and M. Burford, "Nitrogen budget and effluent nitrogen components at an intensive shrimp farm," Aquaculture, vol. 218, no. 1–4, pp. 397–411, 2003.

32. O. Schneider, V. Sereti, E. H. Eding, and J. A. J. Verreth, "Analysis of nutrient flows in integrated intensive aquaculture systems," Aquacultural Engineering, vol. 32, no. 3-4, pp. 379–401, 2005.·

33. R. Alonso-Rodríguez and F. Páez-Osuna, "Nutrients, phytoplankton and harmful algal blooms in shrimp ponds: a review with special reference to the situation in the Gulf of California," Aquaculture, vol. 219, no. 1–4, pp. 317–336, 2003.

34. L. R. Martínez-Córdova and F. Enriquez-Ocaña, "Study of the benthic fauna in a discharge lagoon of a shrimp faro with special emphasis on polychaeta," Journal of Biological Sciences, vol. 7, pp. 12–17, 2007.

35. L. Deutsch, S. Gräslund, C. Folke et al., "Feeding aquaculture growth through globalization: exploitation of marine ecosystems for fishmeal," Global Environmental Change, vol. 17, no. 2, pp. 238–249, 2007.

36. E. A. H. Priya and S. J. Davies, "Growth and feed conversion ratio of juvenile Oreochromis niloticus fed with replacement of fishmeal diets by animal by-products," Indian Journal of Fisheries, vol. 54, pp. 51–58, 2007.

37. R. Casillas-Hernández, H. Nolasco-Soria, T. García-Galano, O. Carrillo-Farnes, and F. Páez-Osuna, "Water quality, chemical fluxes and production in semi-intensive Pacific white shrimp (Litopenaeus vannamei) culture ponds utilizing two different feeding strategies," Aquacultural Engineering, vol. 36, no. 2, pp. 105–114, 2007.

38. H. A. González-Ocampo, L. F. Beltrán Morales, C. Cáceres-Martínez et al., "Shrimp aquaculture environmental diagnosis in the semiarid coastal zone in Mexico," Fresenius Environmental Bulletin, vol. 15, no. 7, pp. 659–669, 2006.

39. W. L. Shelton and S. Rothbard, "Exotic species in global aquaculture—a review," Israeli Journal of Aquaculture-Bamidgeh, vol. 58, no. 1, pp. 3–28, 2006.

40. M. Krkošek, J. S. Ford, A. Morton, S. Lele, R. A. Myers, and M. A. Lewis, "Declining wild salmon populations in relation to parasites from farm salmon," Science, vol. 318, no. 5857, pp. 1772–1775, 2007.

41. A. B. A. Boxall, "The environmental side effects of medication," EMBO Reports, vol. 5, no. 12, pp. 1110–1116, 2004.

42. C. A. Berlanga-Robles and A. Ruiz-Luna, "Assessment of landscape changes and their effects on the San Blas estuarine system, Nayarit (Mexico), through Landsat imagery analysis," Ciencias Marinas, vol. 32, no. 3, pp. 523–538, 2006.

43. F. Páez-Osuna, "The environmental impact of shrimp aquaculture: a global perspective," Environmental Pollution, vol. 112, no. 2, pp. 229–231, 2001.

44. S. C. Johnson, R. B. Blaylock, J. Elphick, and K. D. Hyatt, "Disease induced by the sea louse (Lepeophtheirus salmonis) (Copepoda: Caligidae) in wild sockeye salmon (Oncorhynchus nerka) stocks of Alberni Inlet, British Columbia," Canadian Journal of Fisheries and Aquatic Sciences, vol. 53, no. 12, pp. 2888–2897, 1996.

45. G. M. Kruzynski, "Cadmium in oysters and scallops: the BC experience," Toxicology Letters, vol. 148, no. 3, pp. 159–169, 2004.

46. C. E. Boyd, "Guidelines for aquaculture effluent management at the farm-level," Aquaculture, vol. 226, no. 1–4, pp. 101–112, 2003.

47. P. Read and T. Fernandes, "Management of environmental impacts of marine aquaculture in Europe,"Aquaculture, vol. 226, no. 1–4, pp. 139–163, 2003.

48. F. J. Magallón-Barajas, A. Arreola, G. Portillo-Clark, et al., "Capacidad de carga y capacidad ambiental en la camaronicultura," in Camaronicultura Sustentable, L. M. Córdova, Ed., pp. 37–80, Trillas, D.F., Mexico, 2009.

49. L. R. Martínez-Córdova, M. M. Porchas, and E. Cortés-Jacinto, "Camaronicultura mexicana y mundial: Actividad sustentable o industria contaminante?" Revista Internacional de Contaminacion Ambiental, vol. 25, no. 3, pp. 181–196, 2009.

50. M. J. Costello, A. Grant, I. M. Davies, et al., "The control of chemicals used in aquaculture in Europe,"Journal of Applied Ichthyology, vol. 17, no. 4, pp. 173–180, 2001.

51. M. Emerenciano, E. L. C. Ballester, R. O. Cavalli, and W. Wasielesky, "Biofloc technology application as a food source in a limited water exchange nursery system for pink shrimp Farfantepenaeus brasiliensis(Latreille, 1817)," Aquaculture Research. In press.

52. C. E. Boyd, J. A. Hargreaves, and J. W. Clay, "Codes of conduct for marine shrimp aquaculture," in The New Wave, Proceedings of Special Session on Sustainable Shrimp Culture. Aquaculture, L.C. Browdy and D. E. Jory, Eds., pp. 303–321, The World Aquaculture Society, Baton Rouge, La, USA, 2001.

53. T. Chopin, A. H. Buschmann, C. Halling et al., "Integrating seaweeds into marine aquaculture systems: a key toward sustainability," Journal of Phycology, vol. 37, no. 6, pp. 975–986, 2001.·

54. L. R. Martínez-Córdova, J. A. López-Elías, G. Leyva-Miranda, L. Armenta-Ayón, and M. Martínez-Porchas, "Bioremediation and reuse of shrimp aquaculture effluents to farm whiteleg shrimp,Litopenaeus vannamei: a first approach," Aquaculture Research, vol. 42, no. 10, pp. 1415–1423, 2011.

55. M. Martínez-Porchas, L. R. Martínez-Córdova, M. A. Porchas-Cornejo, and J. A. López-Elías, "Shrimp polyculture: a potentially profitable, sustainable, but uncommon aquacultural practice," Reviews in Aquaculture, vol. 2, no. 2, pp. 73–85, 2010.

56. A. Sapkota, A. R. Sapkota, M. Kucharski et al., "Aquaculture practices and potential human health risks: current knowledge and future priorities," Environment International, vol. 34, no. 8, pp. 1215–1226, 2008.

57. L. Martínez-Córdova, M. Porchas, and H. Villarreal, "Efecto de tres diferentes estrategias de alimentaciön sobre el fitoplancton, zooplancton y bentos en estanques de cultivo de camarön café penaeus californiens› (holmes 1900)," Ciencias Marinas, vol. 24, no. 3, pp. 267–281, 1998.

58. L. E. Cruz-Suárez, M. Nieto-López, C. Guajardo-Barbosa, M. Tapia-Salazar, U. Scholz, and D. Ricque-Marie, "Replacement of fish meal with poultry by-product meal in practical diets for Litopenaeus vannamei, and digestibility of the tested ingredients and diets," Aquaculture, vol. 272, no. 1–4, pp. 466–476, 2007.

59. C. J. Simon, "The effect of carbohydrate source, inclusion level of gelatinised starch, feed binder and fishmeal particle size on the apparent digestibility of formulated diets for spiny lobster juveniles, Jasus edwardsii," Aquaculture, vol. 296, no. 3-4, pp. 329–336, 2009.

60. T. Ho, Feed attractants for juvenile Chinook salmon (Oncorhynchus tshawytscha) prepared from hydrolisates of Pacific hake (Merluccius productus), M.S. thesis, University of British Columbia, 2009.

61. L. R. Martínez-Córdova and M. A. Porchas-Cornejo, "Manejo del alimento y la alimentación en estanques camaronícolas: una estrategia de acuicultura sustentable," in Camaronicultura Sustentable, L. R. Martínez-Córdova, Ed., chapter III, p. 176, Trillas, D.F., Mexico, 2009.

62. A. Campaña-Torres, L. R. Martínez-Córdova, H. Villarreal-Colmenares, and E. Cortés-Jacinto, "Evaluation of different concentrations of adult live Artemia (Artemia franciscana, Kellogs 1906) as natural exogenous feed on the water quality and production parameters of Litopenaeus vannamei (Boone 1931) pre-grown intensively," Aquaculture Research, vol. 42, no. 1, pp. 40–46, 2010.·

63. M. A. Porchas-Cornejo, L. R. Martínez-Córdova, L. Ramos-Trujillo, J. Hernández-López, M. Martínez-Porchas, and F. Mendoza-Cano, "Effect of promoted natural feed on the production, nutritional, and immunological parameters of Litopenaeus vannamei (Boone, 1931) semi-intensively farmed,"Aquaculture Nutrition, vol. 17, no. 2, pp. e622–e628, 2011.

64. M. E. Azim, D. C. Little, and J. E. Bron, "Microbial protein production in activated suspension tanks manipulating C:N ratio in feed and the implications for fish culture," Bioresource Technology, vol. 99, no. 9, pp. 3590–3599, 2008.

65. E. L. C. Ballester, P. C. Abreu, R. O. Cavalli, M. Emerenciano, L. de Abreu, and W. Wasielesky Jr., "Effect of practical diets with different protein levels on the performance of Farfantepenaeus paulensis juveniles nursed in a zero exchange suspended microbial flocs intensive system," Aquaculture Nutrition, vol. 16, no. 2, pp. 163–172, 2010.

66. M. J. Becerra-Dorame, L. R. Martínez-Córdova, P. Martínez-Porchas, and J. A. López-Elías, "Evaluation of zero water exchange autotrophic and heterotrophic microcosm-based systems on the production response of Litopenaeus vannamei intensively nursed without Artemia," Israeli Journal of Aquaculture-Bamidgeh, vol. 63, pp. 1–7, 2011.

67. L. Stumpf, N. S. Calvo, F. C. Díaz, W. C. Valenti, and L. S. L. Greco, "Effect of intermittent feeding on growth in early juveniles of the crayfish Cherax quadricarinatus," Aquaculture, vol. 319, no. 1-2, pp. 98–104, 2011.

68. R. C. Summerfelt and C. R. Penne, "Septic tank treatment of the effluent from a small-scale commercial recycle aquaculture system," North American Journal of Aquaculture, vol. 69, no. 1, pp. 59–68, 2007.

69. C. P. Balasubramanian, S. M. Pillai, and P. Ravichandran, "Zero-water exchange shrimp farming systems (extensive) in the periphery of Chilka lagoon, Orissa, India," Aquaculture International, vol. 12, pp. 555–572, 2005.

70. C. Lezama-Cervantes, J. J. Paniagua-Michel, and J. Zamora-Castro, "Bioremediacion of effluents ones of the culture of Litopenaeus vannamei (Boone, 1931) using microbial mats in a recirculating system," Latin American Journal of Aquatic Research, vol. 38, no. 1, pp. 129–142, 2010.

71. M. B. Timmons, J. M. Ebeling, F. W. Wheaton, S. T. Summerfelt, and B. J. Vinci, Sistemas de Recirculación para la Acuicultura, Fundación Chile,Vitacura, Santiago de Chile, Chile, 2002.

72. V. H. Rivera-Monroy, L. A. Torres, N. Bahamon, F. Newmark, and R. R. Twilley, "The potential use of mangrove forests as nitrogen sinks of shrimp aquaculture pond effluents: the role of denitrification,"Journal of the World Aquaculture Society, vol. 30, no. 1, pp. 12–25, 1999.

73. L. R. Martinez-Cordova and M. Martinez-Porchas, "Polyculture of Pacific white shrimp, Litopenaeus vannamei, giant oyster, Crassostrea gigas and black clam, Chione fluctifraga in ponds in Sonora, Mexico,"Aquaculture, vol. 258, no. 1–4, pp. 321–326, 2006.

74. J. Paniagua-Michel and O. Garcia, "Ex-situ bioremediation of shrimp culture effluent using constructed microbial mats," Aquacultural Engineering, vol. 28, no. 3-4, pp. 131–139, 2003.

75. P. Chávez-Crooker and J. Obreque-Contreras, "Bioremediation of aquaculture wastes," Current Opinion in Biotechnology, vol. 21, no. 3, pp. 313–317, 2010.

76. P. De Schryver, R. Crab, T. Defoirdt, N. Boon, and W. Verstraete, "The basics of bio-flocs technology: the added value for aquaculture," Aquaculture, vol. 277, no. 3-4, pp. 125–137, 2008.

77. D. D. Kuhn, G. D. Boardman, S. R. Craig, G. J. Flick, and E. Mc Lean, "Use of microbial flocs generated from tilapia effluent as a nutritional supplement for shrimp, litopenaeus vannamei, in recirculating aquaculture systems," Journal of the World Aquaculture Society, vol. 39, no. 1, pp. 72–82, 2008.

78. J. M. E. Hussenot, "Emerging effluent management strategies in marine fish-culture farms located in European coastal wetlands," Aquaculture, vol. 226, no. 1–4, pp. 113–128, 2003.

79. J. Liu, Z. Wang, and W. Lin, "De-eutrophication of effluent wastewater from fish aquaculture by using marine green alga Ulva pertusa," Chinese Journal of Oceanology and Limnology, vol. 28, no. 2, pp. 201–208, 2010.

80. Y. Zhou, H. Yang, H. Hu et al., "Bioremediation potential of the macroalga Gracilaria lemaneiformis(Rhodophyta) integrated into fed fish culture in coastal waters of north China," Aquaculture, vol. 252, no. 2-4, pp. 264–276, 2006.

81. F. J. Martinez-Cordero and P. Leung, "Sustainable aquaculture and producer performance: measurement of environmentally adjusted productivity and efficiency of a sample of shrimp farms in Mexico,"Aquaculture, vol. 241, no. 1–4, pp. 249–268, 2004.

82. F. Corsin, S. Funge-Smith, and J. Clausen, A qualitative assessment of standards and certification schemes applicable to aquaculture in the Asia-Pacific region, RAP Publication 2007/25. Food and Agriculture Organization for the United Nations. Regional Office for the Asia and the Pacific, Bangkok, Thailand, 2007.

83. S. E. Siddall, J. A. Atchue III, and P. L. Murray Jr., "Mariculture development in mangroves: a case study of the Philippines, Panama and Ecuador," in Coastal Resources Management: Development Case Studies. Renewable Resources Information Series, Coastal Management, J. R. Clark, Ed., National Park Service, U.S. Dept. of the Interior, and the U.S. Agency for International Development. Research Planning Institute, Columbia, SC, USA, 1985.

84. W. J. M. Verheugt, A. Purwoko, F. Danielsen, H. Skov, and R. Kadarisman, "Integrating mangrove and swamp forests conservation with coastal lowland development; the Banyuasin Sembilang swamps case study, South Sumatra Province, Indonesia," Landscape and Urban Planning, vol. 20, no. 1–3, pp. 85–94, 1991

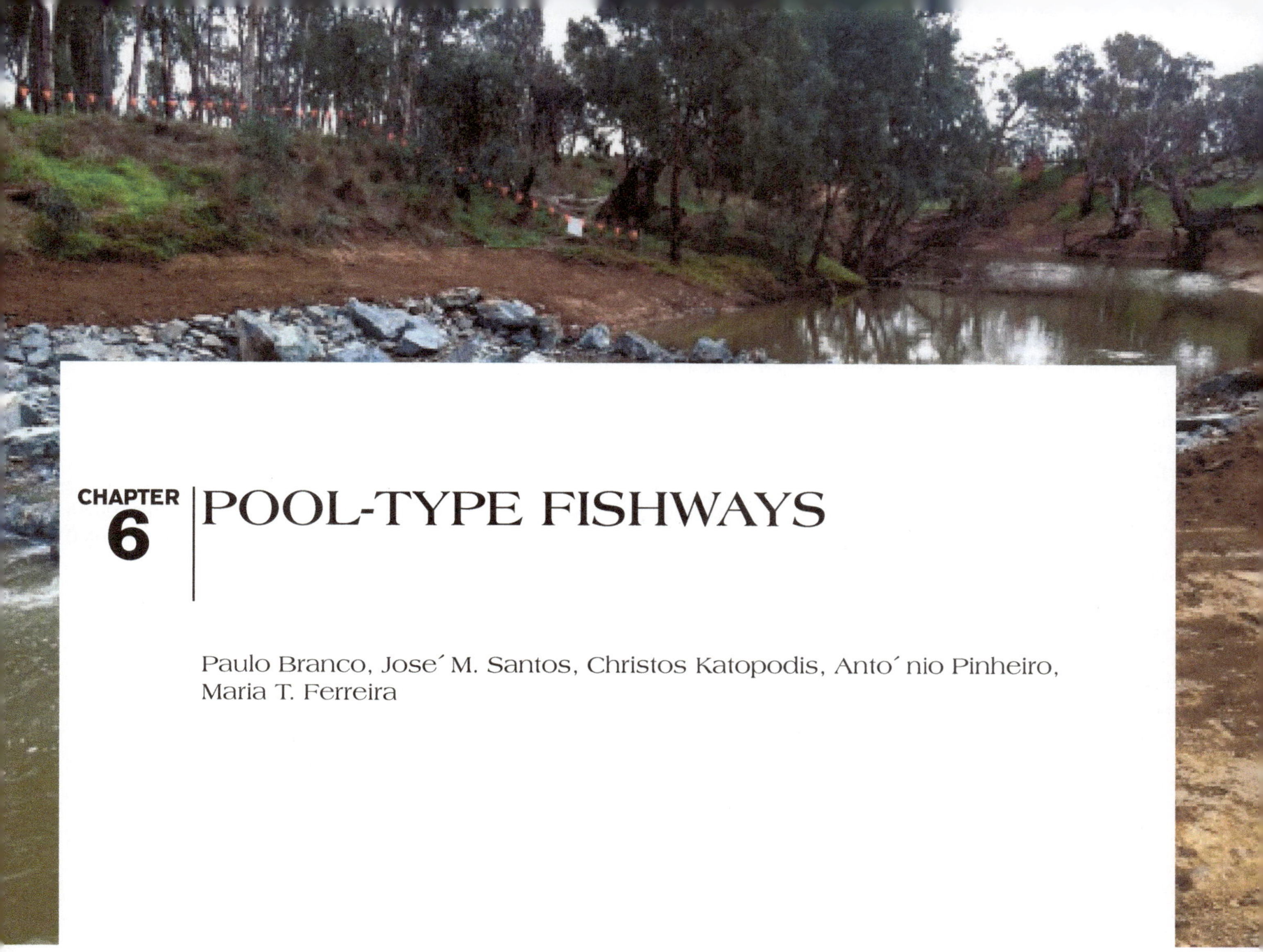

CHAPTER 6 | POOL-TYPE FISHWAYS

Paulo Branco, Jose´ M. Santos, Christos Katopodis, Anto´ nio Pinheiro, Maria T. Ferreira

INTRODUCTION

The increasing demand of water for human consumption propels the construction of dams and weirs that become insurmountable barriers to the migration of fish, thus imperiling the completion of the life cycle of several fish species. Indeed, more than half of the world's largest rivers are currently negatively impacted by dams and weirs [1], which have caused serious declines of both resident and migratory fish populations [2] by affecting upstream adult migration, reproduction, feeding and colonization movements, while promoting genetic impoverishment and dispersion of exotic species [3].

To overcome this problem, barriers must become negotiable by fish. To achieve this goal, the construction of fish transfer devices, commonly known as fishways, has been considered the most feasible measure to improve and restore connectivity. The problem with this measure is the often low efficiency of such devices for weak swimming species [4]. Fishways have traditionally been constructed based on guidelines developed and tested for salmonids, known strong swimmers with great leaping abilities, while studies of fishway performance focused on species with low economic and recreational value (e.g. cyprinids) continue to be neglected [5]. This is rather unfortunate, since these species are an important biological component of fish assemblages and free instream movement is indispensable for their survival [6]. In addition, recently, several non-salmonid species have acquired greater legislative protection (e.g. under the EU Habitats Directive [7]).

It is a complex problem to assure that flow and turbulence conditions within a fishway provide flow patterns suitable for an array of species. Pool-type fishways are a very common type of fish transfer device that has been built since the nineteenth century [8]. They consist of a rectangular flume divided by cross-walls that create a series of consecutive pools arranged in a stepped pattern each upstream from the preceding one. The purpose of these pools is to divide the height to be negotiated by fish, while ensuring no kinetic energy of the jets coming from the bottom orifices or the notches is transmitted to the downstream basins. This creates similar flow patterns in each pool [8]. Additionally, these pools offer resting areas for fish to recover after negotiating the cross-walls. The cross-walls between the pools may be equipped with different opening types - surface notches and submerged orifices at the bottom – that are used by the fish to move from pool to pool. The selection of an opening type by a fish depends on the species swimming ability and on the flow regime passing through the device [9] plunging or streaming [10]. In the plunging flow regime, the water level in the pool immediately below the cross-wall (producing the plunging flow) is below the crest of the notch;

Citation: Branco P, Santos JM, Katopodis C, Pinheiro A, Ferreira MT (2013) Pool-Type Fishways. doi:10.1371/journal.pone.0065089

in the streaming flow mode, a surface stream appears to flow over the crest of the notches, skimming over the water surface of the pools in between (further details in: [11], [12]). Pool-type fishways, which include pool and weir and vertical slot, are the most efficient conventional or technical fishway types constructed, either for salmonids or for non-salmonids [4]. Only fishways with optimal design can be of high efficiency as their success varies according to swimming ability, size [13]–[15] and physiologic state [16], [17] of different fish species, as well as hydraulics [18] and turbulence [19]. The design criteria for pool-type fishways are quite well understood for diadromous species [9], [11], [18]. There is still a knowledge gap though on how to improve and make these passes more efficient and more capable to accommodate a wider range of species and size-classes. This is particularly important for potamodromous and resident cyprinid species, as international environmental legislation requirements are increasingly more stringent for species [18], [20].

Fish species evolved differently, to be adapted to different riverine environments. These specializations can be grouped into different morpho-ecological guilds that introduce differences in swimming abilities, behaviour and niche occupancy. The best method to understand the influence of these differences on fishway negotiation ability is to test different species, representative of different morpho-ecological guilds, on fishways with controlled conditions. Laboratory trials, where conditions found in the field are easily reproduced while manipulating variables and monitoring confounding factors are always preferable [21], and have been proposed as the starting point of successful fishway designs [18], [22].

The purpose of this work is to study the behaviour and performance of cyprinid species with different ecological characteristics, in a full-scale indoor model of a pool-type fishway, fitted with bottom orifices and surface notches while subjected to one of two flow regimes - plunging and streaming. Hence, two species were used, Luciobarbus bocagei (Steindachner, 1864) a large-bodied potamodromous benthic fish, and Squalius pyrenaicus (Günther, 1868) a small-bodied water-column resident fish. It is thus hypothesized that: (1) the proportion of upstream movements through surface notches and submerged orifices will vary depending on the species and flow regime type; (2) both species will use the submerged orifices in greater proportions during the plunging flow regime, and (3) fish upstream movements and successes (when a fish reaches the top of the fishway by negotiating the fifth cross-wall) will be higher for both species during streaming flow conditions.

MATERIALS AND METHODS

Ethics Statement

Animal trials and sampling were conducted in agreement with national and international guidelines to maintain wellfare of the individuals [23], [24]. Fish sampling permits were obtained from the National Forest Authority. The experiments were carried out in strict accordance with the recommendations of the "protection of animal use for experimental and scientific work" of the Department for Health and Animal Protection (Direcção de Serviços de Saúde e Protecção Animal) that authorized animal experiments to be conducted in this experimental facility, and fish to be maintained in the laboratory (permit number: 16546–7/10/2011). All efforts were made to minimize stress and no fish were sacrificed to complete this study.

Fishway Facility

The laboratory experiments were conducted in a full-scale experimental model of a pool-type fishway (Fig. 1). The model structure was comprised of a steel frame with side panels of acrylic glass, allowing visualization of the fish movements occurring within the fishway. The flume was composed of six pools (1.9 m long×1.0 m wide×1.2 m high) divided by compact polypropylene cross-walls equipped with bottom orifices and surface notches. These were placed in a double offset arrangement, i.e., each cross-wall had a bottom orifice on one side and a surface notch on the other, and this pattern alternated between pools. The flume tilted portion (10 m long) was set on an 8.5% slope, which is within the range of those commonly used for this type of fishway [25]. The flow entering the fishway was measured and regulated by a flow meter in the supply pipe. Finer adjustments were made by a valve located at the upstream end of the fishway, where a tank (1.5 m long×1.0 m wide×1.2 m high), separated from the flume by a mesh screen, allowed water to flow smoothly into the flume. The water level on the other hand was regulated by a slot gate located in the downstream end of the flume at a tank (4 m long×3 m wide×4 m high) where the flume starts. This tank is separated from the flume by a mesh panel and mimics the hydraulic environment found downstream of the entrance to the fishway allowing for better fish acclimation. By controlling the flow discharge and the level of water in each pool, either plunging or streaming flow could be produced (Table 1).

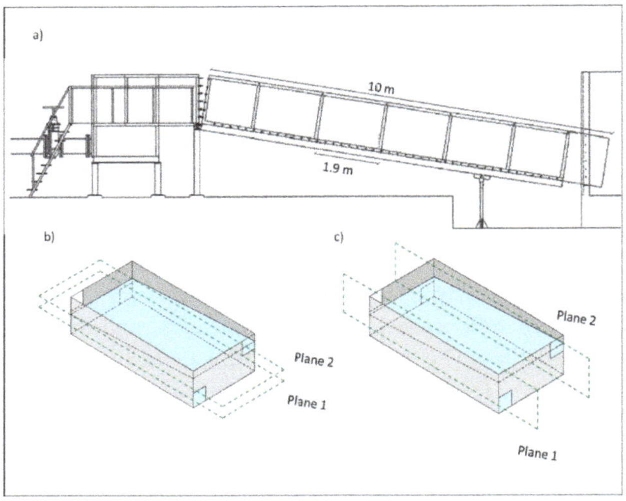

Figure 1. Schematic view of the experimental prototype and measurement planes.

Side view diagram of the pool-type fishway experimental prototype on a slope of 8.5%; b) Three dimensional representation of a pool, showing orifice arrangements and the horizontal planes (dashed green lines) where hydraulic measurements were conducted; c) Three dimensional representation of a pool, showing orifice arrangements and the vertical planes (dashed green lines) where hydraulic measurements were conducted.

Table 1. Description of the two tested flow regimes

Flow Regime	Q (L.s^{-1})	Orifice	Notch (width)	*hm*
Plunging	59.3	0.2 m×0.2 m	0.2 m	0.9 m
Streaming	78.5	0.2 m×0.2 m	0.2 m	1.0 m

Q – fishway discharge; *hm* – pool mean depth.
doi:10.1371/journal.pone.0065089.t001

Fishway hydraulic measurements.

Hydraulic conditions present at each pool have direct influence on the movements of fish [22], especially flow velocity and different turbulence parameters [19], [26], [27]. Therefore, in order to adequately characterize the hydraulics of the fishway, the three components of flow velocity (X, Y and Z) were measured with a Vectrino 3D ADV (Nortek AS) oriented downwards following a point-wise methodology. For this, a sampling point grid was created to cover the entire pool area with tighter spacing near the openings (5 cm), as these are areas of more pronounced velocity fluctuations, and wider spacing in the middle of the grid (10–15 cm). Velocity measurements were performed at a) two horizontal planes parallel to the flume bed (plane 1 at a height of 10 cm from the bottom, i.e. half the height of the bottom orifice and plane 2 at a height of ca. 78% of pool mean depth (hm), the plane directly affected by the surface notch), and b) two vertical planes, at the mid-width of each opening, parallel to the sidewalls (Fig. 1,Table 2). Measurements were performed at a 25 Hz rate for 90 s as this frequency and period were previously defined to be representative [28], at the second upstream pool. The hydraulic balance attained in the fishway allowed for any pool to be representative of every pool in the fishway. Instantaneous measures of velocity were filtered afterwards using the Goring and Nikora [29] phase-space threshold despiking method, modified by Wahl [30], using the WINADV freeware program [31]. Turbulence was characterized through Reynolds shear stress (RSS), which was calculated for the horizontal plane XY and for the vertical plane XZ, using the following formulas respectively:

$$-\rho\overline{\mu' v'} \quad (1)$$

$$-\rho\overline{\mu' w'} \quad (2)$$

where ρ=fluid density, μ′=fluctuating component of the velocity in the x direction, v′=fluctuating component of the velocity in the y direction and w′=fluctuating component of the velocity in the z direction.

Table 2. Point-grid information for the measurements performed with the Acoustic Doppler Velocimeter (ADV) at the 4 planes for each flow regime

Flow Regime	Orientation	Plane	# points in the grid	Measured at
Plunging	Horizontal	1	97	10 cm from the bottom
Plunging	Horizontal	2	97	70.5 cm from the bottom
Plunging	Vertical	1	104	10 cm from the lateral wall adjacent to the notch inlet
Plunging	Vertical	2	104	10 cm from the lateral wall adjacent to the orifice inlet
Streaming	Horizontal	1	97	10 cm from the bottom
Streaming	Horizontal	2	97	78.2 cm from the bottom
Streaming	Vertical	1	105	10 cm from the lateral wall adjacent to the notch inlet
Streaming	Vertical	2	106	10 cm from the lateral wall adjacent to the orifice inlet

Species Selection

The vast majority of studies on negotiating fishways use salmonids as target species. Advancing the scientific knowledge on navigating fishways requires the study of different non-salmonid species. Cyprinids are the dominant group of autochthonous freshwater fish in Mediterranean rivers ranging from large-size benthic potamodromous species to small-size pelagic ones, thus presenting different body shapes and occupying a range of ecological niches [32]. As studying the movements of all species would be time consuming, budget prohibitive and ultimately unfeasible, two species were chosen as representative of two morpho-ecological guilds [33]: the Iberian barbel (L. bocageias the representative of large-bodied potamodromous benthic cyprinids, and the Southern Iberian chub (S.pyrenaicus), as the representative of small-bodied water-column resident cyprinids. The use of guilds, which represent groups of organisms independent of taxonomic envelopes that use the same range of resources [34], has been proposed as a tool for multi-specific approaches [35].

Experiments

Two different flow regimes - plunging and streaming - were tested (Table 1). Three replicates of individual schools of 5 fish for each species were studied independently for each flow regime. Barbel and chub were captured on small coastal rivers from the Tagus river basin, central Portugal, during the migration season (Spring), by means of electrofishing using the procedures adopted by the European Committee for Standardization [23]. Fish were collected using a dip net and promptly placed in a mesh container fixed to the river bed at a location away from the influence of the electric field, allowing fish to be maintained in natural temperature and oxygenation conditions for the duration of the capture period. Fish were chosen to be of comparable length, and were transported to the laboratory in aerated containers filled with river water. Care was taken to reduce stress and expedite the procedure. At the laboratory, fish were stabilized in acclimation tanks (700 L) for at least 48 h before they were tested. Tests were performed in spring during the migration season the quality of the water was examined on a daily basis and changed at a ratio of 150 L per day. Feeding stopped 24 h before each experiment. Fish were acclimatized to the fishway environment as for each experiment they were introduced in the tailwater-pool where the fishway openings were obstructed by a mesh panel that was removed after a period of 30 minutes. Each experiment lasted 90 minutes and was performed during the period of dusk and early night (17 h–01 h) to reflect the natural period of migration [36], [37]. Each fish was only used once and was allowed to ascend the fishway on its own volition. Two independent observers monitored the fish movements within the fishway. It was considered a "movement" whenever a fish negotiated a cross-wall in the upstream direction (a pool-to-pool displacement by one fish); it was considered a "success" when a fish reached the top of the fishway by negotiating the fifth cross-wall. All upstream pool-to-pool displacements performed by any fish in the school during the full length of the experiment (90 min) were registered. After each experiment fish were observed carefully to see if any injury, tissue damage, bruising and direct or delayed mortality was induced by excessive turbulence [38], [39], [40].

Data Analysis

Differences in the number of upstream movements performed by the species during each flow regime - plunging and streaming - and through both opening types - bottom orifice and surface notch - were tested using proportion tests based on Chi-square distribution. The same procedure was applied to test for differences in the number of successful fishway negotiations attained individually and by both species in each of the flow regimes. To test the influence of flow regime, opening type and their interaction (flow regime×opening type) on the upstream movements of both species, a PERMANOVA test (permutation of residuals under a reduced model [41]) was applied. This statistical analysis is a powerful non-parametric approach that uses a permutational technique to enable significance tests for small sample sizes to be conducted [42] and was used to test the null hypotheses: (1) flow regime had no effect on the upstream movements of fish; (2) opening type had no effect on the upstream movements of fish; (3) effects of flow regime and opening types did not interact. Additionally, Mann-Whitney U tests and Sign tests were applied to the hydraulic data, to test for differences in flow velocities and turbulences (i.e. the Reynolds shear stress) between the two flow regimes. Chi-square proportion tests were performed in MedCalc software (MedCalc Software bvba). Mann-Whitney U tests and Sign tests were carried out in the software STATISTICA [43]. PERMANOVA tests were performed with the package PERMANOVA for PRIMER+v6.0 [44],[45].

RESULTS

Hydraulics

Figure 2 shows the flow velocity contours and vectors of both tested flow regimes at the two horizontal planes measured. On horizontal plane 1 of the plunging flow configuration, a jet-flow region could be distinguished from the water inlet to the opposite cross-wall. Horizontal plane 2 of this flow regime was characterized by a velocity peak close to the inlet of the water plunge that shaped a short downstream jet-flow region, interrupted by the main upstream flow. At horizontal plane 1 of the streaming flow regime, a large jet-flow region going from the water inlet to the opposite cross-wall could be observed. Horizontal plane 2 of this flow regime was characterized by a jet-flow from the water inlet to the opposite cross-wall. When compared, through a Sign test, the flow velocity pattern (XYZ resultant) in the two tested flow configurations differed on horizontal plane 2 (Sign test $Z=4.67$, Non-ties=97, $p<0.001$), closest to the water surface. Differences between flow regimes also arose in both horizontal planes, when the patterns of the horizontal and vertical components of Reynolds shear stress, were compared through Sign tests (Plane 1: RSS_{xy}–$Z=3.66$, Non-ties=97, p-value <0.001; RSS_{xz}–$Z=3.66$, Non-ties=97, p-value <0.001; Plane 2: RSS_{xy}–$Z=2.64$, Non-ties=97, p-value=0.008; RSS_{xz}–$Z=3.45$, Non-ties=97, $p<0.001$).

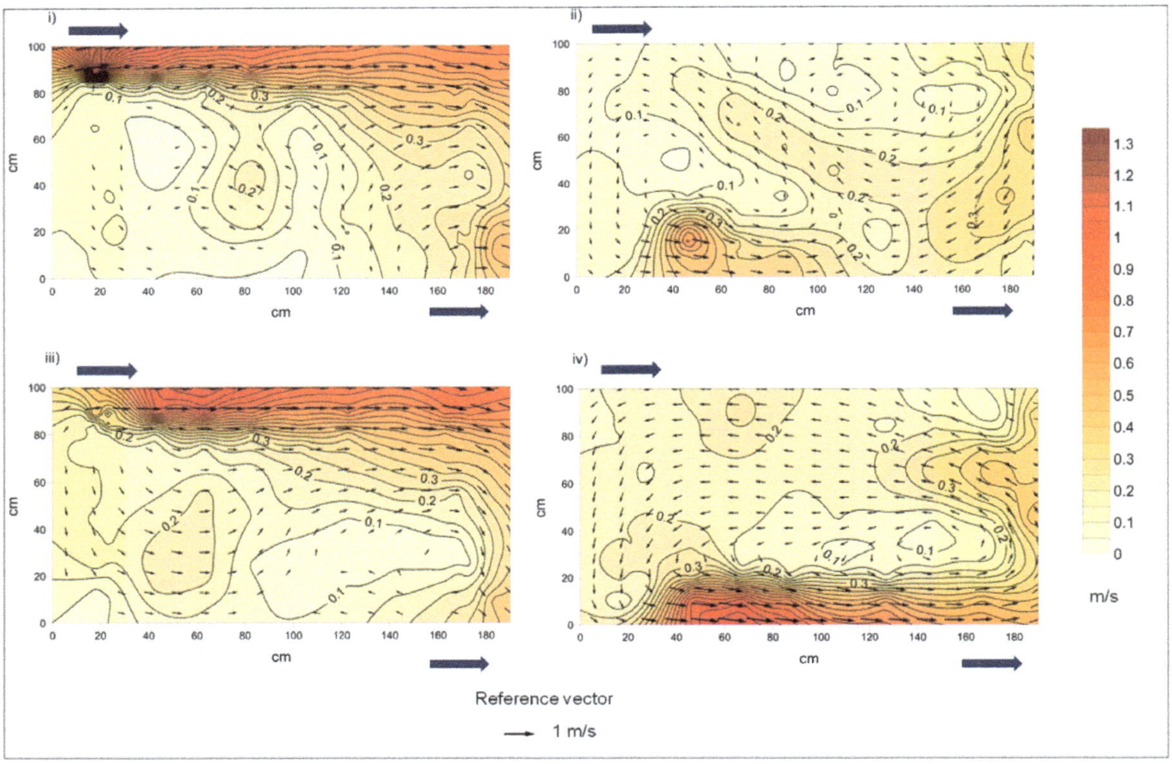

Figure 2. Velocity contour lines and vectors for the two horizontal planes in the two flow regimes.

Plane 1 measured at 10 cm, Plane 2 measured at 78% pool mean depth (hm). Arrows within the diagram show the direction and magnitude of flow. Arrows outside of the diagram show the water inlet and outlet points. i) Plunging flow regime horizontal plane 1; ii) Plunging flow regime horizontal plane 2; iii) Streaming flow regime horizontal plane 1; iv) Streaming flow regime horizontal plane 2.

Figure 3 shows the water velocity contours and vectors of both tested flow regimes at the two vertical planes considered. In the plunging flow, vertical plane 1, there was a noticeably high velocity plunge of water that was interrupted by an upstream flow. Vertical plane 2 of the same regime showed a strong jet stream region t from the water inlet to the opposite cross-wall. The streaming flow regime is characterized by a vertical plane 1 where the water flows, through the entire water column, from the water notch inlet towards the water outlet. Vertical plane 2 of this regime shows a strong jet stream from the water inlet up to the opposite cross-wall. When compared, through a Mann-Whitney U test, the water velocities (XYZ resultant) in the two tested flow regimes differed on vertical plane 1 ($U=4526$, $N_1=104$, $N_2=105$, $p=0.033$). This is the vertical plane which is under the direct influence of the surface notch inlet. In this vertical plane differences also arose when the same comparison was made for the Reynold's Shear Stress horizontal and vertical components (RSS_{xy}–$U=4343$, $N_1=104$, $N_2=105$, p-value=0.011; RSS_{xz}–$U=4139$, $N_1=104$, $N_2=105$, $p=0.0025$).

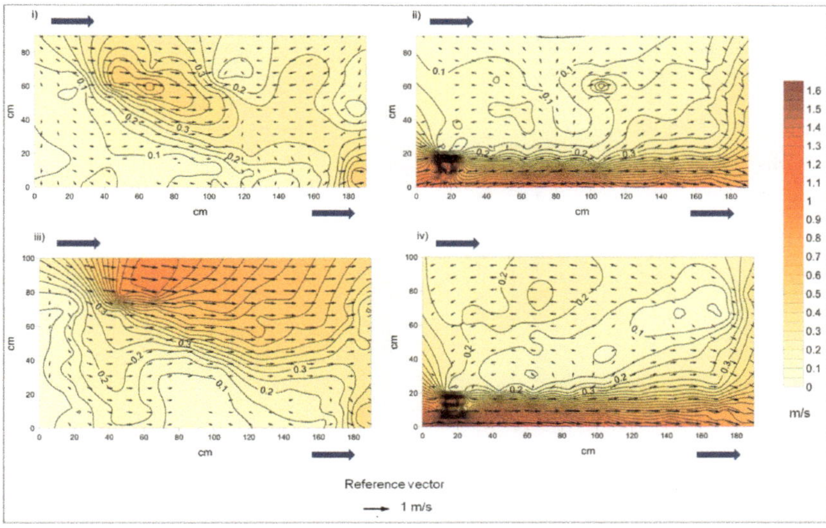

Figure 3. Velocity contour lines and vectors for the two vertical planes in the two flow regimes.

Planes were measured at 50% of the orifice width. Arrows within the diagram show the direction and magnitude of flow. Arrows outside of the diagram show the water inlet and outlet points. i) Plunging flow regime vertical plane 1; ii) Plunging flow regime vertical plane 2; iii) Streaming flow regime vertical plane 1; iv) Streaming flow regime vertical plane 2.

Fish

The fishway experiments showed that fish used the two opening types differently depending on flow regime and were closely related to the differences found in the hydraulic characterization (Table 3). During the plunging flow regime, 94% of all the upstream movements were conveyed through the bottom orifice opening ($\chi^2=163.189$, $p<0.0001$). A different result was observed for the streaming flow where the upstream movements were conveyed preferentially through the surface notch (57%) while 43% occurred through the bottom orifice opening ($\chi^2=10.208$, $p=0.0014$). The overall number of upstream movements between flow regimes was also different, with 74% of the movements occurring during the streaming flow regime ($\chi^2=180.599$, $p<0.0001$). This difference in the number of movements was also reflected in a higher number of successful fishway negotiations, with 74% ($\chi^2=61.725$, $p<0.0001$) occurring in the streaming flow regime.

Table 3. Number of upstream movements (i.e. upstream pool-to-pool displacements of a single individual) performed by all the individuals of each species through the bottom orifice and surface notch in the experimental pool-type fishway in both flow regimes (Plunging and Streaming).

Species	Flow regime	Bottom orifice	Surface notch	Total	# of Successes	Total length	Standard length	Body mass
Barbel	Plunging	37	1	38	15	26±3.9	20.6±3.4	181.2±88.0
	Streaming	71	83	154	48	25.5±2.3	22.7±2.9	239.6±26.0
Chub	Plunging	63	5	68	10	12.9±2.4	10.7±2.0	28.6±16.0
	Streaming	46	98	144	36	12.7±1.5	10.6±1.3	25.5±8.7

The number of successes (i.e. the number of times a fish attained the top of the fishway by the negotiation of the fifth cross-wall) achieved by all the individuals of each species in both flow regimes is also shown. The values of Total length (cm), Standard length (cm) and Body mass (g) are also presented (average ± standard deviation).
doi:10.1371/journal.pone.0065089.t003

The fishway experiments showed that the barbel used the facility differently for each distinct flow regime. Results from the PERMANOVA analysis (Table 4) showed a significant flow regime effect on the number of upstream fish movements within the fishway: overall, experiments conducted during streaming flow conditions revealed a higher proportion of barbel movements (80%) relative to plunging flows (20%). The effect of opening type alone was not significant, as the overall proportion of fish using notches (38%) and orifices (62%) were found to be similar. This produced a non-significant regime-by-opening type interaction. Experiments conducted during plunging flow conditions revealed an unequal proportion of movements through the orifices (97%) and notches (3%) (Fig. 4) (χ^2=64.474, p<0.0001). During the streaming flow conditions, barbel tended to use the two opening types an almost equal percentage of times (χ^2=1.571, p=0.21), as 46% of the upstream movements were undertaken through the surface notch and 54% through the bottom orifice (Fig. 4). In fact, 99% of all barbel's upstream movements through the surface notch were fulfilled in the streaming flow regime (χ^2=132.250, p<0.0001). When analyzing the number of successes registered in each of the flow regimes there was evidence of higher success in the streaming flow regime (χ^2=32.508, p<0.0001), which had over three times more successes (76%) than the plunging flow.

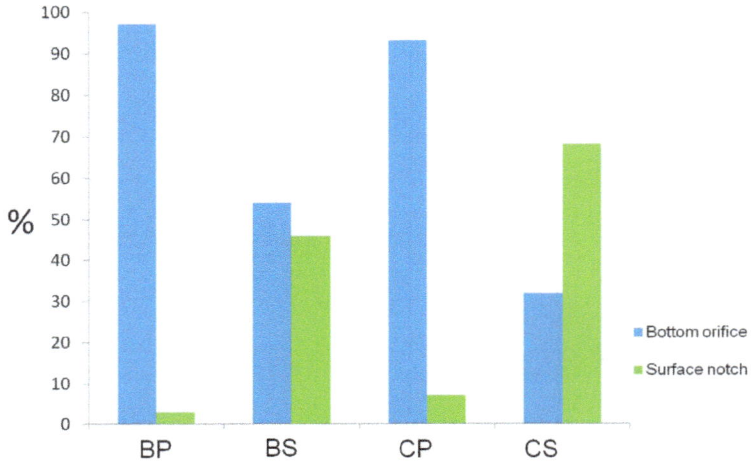

Figure 4. Proportions of upstream movements through both opening types by each species for both flow regimes.

BP – Barbel in plunging flow regime; BS – Barbel in streaming flow regime; CP – Chub in plunging flow regime; CS – Chub in streaming flow regime.

Table 4. Levels of significance of the two-factor PERMANOVA – permutation of residuals under a reduced model – testing the effects of flow regime plunging/streaming (Regime) and surface notch/bottom orifice (opening type) on the upstream movements of barbel and chub.

		Sum of Squares	Degrees of freedom	Mean Squares	Pseudo-F	p	Unique permutations
Barbel	Regime	1121.3	1	1121.3	11.317	0.01*	891
	Opening type	192	1	192	1.9378	0.272	917
	Regime×Opening type	48	1	48	0.48444	0.536	936
	Residuals	792.67	8	99.083			
Chub	Regime	481.33	1	481.33	11.46	0.008*	910
	Opening type	3	1	3	0.00714	0.792	927
	Regime×Opening type	1008.3	1	1008.3	24.008	0.006*	921
	Residuals	336	8	42			

*Significant with an α=0.05.
doi:10.1371/journal.pone.0065089.t004

The fishway experiments showed that the chub used the facility differently for each distinct flow regime. Results from the PERMANOVA analysis showed that flow regime had a significant effect on the number of fish upstream movements within the flume: in general, experiments performed in streaming flow regime revealed a higher ratio of chub movements (68%) relative to plunging flow regime conditions (32%). The isolated effect of opening type was not significant, because the overall proportions of fish using the notch opening (49%) and the bottom orifice (51%) were found to be equivalent. Nevertheless, a significant interaction between regime and opening was detected (p<0.01), indicating thereby that the flow regime affected the number of upstream movements differently, depending on the type of cross-wall opening. Accordingly, tests conducted in plunging flow regime revealed an uneven proportion of movements occurring through the bottom orifice (93%) and surface notch (7%) (Fig. 4) (χ^2=95.559, p<0.0001). As a result, the bulk (58%) of the upstream movements undertaken by chub through the bottom orifice were completed during the plunging flow condition (χ^2=4.697, p=0.0302). In contrast, chub were more likely to use the surface notches (68%) for streaming flows to the detriment of bottom orifices (32%) (χ^2=36.126, P<0.0001), and, as a result, 95% of all upstream movements through the surface notch were accomplished during streaming flow conditions (χ^2=164.349, p<0.0001). When analyzing the successes registered in each of the flow regimes there was evidence of higher success achieved during streaming flow conditions (78%) relative to plunging flows (χ^2=27.174, p<0.0001). After the experiments, fish did not present any signs of injury produced by turbulence.

DISCUSSION

The morphology of a fish species, its body size and shape [15], [46], [47] can determine the swimming ability and hydraulic suitability of the species to a specific environment [33]. This study allowed the assessment of movements and navigation behaviour of cyprinid species representing different morpho-ecological guilds (benthic/ potamodromous vs. water-column/resident) within a full-scale model of a pool-type fishway with different flow regimes. It contrasts with the majority of studies on fish transfer devices that often lack balanced experimental designs. It overcomes study limitation conducted in the wild by enabling the variables of interest to be manipulated while controlling for confounding effects [21]. The importance of this study is highlighted by the fact that it is the first study that examines and compares the movements and navigation behaviour of two morpho-ecologically different fish species in a pool-type fishway with streaming or plunging flow. The vast majority of studies have neglected the streaming flow regime of pool-type fishways and only studied the plunging flow regime [18]. The philosophy behind fishway research, design, and construction has evolved over the years and has been moving towards more holistic fishways, i.e. fishways that can serve a wide spectrum of species with different ecological niches. By focusing on the nature of movements of different species guilds within a fishway, this work moves a step forward towards holistic multi-species fishways.

The upstream movements of any fish are, as those of other animals [48], [49], always non-random and are a consequence of colonization, reproductive or feeding needs. During the experiments both fish species showed to be able to negotiate the fishway in both flow regimes. After the experiments, fish did not present any signs of injury produced by turbulence, indicating that fish probably limited their exposure to high RSS and flow velocities to a minimum.

Recent studies have defined the critical swimming speed of barbel [50] –0.81 $m.s^{-1}$ (3.1 BL. s^{-1}) – and of chub's sister species the bordallo (Squalius carolitertii (Doadrio, 1988)) [51] –0.54 $m.s^{-1}$(4.4 BL. s^{-1}). The flow velocity in the bulk of the pool area for the two flow regimes never surpassed 0.3 $m.s^{-1}$ attaining only higher velocities near the water inlets. This demonstrates that the velocity fields in both flow regimes were acceptable for these species and allowed fish to recover from negotiating the cross-wall where they had to undertake burst movements to offset the high flow velocities found at the cross-wall openings.

The findings of this study showed that during the plunging flow regime, the surface notch was seldom used by either species, which was not surprising since cyprinids are known to have a limited leaping ability [52]. An opposite trend was found for a salmonid species, the Atlantic salmon (Salmo salar), that preferred to use the surface notch opening for its upstream movements in an experimental pool-type fishway during plunging flow regime [53]. Guiny et al.[54] studied salmonids (Atlantic salmon parr and brown trout (Salmo trutta)) and found that both species used almost exclusively the bottom orifice to negotiate a cross-wall. It is important to note though that in the Guiny et al. [54] experiments the water column below the cross-wall was just 20 cm deep, and with this reduced depth the plunging water touched the flume bed and rebounded. This effect created flow vectors moving in different directions and may have reduced fish movements through the surface notch. In the present study, the vast majority of upstream movements during plunging flow by both tested species occurred through the bottom orifice. This result is in line with the findings of Silva et al. [55], where barbel also preferred to use the bottom orifices in contrast to the surface notch opening type for their upstream movements during plunging flows. Ficke et al. [56] also concluded that pool-type fishways that require fish to leap over the crest of the surface notch and instream structures that produce a vertical drop of more than 10 cm will not allow significant upstream passage of small-sized fish species. In the current study, only a few individuals of both species, most likely those with stronger swimming capacity, were able to negotiate the pool through the surface notch during plunging flow. These results confirm the hypothesis that species would not be able to use the surface notch during plunging flow, and will be forced to use the bottom opening. On the other hand, streaming flow facilitated access to the surface notch, allowing species to move upstream more successfully with fewer constraints and possibly with less energy expenditure, increasing the number of movements. By turning the surface notch into a more accessible opening type, the streaming flow regime increased the area available for negotiating cross-walls. However, species displayed different choices of opening types for their upstream movements. These preferences are congruent with their different morpho-ecologic guilds: the chub, a water column species, had a clear preference for the surface notch opening. In contrast, the barbel, a bottom oriented species, divided their upstream movements between opening types, demonstrating that this is a plastic species with the ability to use the entire water column.

The positive rheotactic behaviour during their upstream migrations [22], [52] guides fish upstream by orienting them against the downstream flow. Fish may also display rheotactic responses to turbulence promoting migratory movements [22], [57]. Based on these facts, the hydraulics of the two tested flow regimes allowed a simple explanation of the experimental results. The plunging flow regime provided little or no attraction to the surface notch, since an upstream flow circulation was generated which interrupted the downstream flow possibly miss-orientating fish. The same does not happen for the streaming flow, where there was a continuity of downstream flow allowing fish to be attracted to the surface notch. Looking at vertical plane 1, this pattern becomes even more evident. In streaming flow there was continuity of flow in the entire length of the pool and throughout the pool water column. This is in contrasting with the sudden change in flow direction observed in vertical plane 1 of the plunging flow regime. The fact that the flow regimes differed in their RSS patterns also supports the difference in movements and successes attained in both flow regimes. This proves that not only velocity but also turbulence parameters can explain fish preferences when navigating through a fishway [19], [57].

Streaming flow proved to be a more efficient flow regime by enhancing upstream movements, facilitating the movement of the surface oriented species and increasing the options of the bottom oriented species. Through a more suitable hydraulic environment, streaming flow increased the negotiation success of both species. Thus, the streaming flow configuration is the most suitable for fishways constructed in river systems in which a wide range of fish morpho-ecological traits are found. Existing fishways should be retrofitted to accommodate the behaviour and swimming capabilities of the species present at the site [58]. The present study shows that enabling streaming flows on future and on existing fishways is desirable. This must be considered during fishway design to allow this flow regime to be attained even with low flows. Future studies should concentrate on the behaviour of other non-salmonid species in pool-type fishways. Focusing on the swimming capabilities of different species, finding the best fishway configuration that maximizes attraction [18], [59], [60], keeping velocity gradients[22] and turbulence levels acceptable [19], ensuring negotiation success of all fish species present in the target system, and minimizing fishway selection of species, size-classes and ages [61]–[64] is the best way to move towards the objective of an holistic fishway that accommodates the entire range of species present at a given system.

ACKNOWLEDGMENTS

The authors would like to thank André Fabião for help in collecting fish in the field and Susana Amaral, Miguel Monteiro, Ana Sofia Alves e Gonçalo Duarte for their assistance in the laboratory experiments. The text was improved following consideration of the comments of two anonymous referees.

AUTHOR CONTRIBUTIONS

Conceived and designed the experiments: PB JMS CK AP MTF. Performed the experiments: PB JMS. Analyzed the data: PB JMS AP MTF. Wrote the paper: PB. Significant manuscript editing: JMS CK AP MTF.

REFERENCES

1. Nilsson C, Reidy CA, Dynesius M, Revenga C (2005) Fragmentation and Flow Regulation of the World's Large River Systems. Science 308: 405–408. doi: 10. 1126/science. 1107887
2. Northcote T (1998) Migratory behaviour of fish and its significance to movement through riverine fish passage facilities. In: Jungwirth M, Schmutz S, Weiss S, editors. Migration and Fish Bypasses, Fishing News Books, Cambridge. 3–18.
3. Branco P, Segurado P, Santos JM, Pinheiro P, Ferreira MT (2012) Does longitudinal connectivity loss affect the distribution of freshwater fish? Ecol Eng 48: 70–78 DOI:10. 1016/j. ecoleng. 2011. 05. 008.
4. Noonan MJ, Grand JWA, Jackson CD (2011) A quantitative assessment of fish passage efficiency. Fish Fish. DOI: –10. 1111/j. 1467–2979. 2011. 00445. x.
5. Roscoe DW, Hinch SG (2010) Effectiveness monitoring of fish passage facilities: historical trends, geographic patterns and future directions. Fish Fish 11: 12–33. doi: 10. 1111/j. 1467-2979. 2009. 00333. x
6. Lucas MC, Mercer T, Peirson G, Frear PA (2000) Seasonal movements of coarse fish in lowland rivers and their relevance to fisheries management. In: Cowx IG, editor. Management and ecology of river fisheries. Blackwell Science, Oxford, UK. 87–100.
7. Council Directive 92/43/EEC of 21 May 1992 on the conservation of natural habitats and of wild fauna and flora.
8. Larinier M, Marmulla G (2004) Fish passes: types, principles and geographical distribution an overview. Proceedings of the Second International Symposium on the Management of Large Rivers for Fisheries, 11–14 February 2000, Kingdom of Cambodia.
9. Katopodis C (2005) Developing a toolkit for fish passage, ecological flow management and fish habitat works. IAHR J Hydraul Res 43(5): 451–467. doi: 10. 1080/00221680509500144
10. Rajaratnam N, Katopodis C, Mainali A (1988) Plunging and streaming flows in pool and weir fishway. J Hydraul Eng-ASCE 114: 939–944. doi: 10. 1061/(asce)0733-9429(1988)114:8(939)
11. Clay CH (1995) Design of Fishways and Other Fish Facilities. CRC Press Inc. : Boca Raton, Florida.
12. Ead SA, Katopodis C, Sikora GJ, Rajaratnam N (2004) Flow regimes and structure in pool and weir fishways. J Environ Eng Sci 3: 379–390. doi: 10. 1139/s03-073
13. Hinch SG, Bratty JM (2000) Effects of swim speed and activity pattern on success of adult sockeye salmon migration through an area of difficult passage. Trans Am Fish Soc 129: 604–612. doi: 10. 1577/1548-8659(2000)129<0598:eossaa>2. 0. co;2
14. Castro-Santos T (2005) Optimal swim speeds for traversing velocity barriers: an analysis of volitional high-speed swimming behavior of migratory fishes. J Exp Biol 208: 421–432. doi: 10. 1242/jeb. 01380
15. Katopodis C, Gervais R (2012) Ecohydraulic analysis of fish fatigue data. River Res Appl 28: 444–456 DOI: 10. 1002/rra. 1566.
16. Pon LB, Hinch SG, Cooke SJ, Patterson DA, Farrell AP (2009) Physiological, energetic and behavioural correlates of successful fishway passage of adult sockeye salmon Oncorhynchus nerka in the Seton River, British Columbia. J Fish Biol 74: 1323–1336. doi: 10. 1111/j. 1095-8649. 2009. 02213. x
17. Hasler CT, Donaldson MR, Sunder RPB, Guimond E, Patterson DA, et al. (2011) Osmoregulatory, metabolic, and nutritional condition of summer-run male Chinook salmon in relation to their fate and migratory behavior in a regulated river. Endanger Species Res 14: 79–89. doi: 10. 3354/esr00343
18. Katopodis C, Williams JG (2012) The development of fish passage research in a historical context. Ecol Eng 28: 407–417 DOI: 10. 1002/rra. 1551.
19. Silva AT, Katopodis C, Santos JM, Ferreira MT, Pinheiro AN (2012) Cyprinid swimming behaviour in response to turbulent flow. Ecol Eng 44: 314–328http://dx. doi. org/10. 1016/j. ecoleng. 2012. 04. 015.

20. Wang RW, David L, Larinier M (2010) Contribution of experimental fluid mechanics to the design of vertical slot fish passes. Knowl Manag Aquat Ecosyst 396: 1–21. doi: 10. 1051/kmae/2010002

21. Kondratieff MC, Myrick CA (2005) Two adjustable waterfalls for evaluating fish jumping performance. Trans Am Fish Soc 134(2): 503–508. doi: 10. 1577/t03-174. 1

22. Williams JG, Armstrong G, Katopodis C, Larinie M, Travade F (2012) Thinking like a fish: a key ingredient for development of effective fish passage facilities at river obstructions. River Res Appl 28: 407–417 DOI: 10. 1002/rra. 1551.

23. CEN (European Committee for Standardization) (2003) Water quality: sampling of fish with electricity. Brussels: CEN, European Standard EN 14011: 2003 E.

24. INAG (Instituto da Água, I. P.) (2008) Manual para a avaliação biológica da qualidade da água em sistemas fluviais segundo a Directiva Quadro da Água – Protocolo de amostragem e análise para a fauna piscícola. Lisboa: Ministério do Ambiente, do Ordenamento do Território e do Desenvolvimento Regional. Instituto da Água, I. P.

25. Larinier M (2008) Fish passage experience at small-scale hydro-electric power plants in France. Hydrobiologia 609: 97–108. doi: 10. 1007/s10750-008-9398-9

26. Haro A, Odeh M, Castro-Santos T, Noreika J (1999) Effect of slope and headpond on passage of American shad and blueback herring through simple Denil and deepened Alaska steeppass fishways. N Am J Fish Manage 19: 51–58. doi: 10. 1577/1548-8675(1999)019<0051:eosaho>2. 0. co;2

27. Bunt CM, Katopodis C, McKinley RS (1999) Attraction and passage efficiency of white suckers and smallmouth bass by two Denil fishways. N Am J Fish Manage 19: 793–803. doi: 10. 1577/1548-8675(1999)019<0793:aapeow>2. 0. co;2

28. Santos JM, Silva AT, Katopodis C, Pinheiro PJ, Pinheiro AN, et al. (2012) Ecohydraulics of pool-type fishways: getting past the barriers. Ecol Eng 48: 38–50 DOI:10. 1016/j. ecoleng. 2011. 03. 006.

29. Goring DG, Nikora VI (2002) Despiking acoustic Doppler velocimeter data. J Hydraul Eng 128(1): 117–126. doi: 10. 1061/(asce)0733-9429(2002)128:1(117)

30. Wahl TL (2002) Discussion of "Despiking Acoustic Doppler Velocimeter Data". J Hydraul Eng 129 (6): 484–487. doi: 10. 1061/(asce)0733-9429(2003)129:6(484)

31. Wahl TL (2001) WINADV. A free-ware software program for the analysis of ADV data. Bureau of Reclamation Water Resources Research Laboratory, Denver, Colorado.

32. Ferreira MT, Sousa L, Santos JM, Reino L, Oliveira P, et al. (2007) Regional and local environmental correlates of native Iberian fish fauna. Ecol Freshw Fish 16: 504–514. doi: 10. 1111/j. 1600-0633. 2007. 00241. x

33. Chan MD (2001) Fish ecomorphology: predicting habitat preferences of stream fishes from their body shape. PhD Dissertation, Virginia Tech. Blacksburg, VA. 269 p.

34. Fauth JE, Bernadro J, Camara M, Resetarits Jr WJ, Van Buskrik J, et al. (1996) Simplifying the jargon of community ecology: a conceptual approach. Am Nat 147(2): 282–286. doi: 10. 1086/285850

35. Leonard PM, Orth DJ (1988) Use of habitat guilds of fishes to determine instream flow requirements. N Am J Fish Manage 8: 399–409. doi: 10. 1577/1548-8675(1988)008<0399:uohgof>2. 3. co;2

36. Santos JM, Ferreira MT, Godinho FN, Bochechas J (2002) Performance of fish lift recently built at the Touvedo dam on the Lima River, Portugal. J Appl Ichthyol 18: 118–123. doi: 10. 1046/j. 1439-0426. 2002. 00309. x

37. Santos JM, Ferreira MT, Godinho FN, Bochechas J (2005) Efficacy of a nature-like bypass channel in a Portuguese lowland river. J Appl Ichthyol 21: 381–388. doi: 10. 1111/j. 1439-0426. 2005. 00616. x

38. Cada GF (2001) The development of advanced hydroelectric turbines to improve fish passage survival. Fisheries 26(9): 14–23. doi: 10. 1577/1548-8446(2001)026<0014:tdoaht>2. 0. co;2

39. Odeh M, Noreika JF, Haro A, Maynard A, Castro-Santos T (2002) Evaluation of the effects of turbulence on the behavior of migratory fish. Final Report to the Bonneville Power Administration, Contract 00000022, Project 200005700, Portland, Oregon. 46 p.

40. Neitzel DA, Dauble DD, Cada GF, Richmond MC, Guensch GR, et al. (2004) Survival estimates for juvenile fish subjected to a laboratory-generated shear environment. Trans. Am. Fish. Soc. 133(2): 447–454. doi: 10. 1577/02-021

41. Anderson MJ, Robinson J (2001) Permutation tests for linear models. Aust NZ J Stat 43: 75–88. doi: 10. 1111/1467-842x. 00156

42. Walters K, Coen LD (2006) A comparison of statistical approaches to analyzing community convergence between natural and constructed oyster reefs. J Exp Mar Biol Ecol 330: 81–95. doi: 10. 1016/j. jembe. 2005. 12. 018

43. STATSOFT INC. (2000) Statistica for Windows (Computer Program Manual). Tulsa: Statsoft, Inc.

44. Clarke KR, Gorley RN (2006) PRIMER v6: User manual/tutorial. PRIMER-E, Plymouth, UK. 192 p.

45. Anderson MJ, Gorley RN, Clarke KR (2008) PERMANOVA+ For PRIMER: guide to software and statistical methods. PRIMER-E, Plymouth, UK. 213 p.

46. Videler JJ, Wardle CS (1991) Fish swimming stride by stride: speed limits and endurance. Rev Fish Biol Fish 1: 23–40. doi: 10. 1007/bf00042660

47. Videler JJ (1993) Fish swimming. Chapman and Hall, London, New York. 284 p.

48. Goedmakers A, Pinkster S (1981) Population dynamics of three gammarid species in a French chalk stream. Part III, Migration. Bijdr Dierkd 51: 145–180.

49. Erman NA (1986) Movements of self-marked caddisfly larvae in a Sierran spring stream. Freshw Biol 16: 455–464. doi: 10. 1111/j. 1365-2427. 1986. tb00988. x

50. Mateus CS, Quintella BR, Almeida PR (2008) The critical swimming speed of Iberian barbel Barbus bocagei in relation to size and sex. J Fish Biol 73: 1783–1789. doi: 10. 1111/j. 1095-8649. 2008. 02023. x

51. Romão F, Quintella BR, Pereira TJ, Almeida PR (2012) Swimming performance of two Iberian cyprinids: the Tagus nase Pseudochondrostoma polylepis (Steindachner, 1864) and the bordallo Squalius carolitertii (Doadrio, 1988). J Appl Ichthyol 28(1): 26–30. doi: 10. 1111/j. 1439-0426. 2011. 01882. x

52. Bunt CM (2001) Fishway entrance modifications enhance fish attraction. Fish Manag Ecol 8: 95–105. doi: 10. 1046/j. 1365-2400. 2001. 00238. x

53. Stuart TA (1962) The leaping behaviour of salmon and trout at falls and obstructions. Freshwater and Salmon Research 28: 1–46.

54. Guiny E, Armstrong JD, Ervine DA (2003) Preferences of mature male brown trout and Atlantic salmon parr for orifice and weir fish pass entrances matched for peak velocities and turbulence. Ecol Freshw Fish 12: 190–195. doi: 10. 1034/j. 1600-0633. 2003. 00025. x

55. Silva AT, Santos JM, Franco AC, Ferreira MT, Pinheiro AN (2009) Selection of Iberian barbel Barbus bocagei (Steindachner, 1864) for orifices and notches upon different hydraulic configurations in an experimental pool-type fishway. J Appl Ichthyol 25: 173–177. doi: 10. 1111/j. 1439-0426. 2009. 01237. x

56. Ficke AD, Myrick CA, Jud N (2011) The Swimming and Jumping Ability of Three Small Great Plains Fishes: Implications for Fishway Design. Trans Am Fish Soc 140(6): 1521–1531. doi: 10. 1080/00028487. 2011. 638579

57. Coutant CC (1998) Turbulent attraction flows for juvenile salmonid passage at dams. ORNL/TM-13608. Oak Ridge National Laboratory, Oak Ridge, Tennessee. 28 p.

58. Johnson EL, Caudill CC, Keefer ML, Clabough TS, Peery CA, et al. (2012) Movement of radio-tagged adul pacific lampreys during a large scale fishway velocity experiment. Trans Am Fish Soc 141: 571–579. doi: 10. 1080/00028487. 2012. 683468

59. Francis F (1870) Reports on Salmon Ladders: With Original Drawings, Plans, and Sections. Horace Cox: London. 32 p.

60. Green TM, Lindmark EM, Lundström TS, Gustavsson LH (2011) Flow characterization of an attraction channel as entrance to fishways. River Res Appl 27: 1290–1297. doi: 10. 1002/rra. 1426

61. Agostinho AA, Gomes LC, Fernandez DR, Suzuki HI (2002) Efficiency of fish ladders for neotropical ichthyofauna. River Res Appl 18: 299–306. doi: 10. 1002/rra. 674

62. Agostinho AA, Pelicice FM, Petry AC, Gomes LC, Julio Jr HF (2007) Fish diversity in the upper Parana River basin: habitats, fisheries, management and conservation. Aquat Ecosyst Health Manag 10: 174–186. doi: 10. 1080/14634980701341719

63. Mallen-Cooper M, Brand DA (2007) Non-salmonids in a salmonid fishway: what do 50 years of data tell us about past and future fish passage? Fish Manag Ecol 14: 319–332. doi: 10. 1111/j. 1365-2400. 2007. 00557. x

64. Naughton GP, Caudill CC, Peery CA, Clabough TS, Jepson MA, et al. (2007) Experimental evaluation of fishway modifications on passage behaviour of adult Chinook salmon and steelhead at lower Granite Dam, Snake River, USA. River Res Appl 23: 99–111. doi: 10. 1002/rra. 957

CHAPTER 7

DOES SELECTION IN A CHALLENGING ENVIRONMENT PRODUCE NILE TILAPIA GENOTYPES THAT CAN THRIVE IN A RANGE OF PRODUCTION SYSTEMS?

Ngo Phu Thoa, Nguyen Huu Ninh,Wayne Knibb, and Nguyen Hong Nguyen

Understanding the interactions between genotype and environment (G × E) is of utmost importance in the design of selective breeding programs to develop improved strains that express optimal growth and adaptability in a wide range of environments. In a review of 24 studies in both farmed animals and model species, Falconer[1] proposed that selection in a less favourable environment (i.e. antagonistic selection) may produce genotypes that perform well across a range of production systems. In contrast, selection under favourable environmental conditions (i.e. synergistic selection) may result in sensitive genotypes which do not thrive under stressed conditions. We have developed a genetic line of Nile tilapia under sub-optimal culture environment (i.e. moderately saline water of 10–20 ppt). Genetic evaluation of the line selected over five generations from 2008 to 2012 showed an increase in body weight in saline water of approximately 7% per generation (one generation was completed each year)[2]. However, it is not known if these increases in saline water are expressed in the freshwater production systems common in tilapia aquaculture. This information is necessary before the improved strain can be released into commercial production.

Several studies have examined the effect of G × E interaction by testing full- and half-sib families in varying conditions, such as between culture systems[3,4,5], dietary protein levels[6], animal *vs.* plant based protein diets[7,8,9], feeding regimes and rearing schemes (high *vs.* low stocking density)and between diverse environmental parameters[10]. A synthesised results from the literature show that when the selection environments differ remarkedly from production systems, the G × E effect is statistically significant[5]. On the other hand, when the selected environment in the nucleus is close to commercial production, the interaction between genotype and environment may not be of biological importance[11].

While many studies have investigated the traits of growth and survival, and their respective G × E interactions, as it relates to profit and income for aquatic species, less well understood are the traits of feed conversion ratio[12], sexual maturity, fish morphology[13,14] and their GxE effects. Further, morphological deformities generally have negative impacts on marketability and these characters (fitness related traits), but typically have not been included as part of the breeding objectives in most previous selection programmes. Abnormalities may be caused by inbreeding[15], non-heritable factors such as diseases[16], environmental disturbances[17,18] as well as sub-optimal culture conditions[19,20,21]. Only a few studies have reported the existence of G × E effect for sexual maturity in aquatic animal species[22,23] which is curious the important possibility that G × E interactions may cause a decline in reproduction. Similarly

Citation: Ngo Phu Thoa, Nguyen Huu Ninh, Wayne Knib and Nguyen Hong Nguyen, Does Selection In A Challenging Environment Produce Nile Tilapia Genotypes That Can Thrive In A Range Of Production Systems?, doi: 10.1038/srep21486.

little is known about G × E interactions for fitness and resistance against diseases. Indeed, there is no information regarding quantitative genetic basis of gill diseases, morphological deformity and food conversion ratio (FCR) in tilapia. The question of whether selection in harsh environments permitted fast growth in benign conditions remains unanswered.

The principal objective of this study was to investigate whether selection for high growth rate in a challenging environment of medium salinity produces tilapia genotypes that can perform well in freshwater systems. In resolving this, we (i) examined the effect of genotype by environment interaction for new traits (sexual maturity, deformity, survival and FCR) in a Nile tilapia population undergoing five generations of selection for high growth in moderate saline environment (15–20 ppt); ii) estimated the genetic parameters for these traits together with gill related diseases and body weight; and (iii) deciphered phenotypic and genetic associations among traits recorded during the selection process (2007 to 2014).

MATERIALS AND METHODS

Experimental Animals

The fish used in this study originated from a genetic line which has undergone five generations of selection (2008–2012) for increased harvest body weight in moderately saline water (15–20 ppt) in Vietnam. During this period, a total number of 922 breeders (525 females and 397 males) were used to produce progeny with the average selected proportion of 4.43% in females and 3.48% in males each generation. Within- and between-family selection was practised based on estimated breeding values (EBVs) for body weight. The EBVs were estimated for all individuals in the pedigree using a statistical model that included the fixed effects of generation, sex and age and the random additive genetic effect of individual animal[2]. The number of selected individuals contributed by each family to subsequent generations was restricted and closely related matings among full- and half-sibs was avoided to control realized inbreeding. The accumulative level of inbreeding in the present population (0.51% per generation) was within the acceptable level (which is less than 1% per generation) in closed breeding nucleus[24]. The selection program resulted in a significant improvement in body weight withan average gain of approximately 7% per generation (one generation per year)[2].

Pond Preparation and Family Production

Earthen ponds of various sizes (e. g. 50 × 40 × 1.2 m) were used for conditioning, breeding and nursing and were prepared following a standard protocol[25]. Briefly, ten days before the experiment, an inorganic fertilizer (triple super phosphate) was applied to each pond at a rate of 50 kg per hectare to stimulate the production of natural food sources (e.g. phytoplankton and zooplankton). Breeding to produce offspring for this study began in March 2014 and followed a period of conditioning of the selected female and male breeders in hapas (small enclosed semi rigid nets) (20 × 5 × 2 m). A total of 103 breeding hapas (2 × 1 × 1.5 m) were suspended in moderately saline water ponds (8–10 ppt) and in each hapa, one female was kept with one male. After seven days, fertilized eggs were collected from the mouth of the female and immediately transferred to hatching trays for artificial incubation. Each family was incubated in a separate hatching tray (40 × 20 × 10 cm) in low saline water (3–5 ppt) at 25–30 °C and pH 7.5–8.2. Once a full-sib family was successfully produced and a sufficiently large quantity of progeny (>600 fry per family) was attained, the male was then paired with a second female to produce paternal half-sib families. The mating pairs were mated again if they produced less 600 fry. A hierarchical nested mating design was applied, targeting one male mated to two females; not all the mate pairs were successful. As a result, there were 49 full-sib and 28 half-sib families. In the 2014 breeding season, a total of 77 families were produced from 63 male and 77 female brooders.

Fry rearing was conducted in nursery hapas (3 × 1.2 × 1.5 m with 2 mm mesh size) in a 1000 m^2 low saline water (5–10 ppt) pond. The fry of each family were reared in three replicate hapas at an initial stocking density of 200 fry per hapa (or about 55 fry per square meter of surface water). For the first nursing period from hatching to 20 d, a commercial powder feed (45% protein AQUAXCEL®, Cargill Company) was provided four times daily with a feeding rate at 10% of the total body mass in hapas. The fry were transferred to a larger hapa (3 × 2 × 1.5 m) for further rearing from 21 d post hatching to tagging at 62 d post hatching. The feeding rate was reduced to 5% of total body weight and fry were fed twice a day (8.00 and 16.00) with a commercial pellet (35% protein). When the fingerlings reached 5–8 g, about 100 individuals were randomly sampled from each family (30–35 fingerlings per nursing hapa) and physically tagged with passive integrated transponder tags (PIT). The identification number, body weight (BW) and standard length (L) were recorded. The tagged fingerlings were then conditioned for 4–5 d in fibreglass tanks without feeding. During the conditioning period, fish that lost their tags or died were replaced by other individuals from the same family.

Testing Environments

After conditioning, the tagged fingerlings from 77 families were split into two groups (50 individuals per family in each group) and were randomly assigned to freshwater or moderate saline water (15–20 ppt) earthen ponds (1,500 m^2 30 × 50 × 1.5 m) located at Quy Kim Station in Hai Phong, 120 km from Hanoi. The salinity level in experimental ponds was controlled by adding freshwater (0 ppt) and sea water (30 ppt). The temperature and dissolved oxygen ranged from 17.3–34.5 °C and 3.5–6.4 mg/l in freshwater environment and from 17.8–34.7 °C and 3.7–6.5 mg/l in the moderately saline water environment. In both culture systems, the initial stocking density was 2.5 fish per square meter of surface water. During the grow-out period, the fish were fed a commercial dry pellet with 35% protein content twice a day (8.00 and 16.00) at the average feeding rate of 5% of the total body mass. The feeding rate was adjusted monthly according to the growth rate of the fish, the latter being determined from a random sample of 100 animals per environment. The same feeding, culturing and management practices were applied in both environments.

Measurements

At the end of the grow-out period of approximately 120 d, all the experimental fish were harvested to measure body traits, sexual maturity, morphological deformity, gill condition and survival. Individual fish were weighted using a digital scale (nearest to 0.1 g). Standard length was also measured with a ruler (with precision of 1 mm). Body depth and width was measured at the highest vertical distance and thickness position, respectively, with a calliper (±0.1 mm). Survival was recorded as a binary trait and also computed from the difference between the number of fish at stocking and at harvest. Fish alive at harvest were coded as 1, otherwise as 0 for those that was dead (missing) at harvest. Maturity status of all individual fish was recorded as either: (i) immature and not ready to spawn with white/clear and flat genital papilla and normal abdomen (=0); and (ii) mature and ready to spawn with pink to red and protruding genital papilla, fully opened genital pore, and distended abdomen (=1). Morphological deformities commonly found in tilapia include: scale disorientation, gill, eyes, skeletal, soft tissue, and split fins deformity[18]. They were treated in the form of presence (abnormal) or absence (normal) and were coded as 0 = normal fish and 1 = morphological deformity (abnormal) fish (e.g. fin, body, operculum, eye and head deformities)[26]. Gill condition (red = 0; pale = 1 was an indicator of fish health related to parasite, bacteria or virus diseases) was recorded in all individual fish in both fresh and saline water. The total volume of feed which was used in each pond was recorded to calculate the FCR for individual fish with an assumption that all the fish in each environment had an equal chance to get the feed. FCR of individual fish in each environment was calculated:

$$FCR_{ij} = \frac{FI_i}{N_i} \times (Wtgain_j)^{-1} \quad (1)$$

where $\boldsymbol{FCR_{ij}}$ is feed conversion ratio of fish j^{th} in environment i^{th}; $\boldsymbol{FI_i}$ is total volume of feed in kilogram unit used in environment i^{th} ($i = 1, 2$); N_i is the average number of fish during the grow-out period in environment i^{th}. $\boldsymbol{Wtgain_j}$ is weight gain in kilogram unit of fish j^{th}.

In addition, body condition or Fulton condition factor (**K**), which is the most common measurement to describe body shape of several fish species, was derived from measurements of body weight and standard length[4]. The K factor of each individual fish was calculated as described in the below equation by Casselman[27]:

$$K_i = 1000 \times \frac{Wt_i}{Lth_i^{\,3}} \quad (2)$$

where $\boldsymbol{K_i}$ is Fulton condition factor of fish i^{th}; $\boldsymbol{Wt_i}$ is the harvest weight in gram unit of fish i^{th}; $\boldsymbol{Lth_i}$ is the harvest length in centimetre unit of fish i^{th}.

Statistical Analysis

Statistical analyses were performed on 26,295 data records collected between 2007 and 2014 in saline water together with 7,270 performance data in freshwater in 2014. The pedigree included six generations tracing back to the base population since 2007, comprising 36,145 individuals. The data was evaluated for normality before undertaking further analyses. Logarithmic, arcsine, square root or box-cox transformation was considered where appropriate. Exploratory analyses using a general linear model (GLM) were undertaken using PROC UNIVARIATE in SAS 9.3[28].

After all significant fixed effects were identified (additional file 1); the final model fitted for all traits included the systematic effects of generation, testing environment, sex and their two-way interaction. Stocking weight and age from birth to harvest within sex and generation were fitted as linear covariates for body traits. The random effects are the additive genetics effects of individual fish and the maternal and common environment of dam. In matrix notation the mixed model was written as:

$$\boldsymbol{y = Xb + Za + Wc + e} \quad (3)$$

where **y** is the vector of observations for body traits, survival rate, maturity, FCR, deformity and gill condition, **b** is the vector of the fixed effects of generation (six generations and a base population pedigree), sex (female or male) and testing environments (saline- and fresh- water). Linear covariates of age from birth to harvest and stocking weight were fitted within sex and generation in the model. Vector **a** is the random animal additive genetic effects ~ (0, $\mathbf{A}\sigma_a^2$) where **A** is the additive genetic (numerator) relationship matrix among the animals, **c** is the vector of dam effects (or maternal effects in addition to the additive genetics) ~ (0, $\mathbf{I}\sigma_c^2$) and ***e*** is the vector of residual effects ~ (0, $\mathbf{I}\sigma_e^2$). The dam component (σ_D^2) is most likely a combination of maternal and common environmental effects (thus, $\sigma_D^2 = \sigma_{M+CE}^2$, referred to as σ_c^2) caused by the separate rearing of full-sib families until individuals reached a suitable size for physical tagging. **X**, **Z** and **W** are incidence matrices relating observations to fixed effects, additive genetic effect of the individual animal and common full-sib effect included in the model, respectively. Under the model [3], var(**a**) = **G** = $\mathbf{A}\sigma_a^{\,2}$. The remaining effects are assumed to be distributed as var(**e**) = **R** = $\mathbf{I}\sigma_e^2$, var(**c**) = **W** = $\mathbf{I}\sigma_c^2$, where **I** is an identity matrix. The expectations of all random effects are zero, cov (**a,e**) = 0 and cov (**a,c**) = 0 and thus var (**y**) = $\mathbf{ZGZ'}\sigma_a^{\,2} + \mathbf{WI}\sigma_c^2\mathbf{W'} + \mathbf{R}$.

The mixed model equation for the best linear unbiased estimator (BLUE) of estimable functions of **b** and the best linear unbiased prediction of **a** and **c** are:

$$\begin{bmatrix} \hat{b} \\ \hat{a} \\ \hat{c} \end{bmatrix} = \begin{bmatrix} X'X & XZ & X'W \\ Z'X & Z'Z + A^{-1}\alpha_1 & Z'W \\ W'X & W'Z & W'W + I\alpha_2 \end{bmatrix}^{-1} \begin{bmatrix} X'y \\ Z'y \\ W'y \end{bmatrix} \quad (4)$$

where $\alpha_1 = \sigma_e^2/\sigma_a^2$ and $\alpha_2 = \sigma_e^2/\sigma_c^2$.

Heritabilities, phenotypic and genetic correlations: For estimation of heritability and genetic correlations, the ASReml software version 3.0[29] was used. In details: heritabilities were estimated from a single trait model. Phenotypic and genetic correlations were obtained from a series of bi- and tri-variate analyses, involving body weight (the selection criterion recorded on all animals) to avoid selection bias[30].

First, measurements in both environments were treated as one trait, and the model included all the significant fixed and random effects as described in Equation 3. Second, trait measurements in fresh and saline water were treated independently in order to examine whether there are differences associated with the two culture environments. The model is as described above, also note that the fixed effect of environment was omitted from the statistical model. Genetic correlations between expressions of body traits or other characters studied in freshwater and moderate saline water ponds were estimated via numerator genetic relationships matrix (A). Since the traits were measured on animals in different environments, there is no environmental covariance between them. The phenotypic correlations do not exist because any individual fish expressed the trait in only one environment.

Heritability for weight, body shape, FCR, survival and maturity were calculated as $h^2 = \frac{\hat{\sigma}_a^2}{\hat{\sigma}_a^2 + \hat{\sigma}_c^2 + \hat{\sigma}_e^2}$ and the maternal effect as $c^2 = \frac{\hat{\sigma}_c^2}{\hat{\sigma}_a^2 + \hat{\sigma}_c^2 + \hat{\sigma}_e^2}$ where $\sigma_a^2$is the additive genetic variance, the maternal variance (σ_c^2) and the residual variance (σ_e^2). Genetic and phenotypic correlations among traits were calculated as the covariance divided by the product of the standard deviations of traits: $r = \frac{\sigma_{12}}{\sqrt{\sigma_1^2}\sqrt{\sigma_2^2}}$ where σ_{12} is the estimated additive genetic or phenotypic covariance between the two traits, and $\sigma_1^2$and $\sigma_2^2$are the additive genetic or phenotypic variances of traits 1 and 2, respectively.

In addition to linear mixed model as described above (equation 3), survival, maturity and deformity were analysed by generalised linear mixed model (GLMM), assuming that the data followed a binomial distributions with a logit link function. Under the logit model, the link function ($\hat{p} = e^x/(1+e^x)$) was used where p is the probability of fish survival (maturity or deformity) recorded at harvest. The fixed effects (F_i) are the same as defined in Equation 3, except for the omission of sex when survival was analysed because this information was not known for dead or missing fish. The random effects are individual fish (a_k) and dam (d_l) for survival and maturity (Equation 6). Only the additive genetic effect of individual fish was fitted as a random term in the model for deformity. With this logit threshold model, heritability for survival and maturity was calculated using the variance of the logit link function, which implies a correction of the residual variance by factor $\pi^2/3$.

$$h^2 = \frac{\sigma_a^2}{\sigma_a^2 + \sigma_c^2 + \sigma_e^2 \frac{\pi^2}{3}} = \frac{\sigma_a^2}{\sigma_a^2 + \sigma_c^2 + 3.289} \tag{5}$$

where σ_a^2 is the additive genetic variance, σ_c^2 maternal and common environmental variance and $\sigma_e^2 = 1$

$$\log\left(\frac{p_{ijkl}}{1 - p_{ijkl}}\right) = \mu + F_i + a_j + d_k + e_{ijkl} \tag{6}$$

The maternal and common environmental variance σ_c^2 was not estimable for deformity, thus it was omitted from the above formula to calculate heritability for this traits. For binary observations (survival, maturity and deformity), estimates of heritability (h^2) on the liability scales (logit) can be transformed to observed (0/1) scale using the formula of Robertson and Lerner[31] as described by Thoa, *et al.*[25]. Due to only 0.016% of fish that had pale condition, heritability for this trait was not estimable with GLMM procedure as used for other binary traits. Linear sire model was employed to analyse this trait.

To test for significant differences in genetic parameter estimates between fresh and saline environments, we used the z-score as an approximate method for assessing whether heritability and correlation estimates were significantly different from each other or from zero, and whether genetic correlations between the two environments were significantly different from one[32]. The formula to calculate z-scores was as follows:

$$z = \frac{x_i - x_j}{(\sigma_i^2 + \sigma_j^2)^{0.5}} \tag{7}$$

where x_i and x_j are the estimates of heritability, maternal effects, or genetic correlations for the two traits, and ${}_i$and ${}_j$ are their respective standard errors. Both x_j and ${}_j$ were set to zero or one when we tested whether an estimate was significantly different from zero or one, respectively. Resulting z-scores were then tested against a large sample normal distribution.

Ethics Statement

All animal care and experimental procedures were performed in accordance with guidelines and regulations approved by the ethical committee of University of the Sunshine Coast, Queensland, Australia (ethical approval No. AN/S/13/29).

RESULTS

Characteristics of Testing Environments and Basic Statistics for Traits Studied

The main difference between the two environments was the level of salinity (i.e. 0.74 ppt±0.07 in freshwater; 16.87 ppt±0.07 in saline water).The fish were reared at a starting weight of 4.5 g and a steady linear increase in growth over three sampling periods was recorded in both fresh and medium salinity environments (results not presented). At the final harvest (7270 fish measured at 196 d), the mean body weight of tilapia in moderate saline water was greater than those in freshwater (256.9 *vs.* 226.9 g, $P<0.05$). Basic statistics for all traits are given in Table 1. Body traits generally had low standard deviations with one exception for body weight. Over six generations, the survival rate from stocking to harvest was 72.8%. The average maturity rate of this population was high (86.93%), whereas the proportion of fish having 'pale' gill condition was very low (0.016%). Deformity rate in both testing environments was only 0.148%. Feed conversion ratio, FCR (i.e. kg of food used to produce one kg of fish live weight) was 1.45 with high standard deviation (0.57).

Table 1. Pedigree information and descriptive statistics for traits studied across the two testing environments.

Trait differences between moderate saline- and fresh-water environments

Traits	Unit	Generation	N	Sire	Dam	Mean	SD	Min	Max
Weight	G	2007–2014	26,295	335	424	246.6	65.3	20.6	524
Length	cm	2007–2014	26,295	335	424	23.0	1.9	14.5	29.7
Width	cm	2014	6082	335	424	3.3	0.4	1.9	6.8
Depth	cm	2014	6082	335	424	7.4	0.8	4	10.3
Body condition	ratio	2007–2014	26,295	335	424	21.7	3.5	2.2	75.6
FCR	unit	2014	6082	77	77	1.45	0.57	0.48	5.31
Survival	%	2007–2014	36,145	335	424	72.8	44.5	0	1
Maturity	%	2014	7,270	77	77	87.5	33.71	0	1
Gill condition	%	2014	7,270	77	77	0.016	1.28	0	1
Deformity	%	2014	7,270	77	77	0.148	3.84	0	1

Least squares means of traits studied (body weight, fitness, maturity, FCR and gill condition) obtained from a mixed model that included all possible systematic fixed effects (additional file 1) and random terms of individual fish and common full-sibs are shown in Table 2. The experimental fish reared in moderate saline water environment had significantly greater growth and lower FCR than those in freshwater whereas the opposite trend was found for body condition (Fulton condition or K factor) and survival rate. Maturation rate was 6% lower in moderately saline water than in freshwater ($P<0.001$). The difference in deformity rate between the two testing environments was not significant ($P>0.05$).

Table 2. Growth related traits (weight, length, width and depth), body condition, food conversion ratio, survival, sexual maturity and deformity of fish reared in saline water and freshwater environments in the latest generation in 2014 (Least squares means ± S.E).

Traits	Model	Saline water	Freshwater	Significant probability
Weight (g)	LMM	256.9 ± 1.09	226.9 ± 1.05	<0.001
Length (cm)	LMM	23.1 ± 0.03	22.5 ± 0.03	<0.001
Width (cm)	LMM	3.4 ± 0.01	3.2 ± 0.01	<0.001
Depth (cm)	LMM	7.5 ± 0.01	7.2 ± 0.01	<0.001
Body condition (ratio)	LMM	19.4 ± 0.04	20.4 ± 0.04	<0.001
Body condition (ratio)*	LMM	4.4 ± 0.009	4.7 ± 0.003	<0.001
FCR (unit)	LMM	1.08 ± 0.008	1.89 ± 0.008	<0.001
Survival (%)	GLMM	80.7 ± 0.66	86.6 ± 0.56	<0.001
Maturity (%)	GLMM	83.9 ± 0.69	89.5 ± 0.55	<0.001
Deformity (%)	GLMM	0.12 ± 0.06	0.17 ± 0.08	NS

Heritability and Common Environmental Effects

Heritability (h^2) and common environmental effects (c^2) for traits studied (body weight, body condition, FCR, gill related disease, deformity, survival and sexual maturity) were estimated separately for each testing environment and across the two environments, using both animal and animal and dam models (Table 3). Across the environments, the estimates of heritability for harvest weight using the two models were moderate to high and significantly different from zero ($P<0.05$). In general, heritability and common environmental effects for body traits was higher in saline water than in freshwater, but lower for body condition and survival rate. However, the differences in heritability for all traits studied between the two environments were not significant ($P>0.05$). Moderate to high heritabilities for body condition and FCR were estimated by the animal model, whereas they were low and associated with high standard errors when the animal and dam model was used. Similar results were found for heritability of sexual maturity across the models. The heritability for survival was low to moderate (0.03–0.27) across the two environments. In addition, a very low heritability (0.01–0.07) was estimated for gill condition and deformity. For all traits except gill condition and deformity, the c^2 effect ranged from 0.01 to 0.25 in both saline and fresh water. The large range of c^2indicates that by omission of this effect, the heritability for traits studied were overestimated.

Table 3. Heritability ($h^2 \pm$ S.E) and common environmental effects ($c^2 \pm$ S.E) for traits studied in saline water (SW), fresh water (FW) and in both environments.

Traits	Environment	No. of records	Animal model	Animal + Dam model	
			h^2	h^2	c^2
Weight	SW	23,109	0.57 ± 0.02	0.30 ± 0.02	0.22 ± 0.02
	FW	3,186	0.48 ± 0.07	0.19 ± 0.12	0.10 ± 0.05
	Both	26,295	0.62 ± 0.01	0.32 ± 0.04	0.25 ± 0.02
Body condition	SW	23,109	0.17 ± 0.02	0.08 ± 0.02	0.02 ± 0.004
	FW	3,186	0.28 ± 0.05	0.26 ± 0.11	0.01 ± 0.04
	Both	26,295	0.19 ± 0.02	0.09 ± 0.02	0.04 ± 0.005
FCR	SW	2,896	0.59 ± 0.08	0.12 ± 0.11	0.18 ± 0.06
	FW	3,186	0.52 ± 0.07	0.16 ± 0.13	0.13 ± 0.06
	Both	6,082	0.56 ± 0.07	0.12 ± 0.11	0.16 ± 0.05
Survival	SW	28,875	0.09 ± 0.01	0.15 ± 0.02	0.03 ± 0.01
	FW	7,270	0.01 ± 0.01	n.e.	n.e.
	Both	36,145	0.03 ± 0.006	0.17 ± 0.03	0.03 ± 0.01
	Both*	36,145	0.14 ± 0.01	0.04 ± 0.01	0.06 ± 0.01
Maturity	SW	2,896	0.40 ± 0.06	0.13 ± 0.09	0.09 ± 0.04
	FW	3,186	0.32 ± 0.06	0.06 ± 0.07	0.09 ± 0.04
	Both	6,082	0.36 ± 0.05	0.09 ± 0.07	0.08 ± 0.03
	Both*	6,082	0.25 ± 0.03	0.05 ± 0.06	0.18 ± 0.05
Deformity	SW	2,896	0.05 ± 0.03		
	FW	3,186	0.05 ± 0.03		
	Both	3,186	0.07 ± 0.02		
	Both**	6,082	0.53 ± 0.14	n.e.	n.e.
Gill condition	SW***	2,896	0.05 ± 0.02		
	FW***	3,186	0.01 ± 0.02		
	Both***	6,082	0.07 ± 0.02		

Genetic and Phenotypic Correlations Among Traits

Genetic and phenotypic correlations between traits studied were estimated separately in freshwater and moderate saline water (Table 4) and in the both environments (Table 5). Across environments, genetic and phenotypic correlations between body weight and FCR were very high and negative, whereas a positive and close to one genetic correlation was found between body weight and maturity. Genetic associations between body weight and fitness traits (survival and deformity) were not significant (Table 5). Similarly, there were low genetic correlations between body condition and the other traits studied. The genetic correlations of maturity rate with survival and deformity were moderate to high and negative, ranging from −0.60±0.14 to −0.82±0.06. On the other hand, survival was genetically correlated positively with sexual maturity (Table 5). Both genetic and phenotypic correlations among traits studied in saline water and freshwater environments (Table 4) had similar signs and magnitudes to those obtained across the two environment (Table 5).

Table 4. Phenotypic (above the diagonal) and genetic (below the diagonal) correlations (SE in parenthesis) in saline- and fresh- water environments.

Traits	Saline water						Freshwater					
	Weight	BC	FCR	Survival	Maturity	Deformity	Weight	BC	FCR	Survival	Maturity	Deformity
Weight		0.17 (0.01)	−0.91 (.003)	0.05 (0.01)	0.44 (0.020)	−0.04 (0.02)		0.33 (0.03)	−0.92 (.004)	n.e.	0.42 (0.02)	−0.04 (0.02)
BC	0.22 (0.05)		−0.46 (0.03)	0.06 (0.01)	0.15 (0.03)	0.02 (0.02)	0.17 (0.14)		−0.38 (0.03)	−0.45 (0.070)	0.09 (0.03)	−0.03 (0.030)
FCR	−0.99 (0.01)	−0.46 (0.13)		−0.50 (0.02)	−0.47 (0.03)	n.e.	−0.93 (0.02)	−0.24 (0.13)		−0.66 (0.04)	−0.49 (0.02)	0.006 (0.02)
Survival	0.12 (0.11)	0.10 (0.06)	−0.76 (0.06)		−0.26 (0.03)	−0.32 (0.02)	n.e.	−0.10 (0.50)	−0.40 (0.26)		0.001 (0.03)	−0.33 (0.03)
Maturity	0.78 (0.12)	0.48 (0.14)	−0.77 (0.06)	0.38 (0.15)		−0.01 (0.02)	0.84 (0.05)	0.34 (0.13)	−0.85 (0.05)	0.10 (0.30)		−0.02 (0.02)
Deformity	−0.17 (0.28)	0.09 (0.34)	0.14 (0.28)	−0.82 (0.12)	−0.01 (0.29)		−0.23 (0.29)	−0.61 (0.15)	0.26 (0.29)	−0.97 (0.05)	0.05 (0.29)	

Table 5. Phenotypic (above the diagonal) and genetic (below the diagonal) correlations in both environments.

Genetic correlations between trait expressions in freshwater and moderate saline water

Traits	Weight	BC	FCR	Survival	Maturity	Deformity
Weight		0.21 ± 0.01	−0.90 ± 0.01	0.08 ± 0.01	0.50 ± 0.02	−0.04 ± 0.02
BC	0.19 ± 0.09		−0.50 ± 0.02	−0.43 ± 0.01	0.13 ± 0.03	−0.006 ± 0.02
FCR	−0.98 ± 0.04	−0.55 ± 0.10		−0.16 ± 0.03	−0.41 ± 0.01	0.05 ± 0.02
Survival	0.19 ± 0.04	−0.09 ± 0.11	−0.81 ± 0.05			0.05 ± 0.02
Maturity	0.84 ± 0.08	0.20 ± 0.15	−0.74 ± 0.14	0.58 ± 0.11		−0.34 ± 0.02
Deformity	−0.13 ± 0.18	0.28 ± 0.19	0.15 ± 0.17	−0.82 ± 0.06	−0.03 ± 0.17	

Table 6 shows genetic and common environmental correlations between the trait expressions in the two testing environments. The genetic correlations estimated for body weight, FCR, maturity and survival traits between freshwater and moderate saline water environments were very high and positive (r_g=0.78 to 0.99, P<0.05). The genetic correlation estimate for morphological deformity trait had a high standard error (0.99±0.62). Similarly, common environmental correlations estimated for body weight, FCR and maturity were very high (0.87–0.99) (results not tabulated). The close to one genetic correlations for traits studied indicated that genotype by environment interaction was likely to be unimportant for many of these characteristics in the present population.

Table 6. Genetic (r_G) correlations between the traits expressed in saline- (SW) and fresh-water (FW) environments.

Traits	Environment	Genetic correlation
Weight	Between SW and FW	0.92 ± 0.04
Body condition	Between SW and FW	0.97 ± 0.02
FCR	Between SW and FW	0.96 ± 0.08
Survival	Between SW and FW	0.77 ± 0.03
Maturity	Between SW and FW	0.96 ± 0.02
Deformity	Between SW and FW	0.99 ± 0.62^{NS}

DISCUSSION

The principal aim of our study was to understand if the hypothesis proposed by Falconer[1] is applicable to a tilapia population selected under medium salinity (i.e. less conducive for Nile tilapia), will produce genotypes which can perform well in both freshwater and saline water culture systems. We tested this by treating trait expressions in saline- and fresh-water as different characteristics and using multivariate analysis to estimate the genetic correlations for new traits, including sexual maturity, deformity and feed conversion ratio (FCR). The high and close to one estimates of genetic correlations obtained across trait expressions in saline- and fresh-water showed that the genotype by environment interaction (G×E) was not important for fitness (i.e. sexual maturity, deformity and survival) and growth related traits (i.e. feed conversion ratio and body weight) in the present population of Nile tilapia after six generations of selection. Our results indicate that the genetically improved line selected in saline water for high growth can be cultured successfully in freshwater systems. To the best of our knowledge, performance of the present salinity tolerance line has not been experimentally tested in prevailing environments that are conducive to tilapia culture, i.e. freshwater pond. Almost all studies in aquaculture species examined the G×E effect on body traits, with a few exceptions for fitness related traits, such as sexual maturity or deformity[22,23]. The genetic correlation for sexual maturity between freshwater and saline water environments was consistent with that in Rainbow trout (*Oncorhynchus mykiss*) shown by Kause, *et al.*[23]. In contrast, Wild, *et al.*[22]reported a genotype by salt water interaction for early maturity in Atlantic salmon (*Salmo salar*). Across species, there are no published information on the G×E effect on feed conversion ratio (FCR) in aquaculture to compare with our current study.

In a comparison with the present study, in which the experimental fish were selected in a sub-optimal culture environment (i.e. moderate saline water of 15–20 ppt), other researchers used genetic lines selected under favourable environmental conditions and then conducted performance testing under culture systems that are generally close to the selection environment used in the nucleus. The literature[11] suggest that the effect of genotype by environment interaction for production traits (body weight and survival) is unimportant when the selection and production environments are similar, such as freshwater ponds and cages[3,4,33,34,35]. These results are in good agreement with our high genetic correlations between the two environments for body weight (r_g=0.92±0.04) and for survival (r_g=0.77±0.03). By contrast, a significant G×E interaction occurred for harvest body weight when the selection environment in the nucleus differed greatly from production systems, such as freshwater *vs.* moderate saline water in Nile tilapia[5], Arctic char (*Salvelinus alpinus*)[36] and/or recirculating systems compared with pond in Sole (*Solea solea*)[37]. Similarly, Sylvén, *et al.*[38] reported low genetic correlations in slaughter weight of Rainbow trout between fresh- and salt- (r_g=0.72) and brackish water (r_g=0.58). These results suggest that selection under favourable well-controlled environments in the nucleus not only result in re-ranking of breeding candidates but also a fraction of genetic gain achieved from the selection program can be captured in production systems. This contrasts with our findings showing that the present selected line under moderate saline water can capture about 77% and 92% genetic gain in survival and body weight when its offspring were grown under freshwater ponds. Our breeding program to improve growth performance under moderate saline water provides tilapia seed for both freshwater and saline water production systems. Conducting separate breeding programs may not be justified when the genetic correlation of the trait expressions between the two environments is greater than 0.7–0.8[39].

In addition to the genetic correlations, we estimated heritability for a range of traits in saline- and fresh-water, separately. The non-significant difference in magnitude of heritability for these traits indicates that the G×E interaction was mainly due to scaling effect, i.e. there was no re-ranking of selected candidates between saline- and fresh-water. We also aimed to estimate heritability for new traits in tilapia, including deformity, sexual maturity, gill condition and FCR. Our first estimates of genetic parameters for fitness related traits show that there is an additive genetic component especially for sexual maturity (h^2=0.32–0.4). This suggests that sexual maturity can respond to selection as reported by Wild, *et al.*[22] in Atlantic salmon and Kause, *et al.*[23] in Rainbow trout. Due to the very low proportion of animals that had deformity (0.15%) and 'pale' gill condition (0.016%) in this population, the heritability for these traits are not significantly different from zero, indicating that continuing collection of the data in future generations is needed in order to obtain reliable estimates for binary characteristics. Other studies show existence of useful additive genetic component for a range of deformity measures but the heritability for these traits was low to moderate and associated with high standard errors: 0.16–0.29 in European sea bass *Dicentrarchus labrax*[40] and Yellowtail kingfish *Seriola lalandi*[41].

Improving efficiency of food utilisation is crucial to reduce production costs of aquaculture enterprises. In aquatic species, it is costly and practically infeasible to record the feed intake of individual animal. Our calculation of feed conversion ratio

(FCR) can be used as an approximate measure although it does not reflect a 'real' variation in feed efficiency among animals. However, this measure of FCR is practical under the conditions of commercial production. The present estimate of heritability indicates that FCR is heritable (h^2=0.52–0.59 with animal model and 0.12–0.16 with animal and dam model). Using animal mixed model, Henryon, *et al.*[42] studying rainbow trout reported a low additive genetic variation for feed conversion efficiency with a coefficient of additive genetic variation from 4.0–13.9%. Henryon, *et al.*[42] reared each family separately in different tanks during the grow-out period whereas a communal rearing method was applied in our study. Similarly, low heritability for feed efficiency (0.03±0.10–0.07±0.11) was reported in European whitefish *Coregonus lavaretus*[43] and Rainbow trout[44].

After six generations of selection, our population continues to display large genetic variation for body weight in saline and freshwater environments with the estimate of heritability from 19 to 62%. The heritability obtained for body weight across the testing environments indicated that this population will continue responding to future selection. Heritability for body weight achieved in our study was higher than that reported previously in Nile tilapia[3,4,32,33,45,46,47]. In addition, estimates of heritability for survival (0.15 to 0.32) and body condition (0.06–0.25) in our study was in good agreement with those reported for survival[5,48,49] and body shape of Nile tilapia[4]. It is important to note that, the above studies were conducted under freshwater condition with an exception in moderate saline water system by Luan, *et al.*[5].

The third objective of our study was to test if the covariance between traits varies with culture environment. The difference in size and magnitude of genetic correlations are not significant between saline- and fresh-water environments. Furthermore, in tilapia, the genetic relationships among body traits, FCR, sexual maturity and deformity were not available. The negative genetic correlation between body weight and FCR indicates that fast growing animals selected for increased harvest body weight also converted feed more efficiently into growth. The close to one genetic correlation between weight and FCR is consistent with our assumption that animals in the same pond had equal access to feed and had a fixed amount of feed intake. Under a liberal feeding regime, body weight was moderately correlated genetically with FCR for Japanese flounder[50], Rainbow trout[51] and Coho salmon[52].

Another important observation from our study is a high and positive genetic correlation (r_g=0.84±0.08) between body weight and sexual maturity. This suggests that selection for increased growth rate may lead to early maturation in Nile tilapia. Our results are consistent with those reported for other species, including Rainbow trout[23] or Atlantic cod[53]. However, selection experiments for high production performance in aquatic animals generally showed both negative[54,55] and positive[23,53,56] effects on maturation. In a long term, a multi-trait selection approach using a desired gain (or restricted) index, as demonstrated in kingfish[41], can improve traits showing antagonistic genetic correlations.

This is the first time, that genetic and phenotypic correlations between survival and deformity are reported in Nile tilapia. The high and negative genetic and phenotypic correlation (r_g=−0.82±0.06) between survival and deformity is expected because deformed fish usually have lower survival abilities than healthy animal. Consistent with other studies[32], we found that the body traits are controlled by the same set of genes as indicated by the close to one positive genetic correlations among weight, length, width and depth (results not shown). Interestingly, it is predicted that in the present population, selection for increased body weight can improve body condition (r_g between body weight and K factor=0.19±0.04). This is in contrast with the non-significant estimates between the two traits in other studies[4].

The fish cultured in saline water showed better growth performance and efficiency of feed utilisation than those grown in freshwater (P<0.001). The FCR estimate in our research (1.08–1.89) was within the range from 1.5 to 2.5 reported under practical commercial grow-out conditions in freshwater environments when using pelleted Nile tilapia feeds[57]. It is however not rigorous enough to compare the published information with our study because the experimental fish used in the literature are freshwater (e.g. GIFT strain) or high euryhaline capacity lines (e.g. Florida red tilapia), whereas our fish have undergone six generations of selection in a moderate saline water environment (15–20 ppt). The FCR is different among strains, life stages of growth development (larvae, fingerling and adult) and grow-out period/condition. Furthermore, the differences in type of feed and feeding approach could also lead to variation in FCR. In addition to the higher growth efficiency (high growth and low FCR) in the saline- than in fresh-water, the selected line had a delayed maturity in the former (saline water) relative to the latter (freshwater) environments that is favourable to commercial production. Early maturation has negative effects on growth and thus economic returns for fish producers, owing to early spawning and over density control in culture ponds. On the other hand, survival of the selected strain in this study was lower in moderate saline-than fresh-water environments.

In summary, we report for the first time the additive genetic variation and non-significant genotype by environment interaction effect for novel traits in Nile tilapia (sexual maturity, deformity, gill condition and feed conversion ratio) tested in both saline-and fresh-water systems. The moderate to high heritability for traits studied, especially body weight, indicates that our improved strain will continue responding to future selection. The close to one genetic correlations between trait expressions in the two testing environments indicate that our selected line under moderate salinity water can perform well in both saline-and fresh-water production environments, and that conducting two separate breeding programs for saline- and fresh- water is not justified for this population of Nile tilapia. Objective measurements of gonadal development of tilapia females and gill condition of live animals also merit further studies.

ACKNOWLEDGMENTS

The experimental field work of this study was funded by Research Institute for Aquaculture No. 1 (RIA1), Vietnam as part of the first author's full-time PhD program at the University of the Sunshine Coast (USC). Ngo Phu Thoa was awarded the VIED_USC scholarship. We would like to thank Mr. Tran The Muu, Ms. Ngo Thi Thu Thuy, Mr. Hoang Nhat Son, Mr. Tran Anh Tuan, and Mr. Nguyen Cong Duong; Mr. Cao Van Hanh and Mr. Bui Van Dien; Mr. Le Anh Thuy, Mr. Ngo Ngoc Ninh, Mr. Dao Xuan Hung, Mr. Nguyen Van Dai, Ms. Pham Hong Nhat, Ms. Vu Thi Trang and Ms. Luu Thi Ha Giang at RIA1 for their assistance during the experimental period. We extend our sincere thanks to Prof. Richard Burns at University of the Sunshine Coast for his constructive comments that helped us to improve the manuscript.

REFERENCES

1. Falconer D. Selection in different environments: effects on environmental sensitivity (reaction norm) and on mean performance. Genetical Research56, 57–70 (1990).
2. Ninh N. H., Thoa N. P., Knibb W. & Nguyen N. H. Selection for enhanced growth performance of Nile tilapia (*Oreochromis niloticus*) in brackish water (15–20 ppt) in Vietnam. Aquaculture 428, 1–6 (2014).
3. Bentsen H. B. et al. Genetic improvement of farmed tilapias: Genetic parameters for body weight at harvest in Nile tilapia (*Oreochromis niloticus*) during five generations of testing in multiple environments. Aquaculture 338–341, 56–65 (2012).
4. Trọng T. Q., Mulder H. A., van Arendonk J. A. & Komen H. Heritability and genotype by environment interaction estimates for harvest weight, growth rate and shape of Nile tilapia (*Oreochromis niloticus*) grown in river cage and VAC in Vietnam. Aquaculture 384, 119–127 (2013).
5. Luan T. D., Olesen I., Ødegård J., Kolstad K. & Dan N. C. Genotype by environment interaction for harvest body weight and survival of nile tilapia (*Oreochromis niloticus*) in brackish and Fresh water ponds In the proceedings of the 8th International Symposium on Tilapia in Aquaculture 231–240 (Egypt, 2008).
6. Santos A. I. et al. . Growth and survival rate of three genetic groups fed 28% and 34% protein diets. Aquaculture Research 45, 353–361 (2014).
7. Le Boucher R. et al. . Genotype by diet interactions in European sea bass (*Dicentrarchus labrax* L.): Nutritional challenge with totally plant-based diets. Journal of animal science 91, 44–56 (2013).
8. Quinton C., Kause A., Ruohonen K. & Koskela J. Genetic relationships of body composition and feed utilization traits in European whitefish (*Coregonus lavaretus* L.) and implications for selective breeding in fishmeal- and soybean meal-based diet environments. Journal of animal science85, 3198–3208 (2007).
9. Ibarra, A. M. & Famula R. Genotype by environment interaction for adult body weights of shrimp *Penaeus vannamei* when grown at low and high densitie. Genet. Sel. Evol. 40, 541–551 (2008).
10. Fishback A. G., Danzmann R. G., Ferguson M. M. & Gibson J. P. Estimates of genetic parameters and genotype by environment interactions for growth traits of rainbow trout (*Oncorhynchus mykiss*) as inferred using molecular pedigrees. Aquaculture 206, 137–150 (2002).
11. Nguyen N. H. Genetic improvement for important farmed aquaculture species with a reference to carp, tilapia and prawns in Asia: achievements, lessons and challenges. Fish and Fisheries. (2015).10.1111/faf.12122
12. Kankainen M. et al. . How to measure the economic impacts of changes in growth, feed efficiency and survival in aquaculture. Aquaculture Economics & Management 16, 341–364 (2012).
13. Verhaegen Y., Adriaens D., Wolf T. D., Dhert P. & Sorgeloos P. Deformities in larval gilthead sea bream (*Sparus aurata*): A qualitative and quantitative analysis using geometric morphometrics. Aquaculture 268, 156–168 (2007).
14. Castro J. et al. . Heritability of skeleton abnormalities (lordosis, lack of operculum) in gilthead seabream (*Sparus aurata*) supported by microsatellite family data. Aquaculture 279, 18–22 (2008).
15. Tave D. Genetics for Fish Hatchery Managers . 2nd edn (Van Nostrand Reinhold, 1993).
16. Pasnik D., Evans J. & Klesius P. Development of skeletal deformities in a Streptococcus agalactiae-challenged male Nile tilapia (*Oreochromis niloticus*) broodfish and in its offspring. Bulletin-European Association of Fish Pathologists 27, 169 (2007).
17. Wargelius A., Fjelldal P. G. & Hansen T. Heat shock during early somatogenesis induces caudal vertebral column defects in Atlantic salmon (*Salmo salar*). *Development Genes* Evolution 215, 350–357 (2005).
18. Sun P. E., Hawkins W. E., Overstreet R. M. & Brown-Peterson N. J. Morphological deformities as biomarkers in fish from contaminated rivers in Taiwan. International journal of environmental research and public health 6, 2307–2331 (2009).
19. Leary R. F., Allendorf F. W. & Knudsen K. L. Effects of rearing density on meristics and developmental stability of rainbow trout. Copeia 1991, 44–49 (1991).
20. Krise W. F. & Smith R. A. Eye abnormalities of lake trout exposed to gas supersaturation. Prog Fish-Cult 55, 177–179 (1993).
21. Sawada Y. et al. . Hypoxic conditions induce centrum defects in red sea bream Pagrus major (Temminck and Schlegel). Aquaculture Research 37, 805–812 (2006).
22. Wild V., Simianer H., Gjøen H.-M. & Gjerde B. Genetic parameters and genotype × environment interaction for early sexual maturity in Atlantic salmon (*Salmo salar*). Aquaculture 128, 51–65 (1994).
23. Kause A., Ritola O., Paananen T., Eskelinen U. & Mäntysaari E. Big and beautifu? Quantitative genetic parameters for appearance of large rainbow trout. Journal of Fish Biology 62, 610–622 (2003).
24. Meuwissen T. H. E. & Woolliams J. A. Effective sizes of livestock populations to prevent a decline in fitness. Theoret. Appl. Genetics 89, 1019–1026 (1994).
25. Thoa N. P. et al. . Genetic variation in survival of tilapia (*Oreochromis niloticus*, Linnaeus, 1758) fry during the early phase of rearing in brackish water environment (5–10 ppt). Aquaculture 442, 112–118 (2015).
26. Tave D. & Kim D. S. Review Article: Gross Abnormalities in Tilapia. Fisheries and Aquatic Sciences 14, 148–160 (2011).

27. Casselman J. M. Determination of age and growth In The biology of fish growth (eds Weatherley Alan H. & Gill H. S.) Ch. 7, 209–242 (Academic Press, 1987).

28. SAS Institute. *SAS/STAT® 9.3. User's Guide Release 9.3* . (SAS Institute Inc., NC, USA, 2011).

29. Gilmour A. R., Gogel B., Cullis B., Thompson R. & Butler D. ASReml user guide release 3.0. VSN International Ltd, Hemel Hempstead, UK(2009).

30. Kennedy B. W. Use of mixed model methodology in analysis of designed experiments in *Advances in Statistical Methods for Genetic Improvement of Livestock* (Springer-Verlag, Berlin, 77–97, 1990).

31. Robertson A. & Lerner I. M. The heritability of all-or-none traits; viability of poultry. Genetics 34, 395–411 (1949).

32. Nguyen N. H., Khaw H. L., Ponzoni R. W., Hamzah A. & Kamaruzzaman N. Can sexual dimorphism and body shape be altered in Nile tilapia (*Oreochromis niloticus*) by genetic means? Aquaculture 272, S38–S46 (2007).

33. Eknath A. E. et al. . Genetic improvement of farmed tilapias: Composition and genetic parameters of a synthetic base population of *Oreochromis niloticus* for selective breeding. Aquaculture 273, 1–14 (2007).

34. Thodesen J. et al. . Genetic improvement of tilapias in China: Genetic parameters and selection responses in growth of Nile tilapia (*Oreochromis niloticus*) after six generations of multi-trait selection for growth and fillet yield. Aquaculture 322–323, 51–64 (2011).

35. Khaw H. L., Ponzoni R. W., Hamzah A., Abu-Bakar K. R. & Bijma P. Genotype by production environment interaction in the GIFT strain of Nile tilapia (*Oreochromis niloticus*). Aquaculture 326–329, 53–60 (2012).

36. Chiasson M. A., Quinton C. D., Danzmann R. G. & Ferguson M. M. Comparative analysis of genetic parameters and quantitative trait loci for growth traits in Fraser strain Arctic charr (*Salvelinus alpinus*) reared in freshwater and brackish water environments. Journal of Animal Science 91, 2047–2056 (2013).

37. Mas-Muñoz J., Blonk R., Schrama J. W., van Arendonk J. & Komen H. Genotype by environment interaction for growth of sole (*Solea solea*) reared in an intensive aquaculture system and in a semi-natural environment. Aquaculture 410–411, 230–235 (2013).

38. Sylvén S., Rye M. & Simianer H. Interaction of genotype with production system for slaughter weight in rainbow trout (*Oncorhynchus mykiss*).Livestock Production Science 28, 253–263 (1991).

39. Mulder H., Veerkamp R., Ducro B., Van Arendonk J. & Bijma P. Optimization of dairy cattle breeding programs for different environments with genotype by environment interaction. Journal of dairy science 89, 1740–1752 (2006).

40. Dupont-Nivet M. et al. . Heritabilities and GxE interactions for growth in the European sea bass (*Dicentrarchus labrax L.*) using a marker-based pedigree. Aquaculture 275, 81–87 (2008).

41. Nguyen N., Whatmore P., Miller A. & Knibb W. Quantitative genetic properties of four measures of deformity in yellowtail kingfish *Seriola lalandi*Valenciennes, 1833. Journal of fish diseases. (2015).10.1111/jfd.12348

42. Henryon M. et al. . Genetic variation for growth rate, feed conversion efficiency and disease resistance exists within a farmed population of rainbow trout. Aquaculture 209, 59–76 (2002).

43. Cheryl D., Antti K. & Juha K. Breeding salmonids for feed efficiency in current fishmeal and future plant-based diet environments. Genet. Sel. Evol39, 431–446 (2007).

44. Kinghorn B. Genetic variation in food conversion efficiency and growth in rainbow trout. Aquaculture 32, 141–155 (1983).

45. Nguyen N. H. et al. . Correlated response in fillet weight and yield to selection for increased harvest weight in genetically improved farmed tilapia (GIFT strain), *Oreochromis niloticus*. *Aquaculture* 305, 1–5 (2010).

46. Ponzoni R. W. et al. . Genetic improvement of Nile tilapia (*Oreochromis niloticus*) with special reference to the work conducted by the WorldFish Center with the GIFT strain. *Reviews in* Aquaculture 3, 27–41 (2011).

47. Hamzah A. et al. . Genetic evaluation of the Genetically Improved Farmed Tilapia (GIFT) strain over ten generations of selection in Malaysia.Journal of Tropical Agricultural Science 37, 411–429 (2014).

48. Charo-Karisa H. et al. . Heritability estimates and response to selection for growth of Nile tilapia (*Oreochromis niloticus*) in low-input earthen ponds. Aquaculture 261, 479–486 (2006).

49. Rezk M. A. et al. . Selective breeding for increased body weight in a synthetic breed of Egyptian Nile tilapia, *Oreochromis niloticus*: Response to selection and genetic parameters. Aquaculture 293, 187–194 (2009).

50. Ogata H. Y., Oku H. & Murai T. Growth, feed efficiency and feed intake of offspring from selected and wild Japanese flounder (*Paralichthys olivaceus*). Aquaculture 211, 183–193 (2002).

51. Silverstein J. T., Hostuttler M. & Blemings K. P. Strain differences in feed efficiency measured as residual feed intake in individually reared rainbow trout, *Oncorhynchus mykiss* (Walbaum). Aquaculture research 36, 704–711 (2005).

52. Neely K. G., Myers J. M., Hard J. J. & Shearer K. D. Comparison of growth, feed intake and nutrient efficiency in a selected strain of coho salmon (*Oncorhynchus kisutch*) and its source stock. Aquaculture 283, 134–140 (2008).

53. Kristjánsson T. & Arnason T. Strong phenotypic and genetic correlation between size and first maturity in Atlantic cod *Gadus morhua* L. reared in commercial conditions. Aquaculture Research 46, 2185–2193 (2014).

54. Su G.-S., Liljedahl L.-E. & Gall G. A. E. Estimates of phenotypic and genetic parameters for within-season date and age at spawning of female rainbow trout. Aquaculture 171, 209–220 (1999).

55. Quinton C., Moghadasi S., McKay L. & McMillan I. Genetic parameters of body weight, female spawning

date and age at sexual maturation in rainbow trout. Paper presented at *Seventh World Congress on Genetics Applied to Livestock Production: Session 06. Fish and shellfish breeding*, Montpellier, France: INRA. (2002, August 19-23).

56. Kolstad K., Thorland I., Refstie T. & Gjerde B. Genetic variation and genotype by location interaction in body weight, spinal deformity and sexual maturity in Atlantic cod (*Gadus morhua*) reared at different locations off Norway. Aquaculture 259, 66–73 (2006).

57. El-Sayed A.-F.M. On-farm feed management practices for Nile tilapia (*Oreochromis niloticus*) in Egypt. On-farm feeding and feed management in aquaculture (eds Hasan M.R. & New M.B.). FAO Fisheries and Aquaculture Technical Paper No . 583, Rome, FAO 101–129 (2013).

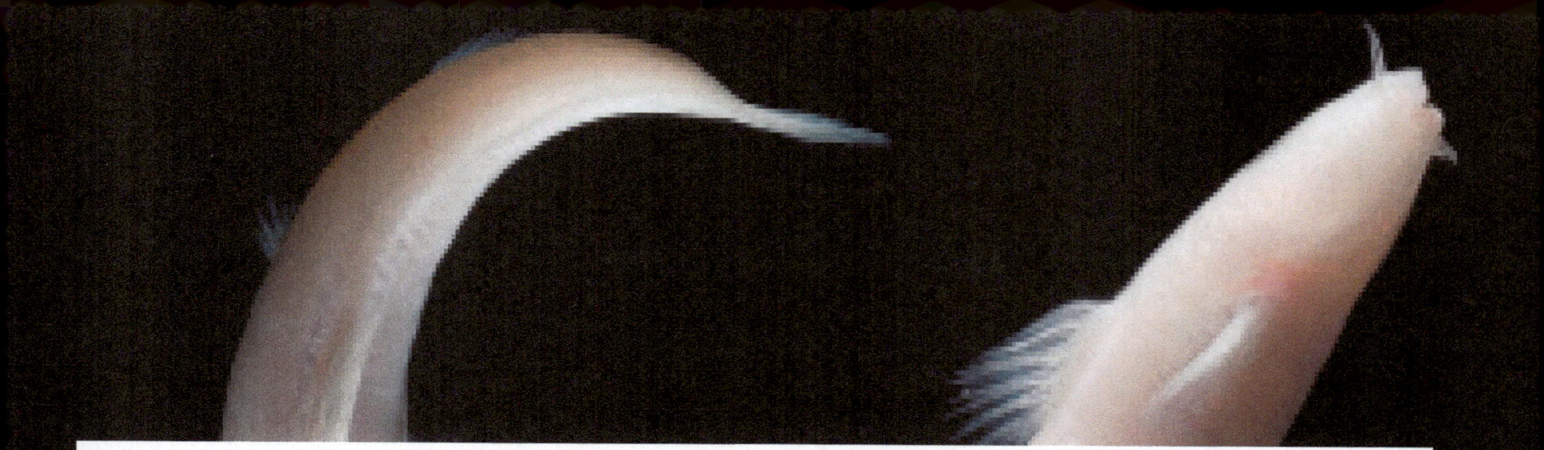

CHAPTER 8

ENERGY COSTS OF CATFISH SPACE USE AS DETERMINED BY BIOTELEMETRY

Ondřej Slavík, Pavel Horký, Libor Závorka

INTRODUCTION

The size of the home range (HR) used by animals during regular day-to-day activities [1] to satisfy their energy needs depends on the quality and availability of resources. Animals search for dispersed resources throughout the HR [2], the size of which can be larger than predicted by energy needs [3], [4] reflecting, for example, low habitat productivity, in which the energy needs of the resident animals are poorly satisfied [5], [6], and/or competition with neighbors for resources within the HR [7]–[9]. An unbalanced distribution of resources within the HR results in a disproportional use of the HR [10]. The preferred area of intensive use [11] or the most concentrated [12], [13] within the HR is referred to as the "core area (CA)" [14] and is where individuals spend most of their time [15]. The CA contains critical resources for species [16], representing the highest quality environment [17] and ensuring the best fitness for residents [18]. The size of the CA and HR are mutually correlated, and both areas display an inverse relationship with available food resources [19]. The CA is frequently reported as the most important part of the HR [17], [15]; thus, we assumed that an important amount of energy spent by an individual would occur within the CA and that the changes in energetics would occur according to the size of the used areas. To test our assumptions, we used the radio biotelemetry method. Specimens of the European catfish *Silurus glanis* (L. 1758), a large freshwater predator, were equipped with physiological sensors measuring energetics of an individual [20]. The correlation between electromyogram (hereafter EMG) records and animal behavior including fish is well documented [20]. EMG records were used to describe the movement activity of fish [21], reproductive behavior [22], a relationship between density and energetic expenditures of individuals in aquaculture [23], parental care [24], stress during transport [25], territorial defense [26] or contact between familiar and unfamiliar individuals [27]. The catfish can achieve a body weight of more than 100 kg [28], hunt aquatic and terrestrial vertebrates [29] and display individually characteristic energy costs of movement activity [30]. The catfish was chosen for the study because they show high site fidelity [31] enabling to determine the HR and CA of specimens [26].

Citation: Slavík O, Horký P, Závorka L (2014) Energy Costs of Catfish Space Use as Determined by Biotelemetry. PLoS ONE 9(6): e98997. doi:10.1371/journal.pone.0098997

MATERIALS AND METHODS

Study Area

The study was conducted on the Berounka and Elbe Rivers in the Czech Republic. The study sites were located in stretches of river characterized as lowland rivers with a gradual slope. These particular sites were chosen because of native catfish occurrence. A detailed description of the study sites can be found in [32] and [26] for the Berounka River (the river stretch studied was 3.5 km long; 49°55′N; 14°14′E; 49°56′N; 14°17′E) and the Elbe River (the river stretch studied was 20 km long; 50°09′N; 15°48′E; 50°02′N; 15°46′E), respectively. No special permit is required for the field work; the species is not protected by legislation in force. The observation was announced to competent authorities that were: Elbe River Authority and Vltava River Authority, River Management Authorities, state enterprises, and to Czech Anglers Club, civic association, Fish Management Authority.

Fish Capture and Tagging

A total of 20 individuals (16 from the Berounka River and four from the Elbe River) were captured through electrofishing (650 V, 4 A, pulsed DC). The surgery was performed under 2-phenoxy-ethanol (0.2 ml l^{-1}; Merck KGaA; www.merck.com) and all efforts were made to minimize suffering. The fish were weighed and measured (Table 1). The mean weight and length of the captured fish were 20.5 kg (range: 5.5–64.5 kg) and 1.34 m (range: 0.83–2.16 m) for the Berounka River and 17.5 kg (range: 4.4–28.5 kg) and 1.29 m (range: 0.85–1.62 m) for the Elbe River Electromyogram (EMG) transmitters (CEMG2–R11–25, mass 12 g in air, 11.49 mm, with an operational life of c. 71 days; Lotek; www.lotek.com) were implanted in the body cavity of each fish through a mid-ventral incision that was closed by three separate stitches using a sterile braided absorbable suture (Ethicon coated VICRYL). The two electrodes of the surgically implanted EMG transmitters were positioned in the red aerobic musculature below the lateral line on the left side of the fish. The distance between the electrodes ranged from 10 to 12 mm [33]. The individuals equipped with the EMG transmitters were kept in cages immersed in the river stretch for c. 60 min after implantation of the transmitters. They were then released at or near the point of capture.

Table 1. The average values of the core area (CA), EMG within CA (EMG_{CA}), Home Range periphery area (HRp) and EMG within HRp (EMG_{HRp}) for particular observed catfish individuals

Fish ID	Weight (kg)	Length (m)	River	CA (m^2)	EMG_{CA}	HRp (m^2)	EMG_{HRp}
25	15.5	1.32	Berounka	755	8.29	2 511	8.19
27	8.5	0.98	Berounka	906	15.99	3 009	15.84
28	9.8	1.05	Berounka	2 433	6.03	8 082	6.24
29	14.6	1.31	Berounka	415	11.98	1 377	11.85
30	7.5	0.87	Berounka	273	7.88	907	7.68
31	9.5	1.07	Berounka	2 054	13.57	6 823	13.17
32	16.5	1.36	Berounka	2 423	12.14	8 048	12.17
33	26.7	1.64	Berounka	302	4.59	1 005	4.92
34	18.5	1.42	Berounka	102	15.68	340	15.63
35	56.5	1.98	Berounka	6 733	11.43	22 365	12.21
36	64.5	2.16	Berounka	4 790	20.95	15 914	20.63
37	11.2	1.12	Berounka	759	7.13	2 523	7.07
38	13.9	1.25	Berounka	287	16.22	954	15.52
39	19.5	1.44	Berounka	6 210	4.55	20 631	4.56
40	29.8	1.71	Berounka	1 409	26.91	4 681	26.41
41	5.5	0.83	Berounka	932	12.21	3 097	12.28
100	4.4	0.85	Elbe	4 014	6.17	13 337	5.27
101	14.8	1.28	Elbe	2 402	6.11	7 982	5.94
102	28.5	1.62	Elbe	46 511	7.27	154 510	6.44
103	22.3	1.39	Elbe	1 487	14.67	4 941	14.93

Sampling Procedures

The fish were monitored from 1 March to 31 May 2006 (Elbe River) and from 3 August to 13 October 2009 (Berounka River). All individuals were tracked weekly to check their positions in the river. In addition, a group of randomly selected individuals (four to eight depending on the tracking conditions) was subsequently tracked during a 48 h cycle with two radio receivers (Lotek SRX_600 W31) and a three-element Yagi antenna equipped with a compass. The positions of the fish were determined during 16 subsequent 3 h intervals using a GPS. The data on individual fish movements were transferred from the GPS to a PC and were then analyzed with the aid of Map Source v5.3 (Garmin Ltd; www.garmin.com). A computer program was developed to obtain the position coordinates of the fish and to subsequently plot the coordinates on a map with the biangulation method proposed by [34]. The EMG-coded transmitters allowed multiple transmitters to be monitored simultaneously at the same frequency. Each transmitter had a unique code to identify each fish. The voltage corresponding to muscle activity was rectified and sampled during a 5 s period. The average value over this period was then determined, and an activity level was assigned that ranged from 0 to 50 [33]; these values are henceforth termed EMG values. A set of ten EMG values (5 s burst rate) was obtained three times for all tagged individuals during each 3 h interval. To obtain information on the resting activity levels for standardizing data, the EMG transmitters were individually calibrated for each fish. The resting values were measured twice: first from the anaesthetized fish and prior to release and second from when the fish were held in cages. The average of all recorded EMG resting signals was subsequently considered the baseline value for the particular fish (the 'baseline EMG value'; [33]). On both rivers, the accuracy of the fish position determination was estimated to be ±3 m according to a calibration procedure performed with a tag located on the riverbed. The calibration was repeated ten times, and the observer did not know the position of the tag.

Data Analysis

Schoener index [35] and Swihart & Slade index [36] were computed to test for autocorrelation of successive catfish locations prior to the HR calculations. Locations were suggested to be autocorrelated in several cases. Such data were treated according to Swihart and Slade [37] and the inappropriate locations were removed. The size of the catfish HR was determined using the fixed kernels with least squares cross validation to estimate the smoothing bandwidth [38]. The 50% contour was used to delineate the CA area [12] and the 95% contour to estimate the entire HR area. The difference between the entire HR and CA areas was termed 'HR periphery'. Bivariate variable describing whether fish was present in the CA or HR periphery was termed 'fish position'. Each final EMG value included in the subsequent analysis was equal to the recorded EMG value minus the baseline EMG value. The final EMG values were used as general descriptors of energy costs of space use. To ensure that the individual EMG analyses were independent of weight, relative EMG/weight ratios were calculated by dividing each final EMG value by fish weight [30], [39]. 'Energy costs of space use' was separately determined in the CA and in the HR periphery as the mean of all recorded EMG/weight ratio values within the catfish 'used area'.

Statistical Analyses

The statistical analyses were performed using the SAS software package (SAS Institute Inc., version 9.2, www.sas.com). The energetic data were analyzed using a linear mixed model (LMM) with random factors (PROC MIXED). The data were transformed for normality prior to the LMM analyses if necessary. The random factors were used to account for the repeated measures collected for the same experimental units (individual fish) across the duration of the experiment in particular rivers. The significance of each exploratory variable was assessed using an F-test. The differences between the classes were tested with a t-test, and a Tukey–Kramer adjustment was used for multiple comparisons. The degrees of freedom were calculated using the Kenward–Roger method [40].

Ethics Statement

The work was prepared and conducted according to valid legislative regulations (Law no. 246/1992, § 19, art. 1, letter c); the permit was subjected to O. Slavík, qualified according to Law no. 246/1992, § 17, art. 1; permit no. CZ00167. All field sampling including EMG transmitter implantation was carried out with the relevant permissions from the Departmental Expert Committee for authorization experimental project of the Ministry of Environment of the Czech Republic (permit no. 26758/ENV/10-1092/620/10-PP6, registered by the Ministry of Environment of the Czech Republic). The study did not involve endangered or protected species.

RESULTS

For the analysis in this study, were used 2 284 records of fish positions and 22 840 EMG values. The average size of the CA was 6784 m^2 (range 25–392 622 m^2) and the average size of the HR periphery was 22 538 m^2 (range 84–1 304 269 m^2). More details according to the individual HR size variation can be found in Table 1.

Final LMM model contained the fixed factors used area and fish position for energy costs of space use analyses. Details of the model are shown in Tables 2 and 3. Differences in energy costs of space use with regard to the size of the used area were observed. The energy spent by catfish decreased with increasing size of the used area (Fig. 1). In other words, energy costs of space use within a small area were higher. As further documented, catfish spent significantly more energy within the CA compared to the HR periphery (Fig. 2) and were present in the CA, delineated by 50% contour of the HR, for 77% of their locations. Thus, the CA was suggested as an important area where catfish spent most of their temporal and energy budget.

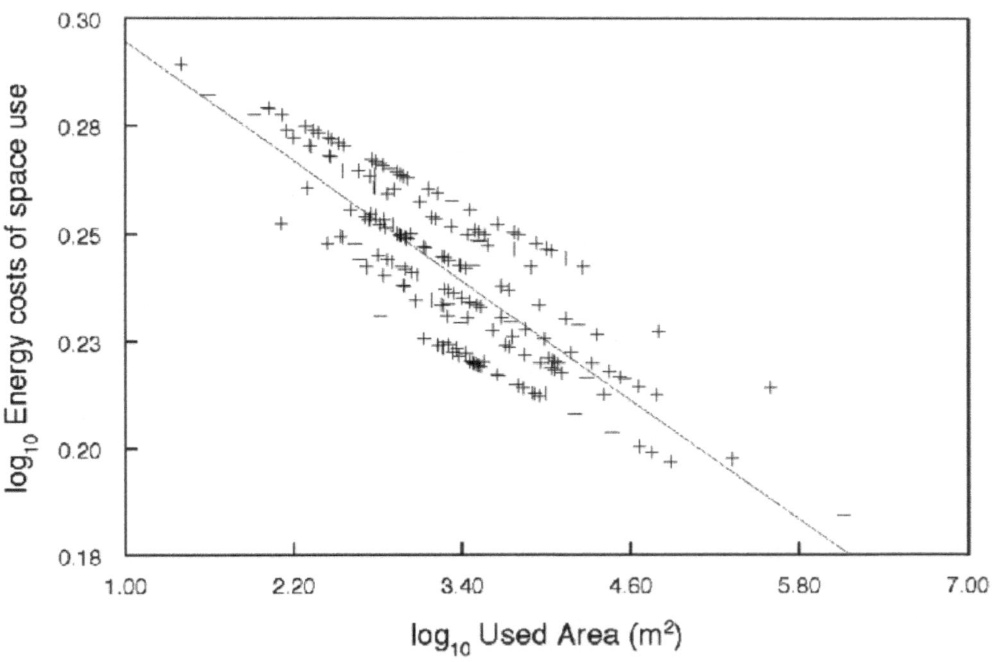

Figure 1. The relationship between energy costs of space use and the size of used area.

Predicted values are from $\log_{10}$ transformed data.

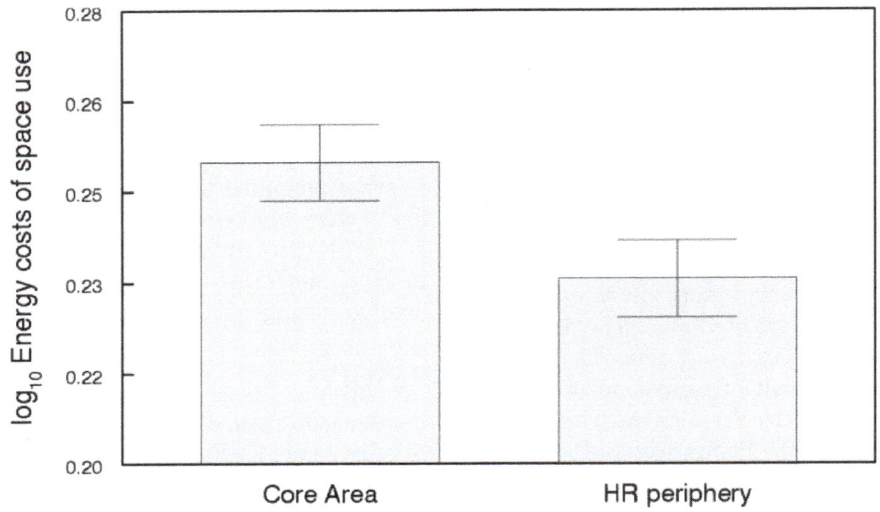

Figure 2. Energy costs of space use within the CA and the HR periphery.

Adjusted means are from $\log_{10}$ transformed data.

Table 2. Solution for fixed effects (parameter estimates with corresponding standard errors and p-values for the final LMM model)

Parameter	Factor level	Estimate	Sd. Error	P<
Intercept	-	0.2853	0.02176	0.0001
Used area	-	−0.01582	0.00579	0.0069
Fish position	CA	0.02038	0.00805	0.0121
	HR periphery	0	0	-

Table 3. Type 3 tests of fixed effects for final LMM model

Effect	Num DF	Den DF	F	P<
Used area	1	201	7.45	0.0069
Fish position	1	214	6.41	0.0121

DISCUSSION

The results of our study revealed that the catfish that occurred within a large area consumed less energy than those that used a small area. It is generally known that animals can satisfy their energy needs within small areas, provided that sufficient resources are available there [41]. For example, a predator occupying a large HR, such as the lynx *Lynx rufus*, reduces size of the area used as food density increases [19]. Similarly, a CA containing a high density and diversity of feeding trees is preferred over non-core areas by the highly mobile spider monkey*Ateles geoffroyi* [17]. Therefore, we can assume that high degree of catfish energy costs in small areas indicates the use of limited resources allocated within that small region. Accordingly, freshwater herbivorous grass carp, *Ctenopharyngodon idella*, were associated with submerged vegetation 50% of time in 16% of the HR area [42]. In addition, collared lizards, *Crotaphytus collaris*, were present in approximately 25% of the HR 66% of the time they were observed [43].

The use of an area and energy costs of an individual are also considered in relation to their defense of a preferred area when the residents exclusively occupy the CA [11]. It is generally known that defended areas where residents pay higher energy costs to maintain exclusivity appeared to be smaller than non-defended areas [44]. The results of our study generally indicate that catfish defended energetically optimal areas rather than a large HR area, as indicated by the inverse relationship between the size of the used area and the associated energy costs. Our findings are in agreement with the previous studies [26], where the catfish energy costs increased when the CAs of conspecifics overlapped and both of them were inside CAs simultaneously. Large catfish aggregations were visually observed in the wild within small areas [45]. Aggregating can provide various benefits as well as costs resulting from sharing resources [46]. One of the costs is interference [47], which does not necessarily mean an overt aggression. In aggregations of another large fish predator, pike *Esox lucius* (L.) was suggested interference operated through intimidation [48]. Stress related to intimidations could also be the reason of increased catfish energy costs within the CA. In addition to feeding and defense, other activities associated with the movement of species within the HR can be considered. For example, low energy costs of movement were reported for fish from the Labridae family moving within a large area, and the activity was expected to represent the testing of a social situation as opposed to active food intake [49]. The movement of animals within the HR appeared to be related to low-energy activities, such as patrolling and searching for new resources and/or partners, whereas energetically demanding movement within small areas appeared to be related to the use of limited resources.

Low amounts and wide dispersal of available resources induce an increase in the HR [2], [19]. Apparently, an increase in movement activity can be expected in the search for these resources [41]; however, these HR relocations can be energetically less demanding than the resources use itself [49]. Similarly, intensive movement does not always imply a large HR for an individual [50]. The aquatic environment allows fish to move with minimal energy expenditures; however, energy expenditures rapidly increases during active movement [33]. Furthermore, our results revealed differences in degree of energy costs with regard to the size of the area used. Hence, we can infer that CA is of the special importance as a source of critical resources. We can conclude that energy costs and the size of the area used are not necessarily positively correlated; sufficient resource availability allows animals to use small areas where the majority of energy is located and consumed.

ACKNOWLEDGMENTS

We thank A. Slavikova for the comments on the earlier versions of the manuscript.

AUTHOR CONTRIBUTIONS

Conceived and designed the experiments: OS PH. Performed the experiments: OS PH LZ. Analyzed the data: OS PH. Contributed reagents/materials/analysis tools: PH. Wrote the paper: OS.

REFERENCES

1. Mace GM, Harvey PH, Clutton-Brock TH (1983) Vertebrate home-range size and energetic requirements. In: Swingland IR, Greenwood PJ, editors. The ecology of animal movement. Oxford: Clarendon. pp 32–53.
2. Harestad AS, Bunnel FL (1979) Home range and body weight – a re-evaluation. Ecology 60: 389–402. doi: 10. 2307/1937667
3. Damuth J (1981) Home range, home range overlap, and species energy use among herbivorous mammals. Biol J Linnaean Soc 15: 185–193. doi: 10. 1111/j. 1095-8312. 1981. tb00758. x
4. Kelt DA, Van Vuren D (2001) The ecology and macroecology of mammalian home range area. Am Nat 157: 637–645. doi: 10. 1086/320621
5. Schoener TW (1968) Sizes of feeding territories among birds. Ecology 49: 123–141. doi: 10. 2307/1933567
6. Kelt DA, Van Vuren D (1999) Energetic constrains and the relationship between body size and home range area in mammals. Ecology 80: 337–340. doi: 10. 2307/177002
7. Schoener TW (1983) Simple models of optimal territory size: a reconciliation. Am. Nat 121: 608–629. doi: 10. 1086/284090
8. Jetz W, Carbone C, Fulford J, Brown JH (2004) the scaling of animal space use. Science 306: 266–268. doi: 10. 1126/science. 1102138
9. Buskirk S (2004) Keeping an eye on the neighbors. Science 306: 238–239. doi: 10. 1126/science. 1104480
10. Hayne DW (1949) Calculation of size of home range. J Mammal 30: 1–18. doi: 10. 2307/1375189
11. Samuel MD, Pierce DJ, Garton EO (1985) Identifying areas of concentrated use within the home range. J Anim Ecol 54: 711–719. doi: 10. 2307/4373
12. Clutton-Brock TH, Guinness FE, Albon SD (1982) Red Deer: Behavior and Ecology of Two Sexes. Edinburgh: Edinburgh University Press.
13. Seaman DE, Powell RA (1990) Identifying patterns and intensity of home range use. Int Conf Bear Res Mgmt 8: 243–249. doi: 10. 2307/3872925
14. Kaufman JH (1962) Ecology and social behavior of the coati, (*Nasua nirica*) on Barro Colorado Island, Panama. Univ Calif Publ Zool 60: 95–222.
15. Vander Wal E, Rodgers AR (2012) An individual-based quantitative approach for delineating core areas of animal space use. Ecol Model 224: 48–53. doi: 10. 1016/j. ecolmodel. 2011. 10. 006
16. Plowman BW, Conner LM, Chamberlain MJ, Leopold BD, Burger LW Jr (2006) Annual dynamics of bobcat (*Lynx rufus*) home range and core use areas in Mississippi. Am Mid Nat 156: 386–393. doi: 10. 1674/0003-0031(2006)156[386:adoblr]2. 0. co;2
17. Asensio N, Lusseau D, Schaffner CM, Aureli F (2012) Spider monkeys use high-quality core areas in a tropical dry forest. J Zool 287: 250–258. doi: 10. 1111/j. 1469-7998. 2012. 00911. x
18. Emery Thompson M, Kahlenberg SM, Gilby IC, Wrangham RW (2007) Core area quality is associated with variance in reproductive success among female in chimpanzees at Kibale National Park. Anim Behav 73: 501–512. doi: 10. 1016/j. anbehav. 2006. 09. 007
19. Benson JF, Chamberlain MJ, Leopold BD (2006) Regulation of space use in a solitary felid: population density or prey availability? Anim Behav 71: 685–693 (doi:10. 1016/j. anbehav. 2005. 08. 005).
20. Cooke SJ, Thorstad EB, Hinch SG (2004a) Activity and energetics of free-swimming fish: insight from electromyogram telemetry. Fish Fisheries 5: 21–52. doi: 10. 1111/j. 1467-2960. 2004. 00136. x
21. Hruska AK, Hinch GS, Healey CM, Farrel AP (2007) Electromyogram telemetry, nondestructive physiological biopsy, and genetic markers: linking recent techniques with behavioral observations for the study of reproductive success in sockeye salmon mating system. Am Fish Soc Sym 54: 17–29.
22. Kaseloo PA, Weatherley AH, Ihssen PE, Anstey DA, Gare MD (1996) Electromyograms from radiotelemetry as indicators of reproductive activity in lake trout. J Fish Biol 48: 664–674. doi: 10. 1111/j. 1095-8649. 1996. tb01462. x

23. Cooke SJ, Chandroo PK, Bedow AT, Moccia DR, McKinley SR (2000) Swimming activity and energetic expenditure of captive rainbow trout *Oncorhynchus mykiss*(Walbaum) estimated by electromyogram telemetry. Aqua Res 31: 495–505. doi: 10. 1046/j. 1365-2109. 2000. 00470. x

24. Cooke SJ, Philipp DP, Wheatherhead PJ (2002) Parental care patterns and energetics of smallmouth bass, *Micropterus dolomieu*, and largemouth bass *M. salmoides*, monitored with activity transmitters. Can J Zool 80: 756–770. doi: 10. 1139/z02-048

25. Chandroo PK, Cooke SJ, McKinley SR, Moccia DR (2005) Use of electromyogram telemetry to assess the behavioral and energetic response of rainbow trout,*Oncorhynchus mykiss* (Walbaum) to transportation stress. Aqua Res 36: 1226–1238 (doi: 10. 1111/j. 1365-2109. 2005. 01347. x).

26. Slavík O, Horký P (2009) When fish meet fish as determined by physiological sensors. Ecol Freshw Fish 18: 501–506. doi: 10. 1111/j. 1600-0633. 2009. 00370. x

27. Slavík O, Pešta M, Horký P (2011) Effect of grading on energy consumption in European catfish *Silurus glanis*. Aquaculture 313: 73–78 (doi: 10. 1016/j. aquaculture. 2011. 01. 002).

28. Copp GH, Britton JR, Cucherousset J, Garcia-Berthou E, Kirk R, et al. (2009) Voracious invader or benign feline? A review of the environmental biology of European catfish *Silurus glanis* in its native and introduced ranges. Fish Fisheries 10: 252–282 (doi: 10. 1111/j. 1467-2979. 200800321. x).

29. Cucherousset J, Boulêtreau S, Azémar F, Compin A, Guillaume M (2012) "Freshwater Killer Whales": beaching behavior of an alien fish to hunt land birds. PLoS ONE 7: e50840 (doi:10. 1371/journal. pone. 0050840).

30. Slavík O, Horký P (2012) Diel dualism in the energy consumption of the European catfish *Silurus glanis*. J Fish Biol 81: 2223–2234. doi: 10. 1111/j. 1095-8649. 2012. 03436. x

31. Carol J, Zamora L, García-Berthou E (2007) Preliminary telemetry data on the movement patterns and habitat use of European catfish (*Silurus glanis*) in a reservoir of the River Ebro, Spain. Ecol Freshw Fish 16: 450–456. doi: 10. 1111/j. 1600-0633. 2007. 00225. x

32. Slavík O, Horký P, Bartoš L, Kolářová J, Randák T (2007) Diurnal and seasonal behavior of adult and juvenile European catfish as determined by radio-telemetry in the River Berounka, Czech Republic. J Fish Biol 71: 101–114. doi: 10. 1111/j. 1095-8649. 2007. 01471. x

33. Cooke SJ, Hinch SG, Wikelski M, Andrews RD, Kuchel LJ, et al. (2004b) Biotelemetry: a mechanistic approach to ecology. Tr Ecol Evol 19: 334–343 (doi: 10. 1016/j. tree. 2004. 04. 003).

34. White GC, Garrott RA (1990) Analysis of wildlife radio-tracking data. New York, USA: Academic Press.

35. Schoener TW (1981) An empirically based estimate of home range. Theor Pop Biol 20: 281–325. doi: 10. 1016/0040-5809(81)90049-6

36. Swihart RK, Slade NA (1985a) Influence of sampling interval on estimates of home-range size. J Wildl Manage 49: 1019–1025. doi: 10. 2307/3801388

37. Swihart RK, Slade NA (1985b) Testing for independence of observations in animal movements. Ecology 66: 1176–1184. doi: 10. 2307/1939170

38. Seaman DE, Powell RA (1996) An evaluation of the accuracy of kernel density estimators for home range analysis. Ecology 77: 2075–2085. doi: 10. 2307/2265701

39. Aarestrup PK, Jepsen N, Koed A, Pedersen S (2005) Movement and mortality of stocked brown trout in a stream. J Fish Biol 66: 721–728. doi: 10. 1111/j. 0022-1112. 2005. 00634. x

40. Kenward MG, Roger JH (1997) Small sample inference for fixed effects from restricted maximum likelihood. Biometrics 53: 83–997. doi: 10. 2307/2533558

41. Marable MK, Belant JL, Godwin D, Wang G (2012) Effects of resource dispersion and site familiarity on movements of translocated wild turkeys on fragmented landscapes. Behav Process 91: 119–124. doi: 10. 1016/j. beproc. 2012. 06. 006

42. Chilton EW, Poarch SM (1997) Distribution and movement behavior of radio-tagged grass carp in two Texas reservoirs. Trans Am Fish Soc 126: 467–476. doi: 10. 1577/1548-8659(1997)126<0467:dambor>2. 3. co;2

43. Stone PA, Baird TA (2002) Estimating lizard home range: the Rose model revisited. J Herpet 36: 427–436. doi: 10. 2307/1566187

44. Grant JWA, Chapman CA, Richardson KS (1992) Defended versus undefended home range size of carnivores, ungulates and primates. Behav Ecol Sociobiol 31: 149–161. doi: 10. 1007/bf00168642

45. Boulêtreau S, Cucherousset J, Villéger S, Masson R, Santoul F (2011) Colossal Aggregations of Giant Alien Freshwater Fish as a Potential Biogeochemical Hotspot. PLoS ONE 6: e25732 (doi:10. 1371/journal. pone. 0025732).

46. Krause J, Ruxton GD (2002) Living in Groups. Oxford University Press, Oxford.

47. Holbrook SJ, Schmitt RJ (2002) Competition for shelter space causes density-dependent predation mortality in damselfishes. Ecology 83: 2855–2868 (doi: [];10. 1890/0012-9658(2002)083[2855:CFSSCD]2. 0. CO;2).

48. Hawkins LA, Armstrong JD, Magurran AE (2005) Aggregation in juvenile pike (Esox lucius): interactions between habitat and density in early winter. Func Ecol 19: 794–799 (doi: 10. 1111/j. 1365-2435. 2005. 01039. x).

49. Bradbury C, Green JM, Bruce-Lockhart M (1995) Home ranges of female cunner, (*Tautogolabrus adspersus*, *Labridae*), as determined by ultrasonic telemetry. Can J Zool 73: 1268–1279. doi: 10. 1139/z95-151

50. Carfagno GLF, Weatherhead JP (2008) Energetics and space use: intraspecific comparison of movements and home ranges of two Colubrid snakes. J Anim Ecol 77: 416–424 (doi: 10. 1111/j. 1365-2656. 2008. 01342x).

CHAPTER 9

ENCOURAGE SUSTAINABILITY BY GIVING CREDIT FOR MARINE PROTECTED AREAS IN SEAFOOD CERTIFICATION

Sarah E. Lester, Christopher Costello, Andrew Rassweiler, Steven D. Gaines, Robert Deacon

Widespread concern over global fish stocks has prompted an increase in research and initiatives aimed at rebuilding ailing fisheries and incentivizing sustainable fishing practices. This promising focus on solutions coincides with a burgeoning consumer and retailer demand for environmentally friendly products (Figure 1). Sustainability certification, labeling, and consumer guides (e.g., Marine Stewardship Council, Fair Trade, Seafood Watch, etc.) are signals that help eco-minded consumers identify products that meet their standards. Accurate signals offer an immense opportunity to incentivize sustainability, increasing demand and profits for sustainable producers. Yet, while the growing number of seafood certification programs and consumer seafood guides fuel and inform demand, the pace of change is slow. One key barrier to progress is the significant lag between the implementation of reforms and the recovery of fish stocks. Without preemptive credits within certification protocols for conservation actions that can be expected to benefit the stock over time, the incentives for reforms may be limited.

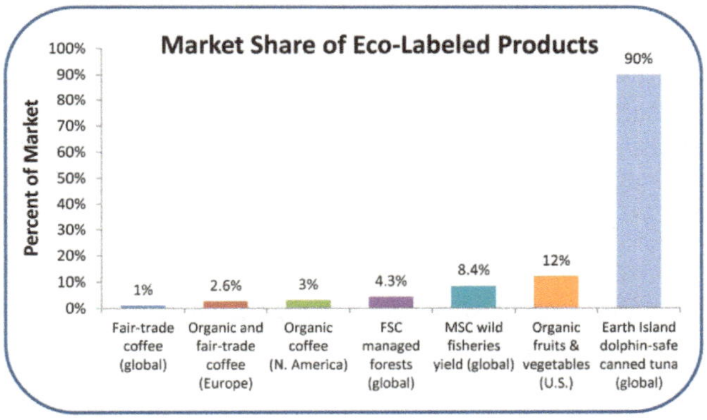

Figure 1. Market share of several certified products throughout various regions of the world.

Included are: the percentage of global wild-capture fisheries yield that is MSC certified[14],[15], the market share of fruits and vegetables in the U.S. that are certified organic[16], the market share of coffee in Europe that is organic and fair-trade certified, the market share of coffee in North America that is certified organic, the global market share of coffee that is fair trade certified [17], the global market share of canned tuna that is Earth Island dolphin-safe certified [18], and the global percentage of forests managed according to Forest Stewardship Council standards [19],[20].

Citation: Lester SE, Costello C, Rassweiler A, Gaines SD, Deacon R (2013) Encourage Sustainability by Giving Credit for Marine Protected Areas in Seafood Certification. PLoS Biol 11(12): e1001730. doi:10.1371/journal.pbio.1001730

Many potential conservation reforms could qualify for preemptive credit, but the most conspicuous absence is the lack of credit for marine protected areas (MPAs). Marine protected areas are one of the most important tools for conserving the ocean›s ecosystems. MPAs, similar to protected areas on land, are locations where regulations prohibit specific human activities. There has been considerable research about the effects of MPAs, particularly focused on a common class of MPAs in which all fishing is prohibited (no-take marine reserves). Studies from more than 120 no-take marine reserves around the world consistently reveal the conservation benefits of this type of protection, including higher total biomass, abundance, and average size of fish within reserves [1] (Figure 2). The data from these many existing MPAs, combined with improving spatial fisheries models, allow us to predict an MPA›s future conservation benefits long before they are realized. Such empirically based model forecasts could be used to assign preemptive credit for MPAs within certification frameworks when the MPA is implemented. Doing so would reward existing MPAs for their contributions to sustainable fisheries and would provide more powerful short-term incentives to create well-designed new MPAs.

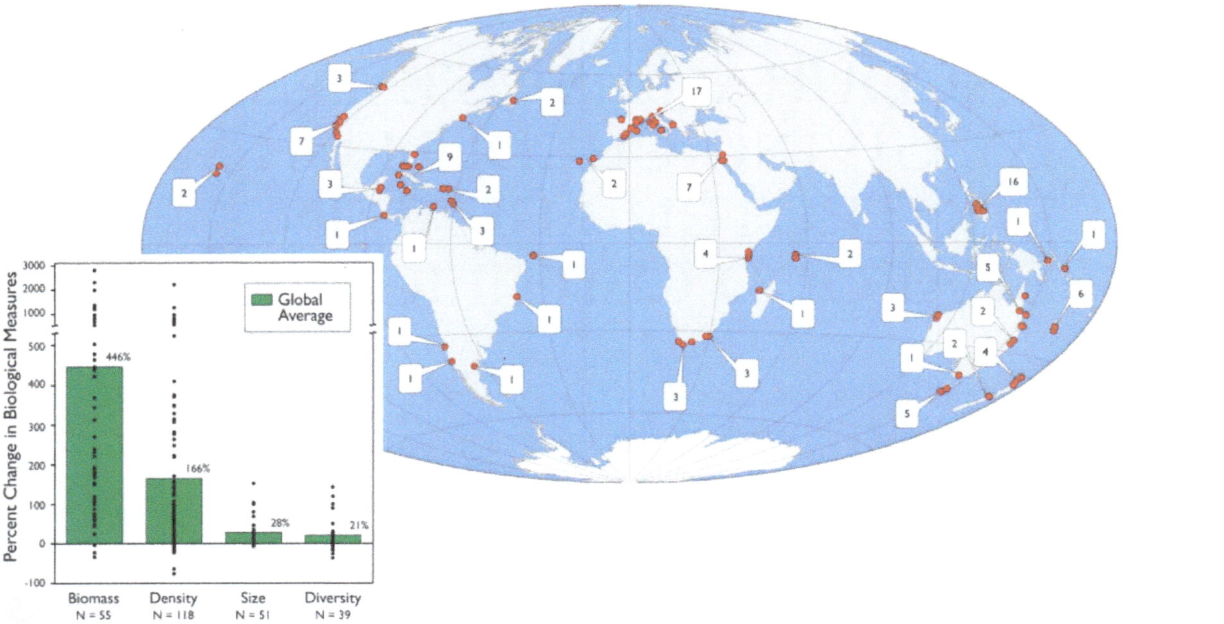

Figure 2. Map showing the locations of 124 marine reserves that have been studied by scientists with the results published in scientific journals.

The graph shows the average changes (green bars) in fishes, invertebrates, and seaweeds within these 124 marine reserves. Although changes varied among reserves (black dots), most reserves had positive changes. Data from [1]. Graphics reproduced with permission from the Partnership for Interdisciplinary Studies of Coastal Oceans[21].

Why We Need to Marry Seafood Certification and MPAs

Given growing green demand, the time is ripe for seafood certification to drive global fishery sustainability. Yet many existing programs are criticized for not being up to this challenge [2],[3]. Consider the example of the most prominent global fishery certifier, the Marine Stewardship Council (MSC). Although MSC has achieved significant market penetration with more than 200 certified fisheries (Figure 1), these tend to be large-scale fisheries in developed nations [4]. The intensive data requirements and high cost of assessment makes MSC certification inaccessible to a much broader range of sustainable fisheries [5].

There are two key consequences of these barriers to certification. First, many fisheries worldwide may already meet sustainability criteria but do not receive market recognition because they are simply too data-poor or cash-poor to seek MSC certification. Second, countless other fisheries might be induced to adopt sustainable practices if some of the practical hurdles to certification could be reduced without compromising sustainability standards. MSC has attempted to lower some of these hurdles recently. Yet neither the MSC, nor any other major seafood certifier, provides explicit credit for MPAs despite their growing global use.

There is strong scientific evidence that MPAs achieve conservation goals, particularly when they close areas to all fishing [1]. There is also evidence that, if designed well, MPAs may be able to increase the resilience of surrounding fisheries and enhance local catches [6],[7]. The magnitude and timing of fisheries benefits are less certain than the conservation benefits, because they depend on both the response of the fish inside the MPA and the responses of fishers outside the MPA to the new regulations. Nonetheless, despite the greater uncertainty, forecasts of fisheries benefits and costs from bioeconomic models have now been used successfully in MPA planning processes [8] and provide a framework for projecting benefits in certification systems. Furthermore, regardless of uncertainty regarding fishery impacts, the scientific evidence for the conservation and sustainability benefits of MPAs are sufficient to justify incorporating credit for MPAs into certification protocols. MPAs, when adequately sized relative to the adult movement potential of the target species, will reliably result in increased biomass inside of the protected area. This can help ensure safe levels of the stock and can mitigate ecosystem impacts of fishing by shielding areas from destructive fishing practices. Nearshore fisheries targeting species that have more limited movement (e.g., reef-associated species) are obvious candidates because these species respond positively to protection even by small MPAs, but all fisheries are potential candidates given a sufficiently sized and well-enforced protected area.

Sustainability standards, including those of MSC and the Monterey Bay Aquarium›s Seafood Watch program, tend to evaluate fisheries based on three core tests: (1) Are harvesting pressure and fish stocks at safe levels? (2) Are fishing practices that have a significant negative impact on the ecosystem prohibited? and (3) Is an effective management system in place [9]? We argue that if explicitly incorporated into certification, MPAs would help fisheries pass each of these tests. First, MPAs protect some of the target stock from fishing. Second, MPAs protect habitat and non-target species from fishing impacts. Third, MPAs can reduce the complexity of required management institutions. If the MPA is well enforced and protects enough of the stock, active management outside may be less critical; effective enforcement of the MPA may be sufficient to ensure stock sustainability. MPAs act as a simple and cost-effective insurance policy against intense fishing and mismanagement in adjacent waters.

If certification schemes were to provide explicit credit for MPAs, certification would become a possibility for a wider range of small-scale, data-poor, and cash-poor fisheries. A stock›s healthy status can be verified and ensured to persist without expensive stock assessments, and developing nations that have invested in fishery conservation measures can be rewarded for their efforts. MPA creation will help associated fisheries gain a competitive advantage from certification, providing further incentives for MPA adoption by regulators and fishing interests. This is critical because many fishermen, particularly in developing countries, lack incentives to support MPAs because they cannot afford to wait for the recovery of fish stocks and the accrual of the associated fisheries benefits from MPAs. Pre-emptive credit for MPAs within certification programs, on the other hand, could provide an immediate reward through higher prices and access to new markets.

How to Provide Credit

Despite the role that MPAs play in sustainability, we are unaware of any sustainability rating that formally and quantitatively assigns credit for establishing MPAs adjacent to fisheries. We highlight three areas of analytical innovation that can improve certification efforts now and allow them to incorporate the benefits of MPAs.

First, the contribution of MPAs to fishery sustainability can be evaluated even in the absence of sufficient information to conduct a stock assessment. Such stock assessments are quantitative analyses of a fished population; they form the backbone of most seafood certifications, but require considerable data and time to conduct. The vast majority of the world's fisheries are too small and/or impoverished to finance such assessments. For these stocks, certification currently cannot provide an incentive for sustainable harvest levels or conservation actions. Of course, MPAs do conserve fish populations in these data-poor situations, provided there is adequate enforcement. In fact, for all stocks, there is an MPA network large enough to guarantee sustainability even in the absence of any other management or detailed information on stock status. And, more commonly, a moderately sized MPA can compensate for deficiencies in management or data. Similar arguments can be made for quantifying and crediting MPAs for their protection of the broader ecosystem.

Second, formal stock assessment methods must be revised to account for fish biomass within MPAs. Traditional stock assessments typically ignore the population of fish inside of the MPAs, substantially undervaluing the role of MPAs in maintaining the fish population as a whole and undermining support for their use as a management tool. For example, consider a hypothetical case where a fishery seeking certification decides to protect 20% of the region in MPAs, prohibiting fishing in these areas. In an assessment that ignores the MPA, a smaller portion of the population is now assessed (encompassing the 80% of the region where fishing is still allowed); the assessment will focus on the fact that populations outside of the MPAs have not increased and will overlook the healthy populations within the MPAs. Fortunately, as the scale and extent of MPA protection increases in many regions of the world's oceans, this shortcoming of stock assessments has been noted [10]. State-of-the-art stock assessment science includes novel approaches to account for the biomass within MPAs [11]–[13]. In data-rich fisheries, where stock assessments are possible, application of these improved methods will inherently give appropriate sustainability credit for MPAs.

Third, we can apply predictive modeling to assess alternative routes to certification. Spatial models can be used to forecast the impacts of alternative pathways to certification and identify the most efficient combination. For example, recently developed spatial models of fish population dynamics, coupled with oceanographic models predicting movement of young fish and fleet models describing fishermen behavior, can forecast the long-term impacts on sustainable stock levels and fishery profit of different management actions, including protecting areas in MPAs. The impacts of MPA networks on stocks and profits have already been demonstrated for several fisheries in southern California (Figure 3; [8]). The same approach can be used to evaluate changes to management regulations (e.g., reduced fishery pressure through a size or season limit or transition to rights-based management) and/or the reduction of uncertainty regarding the status of the stock (e.g., collecting more data to conduct a more sophisticated stock assessment). Such applications can generate a range of viable options for achieving sustainability from which the most efficient action or combination of actions can be selected.

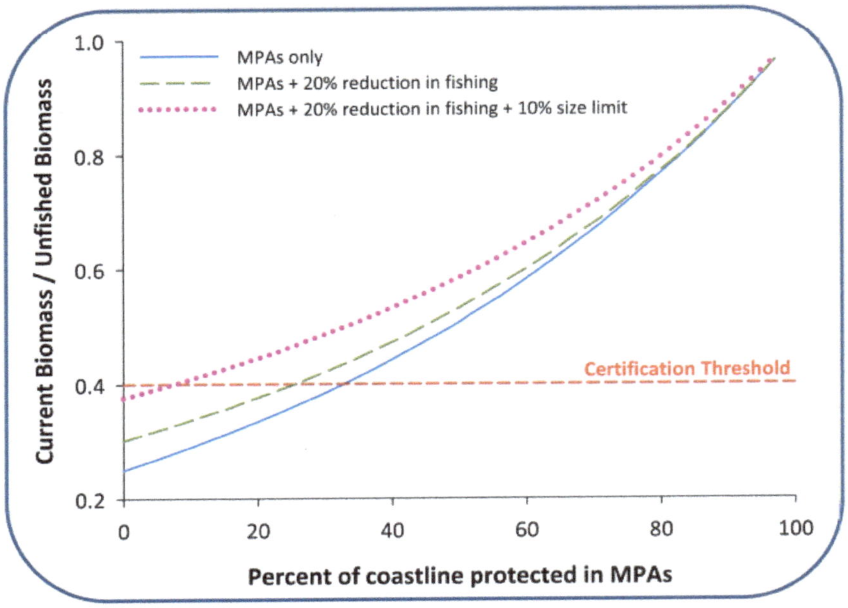

Figure 3. Bioeconomic models can be used to compare different strategies for achieving seafood sustainability.

In this example based on the sheephead fishery in Southern California, we assume that one requirement of the certification scheme is for the biomass of the target species to be at least 40% of unfished biomass. With MPAs alone, biomass is predicted to equilibrate at certifiable levels if one third of the coastline is closed to fishing (blue solid curve). If effort is reduced by 20%, certification can be achieved by closing only one quarter of the coastline to fishing (green dashed curve). If that reduction in effort is accompanied by a 10% increase in the size limit, less than one tenth of the coastline need be closed to raise the population to 40% of its unfished biomass (purple dotted curve). Reducing fishing effort by 20% and raising the size limit is not sufficient for achieving certification without at least some MPA protection, but a mix of more conservative non-spatial regulations could achieve certification without any MPAs (not shown).

The basic approach presented here, which involves evaluating the likely future effects of MPAs on fishery sustainability, could in principle apply to any fishery in the world and could be used to provide sustainability credit. Perhaps the best candidates are unassessed coastal fisheries, many of which occur in the developing world; and these are the very fisheries that are most difficult to certify using conventional stock assessments. These fisheries compose a huge fraction of global fisheries, but a relatively small fraction of global catch. Applying the MPA certification approach to only this class of fisheries could give rise to substantial local benefits in many parts of the world, but would not have dramatic impacts on the global food system. Many high-volume fisheries, composed of migratory species, demersal species, forage fish, and others, can also benefit from MPAs, though larger protected areas would be required.

Of course, not all fisheries supply the developed-world markets in America and Asia that seek sustainability guarantees. For example, many fisheries in the developing world are used for subsistence and local markets, although international fish processing and distribution businesses increasingly target developing-world fisheries and provide the infrastructure for this seafood to enter more lucrative developed-world markets. It should also be noted that even fisheries that are primarily harvested to produce fish meal and fish oils (e.g., anchovy fisheries) are interested in obtaining certification because of consumer demand for farmed fish that are raised on sustainable feed.

LOOKING FORWARD

We have argued that providing credit for MPAs could significantly increase the incentive to implement MPAs. While it is hard to know how many more MPAs would have been implemented had this practice been adopted at the inception of seafood certification, we do note that, until recently, most regions of the world have had negligible area protected from fishing in MPAs. As the conservation and fishery benefits of MPAs become better known, a growing number of MPAs have been established. This trend is expected to continue, given international commitments and NGO initiatives to expand the global network of MPAs, and could accelerate still further if certification programs were to recognize MPAs. Two prominent examples of MPAs that clearly could have received credit in certification schemes are the Great Barrier Reef Marine Park in Australia, which protects about a third of the reef with no-take areas (with an additional third protected from destructive bottom trawling fishing techniques), and the statewide network of MPAs being implemented in California, which protects about 17% of nearshore waters (with about 10% in no-take areas). Both of these examples show a strong investment in conservation. Yet most of the fisheries in these regions get no credit for these management actions under existing certification programs.

A more explicit integration of MPAs into seafood certification is likely to result in significant benefits. This integration will create more widespread incentives for sustainability, particularly in developing-country fisheries that are often data poor but may have adequate local support for MPA networks. Making certification a more readily attainable goal in parts of the world where it is presently infeasible will enhance the adoption of sustainable practices for harvesting seafood and accelerate the spread of sustainability reforms in ocean management. Furthermore, by providing fishermen with the incentive of certification, support for MPA implementation is likely to increase. With broader stakeholder support, protection in MPAs can accelerate to the scale required for more effective global conservation of marine resources

ACKNOWLEDGMENTS

We thank M. Clemence for compiling the data for Figure 1 and the David and Lucile Packard Foundation support to PISCO for the Science of Marine Reserve resources for Figure 2. This paper was improved by comments from N. Baron, M. Wright, E. Goldman, H. Galindo, and two anonymous referees.

REFERENCES

1. Lester SE, Halpern BS, Grorud-Colvert K, Lubchenco J, Ruttenberg BI, et al. (2009) Biological effects within no-take marine reserves: A global synthesis. Mar Ecol Prog Ser 384: 33–46. doi: 10.3354/meps08029
2. Jacquet J, Pauly D, Ainley D, Holt S, Dayton P, et al. (2010) Seafood stewardship in crisis: Opinion. Nature 467: 28–29. doi: 10.1038/467028a
3. Froese R, Proelss A (2012) Evaluation and legal assessment of certified seafood. Mar Policy 36: 1284–1289. doi: 10.1016/j.marpol.2012.03.017
4. Gulbrandsen LH (2009) The emergence and effectiveness of the Marine Stewardship Council. Mar Policy 33: 654–660. doi: 10.1016/j.marpol.2009.01.002
5. Perez-Ramirez M, Phillips BB, Lluch-Belda D, Lluch-Cota S (2012) Perspectives for implementing fisheries certification in developing countries. Mar Policy 36: 297–302. doi: 10.1016/j.marpol.2011.06.013
6. Halpern BS, Lester SE, Kellner JB (2010) Spillover from marine reserves and the replenishment of fished stocks. Environ Conserv 36: 268–276. doi: 10.1017/s0376892910000032
7. Grafton RQ, Kompas T (2005) Uncertainty and the active adaptive management of marine reserves. Mar Policy 29: 471–479. doi: 10.1016/j.marpol.2004.07.006
8. Rassweiler A, Costello C, Siegel D (2012) Marine protected areas and the value of spatially optimized fishery management. Proc Natl Acad Sci U S A 109: 11884–11889. doi: 10.1073/pnas.1116193109
9. MSC (2010) MSC Fishery Standard: Principles and Criteria for Sustainable Fishing: Marine Stewardship Council. doi: 10.4337/9781845428297.00080
10. Field JC, Punt AE, Methot RD, Thomson CJ (2006) Does MPA mean 'Major Problem for Assessments'? Considering the consequences of place-based management systems. Fish and Fisheries 7: 284–302. doi: 10.1111/j.1467-2979.2006.00226.x
11. McGilliard CR, Hilborn R, MacCall A, Punt AE, Field JC (2011) Can information from marine protected areas be used to inform control-rule-based management of small-scale, data-poor stocks? ICES J Mar Sci 68: 201–211. doi: 10.1093/icesjms/fsq151
12. Wilson JR, Prince JD, Lenihan HS (2010) A management strategy for sedentary nearshore species that uses marine protected areas as a reference. Mar Coast Fish 2010: 14–27. doi: 10.1577/c08-026.1
13. Babcock EA, McCall AD (2011) How useful is the ratio of fish density outside versus inside no-take marine reserves as a metric for fishery management control rules? Can J Fish Aquat Sci 68: 343–359. doi: 10.1139/f10-146
14. FAO (2012) FAO Yearbook 2010: Fisheries and Aquaculture Statistics. Rome: Food and Agriculture Organization of the United Nations.
15. MSC (2013) Global impacts report 2013. London, UK: Marine Stewardship Council [MSC]. 56 p.
16. OTA (2011) Organic Industry Survey: 2011: Organic Trade Assocation [OTA]. Available: http://www.ota.com/pics/documents/2011OrganicIndustrySurvey.pdf. Accessed 28 August 2013.
17. Pay E (2009) The market for organic and fair-trade coffee: Study prepared in the framework of FAO project GCP/RAF/404/GER. Rome: Food and Agriculture Organization of the United Nations.
18. EII. Dolphin safe tuna. Earth Island Institute [EII]. Available:http://www.earthisland.org/dolphinSafeTuna/consumer/index.html. Accessed 15 August 2013.
19. FAO (2010) Global forest resources assessment 2010. Rome: Food and Agriculture Organization of the United Nations.
20. FSC (2012) Forest Stewardship Council Annual Report 2012: Available at:https://ic.fsc.org/annual-reports.341.htm. Accessed 28 August 2013.
21. PISCO (2007) The science of marine reserves: Partnership for Interdiscplinary Studies of Coastal Oceans (2nd Edition). Available:http://www.piscoweb.org/publications/outreach-materials/science-of-marine-reserves/smr-booklet-versions. Accessed 28 August 2013. 22 pages.

CHAPTER 10

SURVIVAL, GROWTH AND REPRODUCTION OF NON-NATIVE NILE TILAPIA

Michael R. Lowe, Wei Wu, Mark S. Peterson, Nancy J. Brown-Peterson, William T. Slack, Pamela J. Schofield

INTRODUCTION

Aquaculture is the fastest growing food-production sector in the world and is viewed as a viable solution to global nutritional deficiencies and poverty [1], [2]. Given the declining status of wild fish stocks [3], aquaculture may one day surpass capture fisheries in terms of food-fish production [2]. Despite the economic contributions of aquaculture and potential for mitigating environmental impacts [4], aquaculture does not operate in a vacuum [5] and the same traits that make species desirable for production also lend to their potential as an invasive species[6], [7]. After habitat modification, invasive species are among the largest threats to freshwater and marine fish biodiversity [8], [9]. Despite a growing understanding of the consequences of biological invasions [10]–[13] the number of invasive species [14], [15] resulting from aquaculture suggests that there is a trade-off that favors economic benefit over the potential impacts of culturing non-native species [16].

Nile tilapia (*Oreochromis niloticus* Linnaeus, 1758), a secondary freshwater teleost native to the Nilo-Sudanian ecoregion of Africa [17], is one of the most widely cultured species globally and has been introduced to at least 85 countries [18], [19]. Nile tilapia is capable of rapid adaptation [20] and shows a wide range of biological responses to environmental conditions[21], [22]. Further, many cultured tilapiine fishes have been genetically enhanced for the purpose of increased production; thus their response to local environmental conditions, once escaped, may be less predictable [23], [24]. Therefore, it is invalid to assume that environmental constraints prevent the establishment and dispersal of this highly cultured species. For example, it was assumed that the escaped Nile tilapia from aquaculture facilities in coastal Mississippi would not survive through a temperate winter. However, Nile tilapia has become established in coastal Mississippi [7], [25] and its eradication may not be feasible[26], [27]. Further, the spread of established Nile tilapia does not appear to be limited by salinity as Schofield et al. [22] found Nile tilapia from Mississippi could survive, grow, and reproduce in elevated salinity during summer although overwinter survival occurred up to a salinity of 10. Given these important demographics, it was hypothesized that not only do the waters of the Mississippi Sound (hereafter Sound) provide suitable habitat for Nile tilapia to complete all facets of their life history, but also act as a 'salt-bridge' [28], whereby Nile tilapia can gain access to other freshwater systems by moving through saline waters, potentially facilitating its spread to other freshwater systems throughout the northern Gulf of Mexico (GOM).

Citation: Lowe MR, Wu W, Peterson MS, Brown-Peterson NJ, Slack WT, Schofield PJ (2012) Survival, Growth and Reproduction of Non-Native Nile Tilapia. doi:10.1371/journal.pone.0041580

The invasion process is defined by a progression of stages (e.g., introduction, establishment, and spread) that are both biologically and environmentally filtered at each step [29], [30]. Once an invasive species becomes established, regional patterns in abiotic variables become the principal driver of spread. The growing effort to model the distribution and potential spread of invasive species following establishment [31]–[34] has highlighted the need for a proactive, predictive approach to invasive species management [35]. Species distribution models (SDM) based on ecological niche theory have emerged as a common approach to predicting species› ranges [36]. Though powerful tools, SDMs are based on observational data (e.g., presence/absence) and tend to under-perform with small sample size [37], [38]. As a result, a more prudent approach is to use the fundamental niche of 'data poor' taxa (e.g., recently established invasive species) to mechanistically predict the potential distribution of an organism based on physiological constraints. Models based on the fundamental niche not only help understand the contemporaneous distribution of an organism in the present climate [36], but also allow the prediction of future distributions under climate change scenarios [39].

Current climate predictions from 16 General Circulation Models downscaled to the northern GOM suggest that air temperature will increase from 2.5–5.9°C in summer, and 1.4–4.5°C in winter from 2070–2099 compared to 1961–1990 under the A2 emission scenario in World Climate Research Programme›s Coupled Model Intercomparison Project phase 3 multi-model dataset [40]. Concomitantly, regional precipitation is projected to vary between −28% to 32% in the summer and −48% to 18% in winter. While increasing temperature will likely favor the survival, growth, and reproduction of Nile tilapia [41], [42], the uncertainty surrounding precipitation makes predicting the effect of regional salinity on population spread more problematic.

The objectives of this study are to: 1) develop a predictive model of Nile tilapia survival, growth, and reproduction derived from our earlier research [22]; 2) project the species› fundamental niche in the Sound by integrating spatial environmental data (see Figure 1) with the predictive model; 3) project the potential pathways that an established population of Nile tilapia might use to invade other drainages throughout coastal Mississippi and the GOM; and 4) examine how future climate variability may affect the distribution of Nile tilapia.

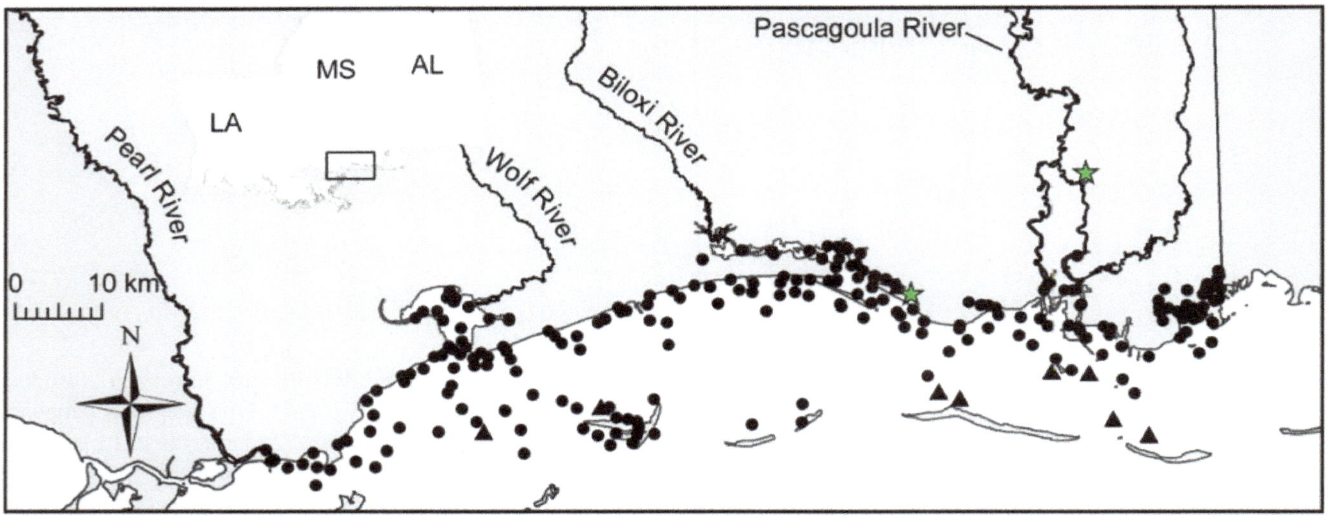

Figure 1. Map of Mississippi Sound and locations of long-term environmental stations used in for salinity projections.

Closed circles represent stations that have both summer and winter data while closed triangles only have summer data. The gray stars represent the location of established populations of *Oreochromis niloticus*.

RESULTS

We were able to find a finite estimate for all of our model parameters using the FLR with penalized maximum likelihood estimation. All intercept and slope estimates were significantly different from zero for all models (Table 1; WALD, all p-values≤0.008) and the probability of a positive outcome for all biological response (i.e., survival, growth, and reproduction) variables was inversely related to salinity (Table 1; negative slope (b) for all models) in both the summer and winter (Figure 2). Further, all models with salinity as a covariate were statistically more reliable than the intercept only models (Table 1; LRT, all p-values≤0.0002) indicating that each model's predictive ability was significantly enhanced by the inclusion of salinity as a covariate.

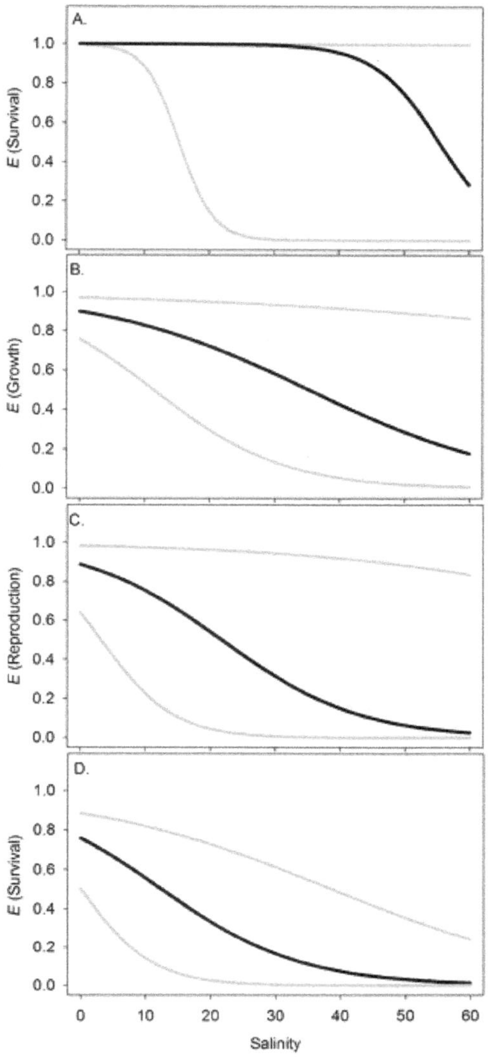

Figure 2. Empirical estimates of the relationships between *Oreochromis niloticus*biological response variables and salinity.

A) summer survival, B) summer growth, C) summer reproduction, and D) winter survival. The black line indicates penalized maximum likelihood estimate of the logistic response function (Eq. 2); gray lines are the 95% confidence intervals based on the profile of the penalized likelihoods.

Table 1. Model estimates and associated statistics from Firth-logistic regression

Parameter	Estimate (SE)	Upper 95% CI	Lower 95% CI	WALD p-value	LRT p-value	R^2
Summer						
Survival					38.3	0.76
a	11.08 (3.28)	19.89	5.84	<0.0001	<0.0001	
b	−0.20 (0.063)	−0.11	−0.38	<0.0001		
Growth					13.48	0.61
a	2.19 (0.59)	3.48	1.14	<0.0001	0.0002	
b	−0.062 (0.019)	−0.027	−0.10	<0.0001		
Reproduction					14.18	0.53
a	2.066 (0.88)	4.10	0.58	0.004	<0.0002	
b	−0.095 (0.033)	−0.041	−0.18	0.001		
Winter						
Survival					23.49	0.55
a	0.89 (0.55)	2.06	0.12	0.008	<0.0001	
b	−0.012 (0.031)	−0.053	0.18	<0.0001		

SE = Standard Error, PML = penalized maximum likelihood, WALD = Wald score, LRT = log ratio test. R^2 = Nagelkerke's R^2.
doi:10.1371/journal.pone.0041580.t001

The final semivariogram models fit to the observed salinity data were chosen based on the lowest sum of squared errors among the exponential, spherical, and Gaussian models (Table 2). Normal salinity years were less variable (shown in total sill) than either wet or dry years during both seasons, due to smoothing of extreme salinity values with high (wet years) or low (dry years) river discharge. Micro-scale variability (i.e., nugget effect) in salinity patterns was generally greatest during periods of high river discharge (i.e., winter and wet years) compared to low discharge periods (i.e., summer and dry years). During normal years, salinity generally increased from west to east throughout the Sound, was lowest in the bays and estuaries near river mouths and greatest in areas with minimal freshwater input (e.g., offshore barrier islands and saltmarshes of eastern Mississippi) and areas associated with major ship channels (Gulfport and Pascagoula). Further, salinity was greater in the summer than in the winter. While these spatial and seasonal patterns were consistent for both wet and dry years, overall salinity decreased and increased, respectively, within the Sound during these periods.

Table 2. Universal kriging results for the predicted mean salinity across 14 years (normal), mean salinity for three extremely wet years (wet) and mean salinity for three extremely dry years (dry) during both summer and winter seasons.

Season	N	Model	Nugget	Partial Sill	Total Sill	Theoretical Range (km)	Effective Range (km)
Summer							
Normal	227	Exponential	2.56	14.5	17.1	9.28	27.8
Wet	172	Exponential	4.14	16.3	20.4	18.7	56.0
Dry	184	Spherical	1.86	19.2	21.0	10.8	10.8
Winter							
Normal	219	Spherical	5.40	23.2	28.6	20.0	20.0
Wet	176	Gaussian	17.4	16.8	34.2	8.51	14.7
Dry	197	Exponential	0.88	30.2	31.1	6.10	18.3

N is the number of long-term stations used in each analysis. Nugget represents discontinuity at the origin due to microscale effects or measurement error. Total sill is the variance estimate. Theoretical and Effective ranges are the distances at which sampling stations are no longer spatially autocorrelated.
doi:10.1371/journal.pone.0041580.t002

During normal summer salinity conditions, the probability of survival was high throughout the Sound. However, the probabilities of positive growth, reproduction, and winter survival decreased from inshore to offshore. Though Nile tilapia can survive throughout the Sound in the summer, growth and reproduction are limited to nearshore and low salinity habitats (Figure 3A), respectively. In the winter, survival is restricted to upper estuaries and rivers where salinity is typically <10. During extremely wet and dry years, the areal coverage of summer survival habitat did not change (Figures 3B, 3C;Table 3) relative to normal years. However, the areal coverage of growth and reproductive habitats increased by 2% and 47%, respectively, (Figures 3B, 3C; Table 3) for wet years. Conversely, growth and reproductive habitats decreased by 92% and 87%, respectively, during dry years due to increased salinity (Figures 3B, 3C; Table 3). Similarly, overwinter survival habitats increased by 368% during wet years (Figure 4B; Table 3) and decreased by 86% during extremely dry years (Figure 4C; Table 3).

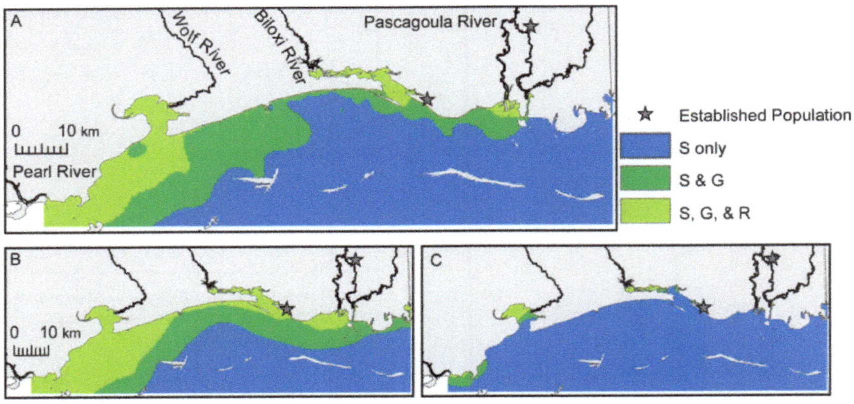

Figure 3. Areas of the Mississippi Sound with the highest probability of*Oreochromis niloticus* **survival (S), growth (G), and reproduction (R) during summer months.**

A) normal years, B) wet years, and C) dry years.

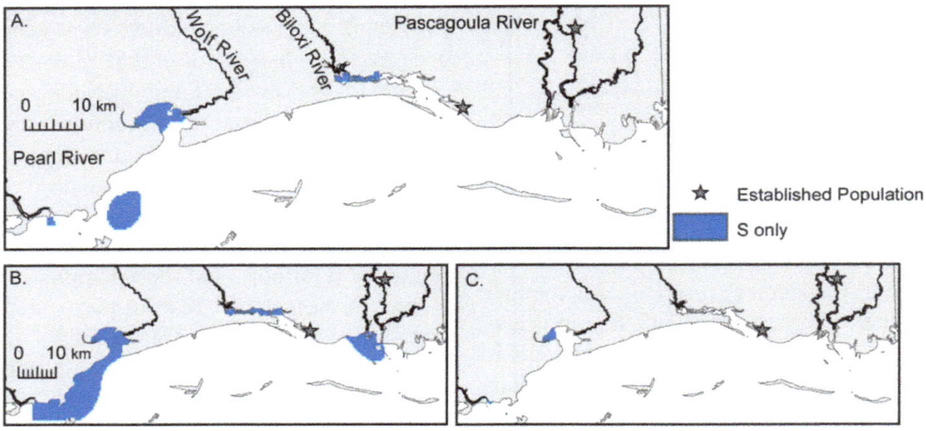

Figure 4. Areas of the Mississippi Sound with the highest probability of*Oreochromis niloticus* **survival (S) during winter months.**

A) normal years, B) wet years, and C) dry years.

Table 3. Areal coverage (km²) of survival, growth, and reproductive habitats for*Oreochromis niloticus* **in the Mississippi Sound for normal salinity** years, wet years, and dry years

	Normal years		Wet years		Dry years	
Parameter	**Summer**	**Winter**	**Summer**	**Winter**	**Summer**	**Winter**
Survival	2.31	0.0763	2.31	0.281	2.31	0.0104
Growth	1.14		1.16		0.0946	
Reproduction	0.329		0.615		0.0438	

All values are $\times 10^4$.

doi:10.1371/journal.pone.0041580.t003

Given the physiological tolerances of Nile tilapia and the range of salinities commonly found in this region, the coastal waters of the Sound do not act as a barrier to dispersal. While seasonal pulses in abundance are likely to occur in the bays and estuaries, establishment of this species is likely limited to freshwater habitats and and low salinity habitats at the river mouths (Figure 5A). Further, under the different scenarios of climate variability (Figure 5B–I), the relative area of these different habitats will likely expand during extremely wet years (e.g.,Figure 5D) and contract during extremely dry years (e.g., Figure 5G).

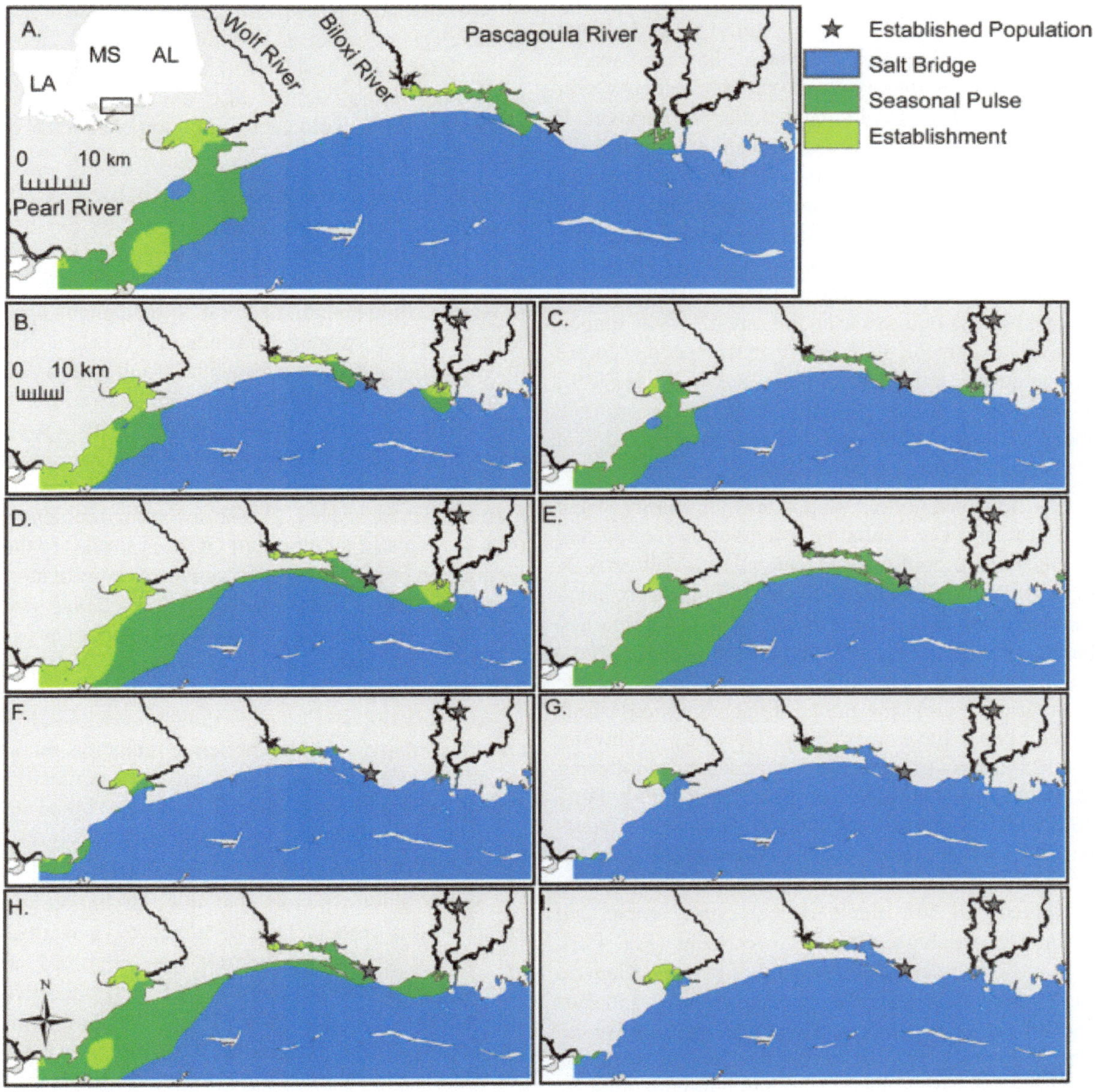

Figure 5. Projected areas of the Mississippi Sound that *Oreochromis niloticus*might use to establish a population (survive, grow, reproduce, and overwinter), pulse seasonally in abundance (survive, grow, reproduce, but not overwinter), or use as a salt bridge between adjacent freshwater systems (Survive and possibly grow, but reproduction and overwintering are unlikely).

Each panel is for the following scenario: A) normal summer and normal winter, B) normal summer and wet winter, C) normal summer and dry winter, D) wet summer and wet winter, E) wet summer and dry winter, F) dry summer and wet winter, G) dry summer and dry winter, H) wet summer and normal winter, and I) dry summer and normal winter.

DISCUSSION

The growing effort to model the spread of invasive species [31]–[34] has highlighted the need for a proactive, predictive approach to invasive species management [35], [43]. The time lag between successive stages of the invasion process is not only unpredictable [44] but also hampers our ability to both monitor the spread and mitigate the impacts of biological invasions[45]. By integrating the empirical results of Schofield et al [22] with a predictive framework, we have expedited processes at the population level that are typically only observable over extended time scales, thus allowing us to proactively predict areas of the Sound that are susceptible to invasion by Nile tilapia. Our fundamental niche projections not only corroborate the 'salt-bridging' hypothesis discussed in Schofield et al. [22] but also highlight that nearshore, low-salinity waters, particularly in the western portion of the Sound, currently provide suitable conditions for establishment.

The present study also highlights the linkage between future precipitation patterns and the potential for Nile tilapia to spread throughout the region. For example, increased precipitation will likely facilitate the spread of Nile tilapia by increasing the areal coverage of habitats where they may establish. Conversely, decreased precipitation may limit the area available for Nile tilapia establishment; however, spread remains possible given that regional salinity patterns will fall within Nile tilapia›s physiological limits for survival. It should be noted that our study conservatively approaches the implications of climate change on the spread of Nile tilapia by only considering changes in salinity and treating temperature as a static variable, particularly in the winter. Given the expectation that predicted increases in global temperature [46] will favor invasive species [41], [42], [47], warming regional temperature will likely have two major impacts on the spread of Nile tilapia. First, northward expansion in river systems becomes more likely as regional air and water temperature increases [48]. Secondly, areas that are not currently projected to promote establishment are likely to become suitable as physiological barriers breakdown during warmer winters.

To date, however, Nile tilapia has primarily been documented in freshwater habitats throughout the region and there have been three confirmed documentations directly in the Sound [49]. While this potentially reflects a discrepancy between fundamental and realized niches by overlooking key habitat characteristics (i.e., sediment type for bower (i.e., nest) formation), we suggest two probable explanations for the lack of observations within the Sound. First, sampling efforts targeting Nile tilapia have been biased towards freshwater habitats [7], [25]and routine fisheries-independent monitoring within the Sound uses sampling gear (e.g., long-lining, large mesh gillnets, trawling) that are not likely to capture Nile tilapia. Second, most invasive species exhibit an extended time lag between initial establishment and the onset of rapid population growth and spread [50]. Nile tilapia life history is characterized by low fecundity and high parental investment [51], [52]. Therefore, it may take several generations for populations to reach a level where local resources become limiting and thus warrant movement to new areas. However, there is wealth of evidence indicating that, as a whole, members of the family Cichlidae are fully capable of both moving through and establishing populations in low salinity, coastal waters both within their natural [53] and introduced ranges[54]–[57]. In Florida, for example, Mayan cichlid (*Cichlasoma urophthalmus*) established in freshwater systems are seeding ephemeral populations in estuarine habitats leading to a general northward expansion of the species into previously un-invaded systems since 1983[55].

The potential for establishment of Nile Tilapia in the western Sound has much broader implications given the proximity to Louisiana's coastal habitats. Lake Pontchartrain, for example, is a large (1,839 km^2), highly altered, shallow, low salinity estuary situated north of New Orleans, Louisiana that receives decreased freshwater inputs from the Mississippi River due to historical modification [58]. Proposed plans to restore freshwater input include periodic opening of flood control structures created by the U.S. Army Corps of Engineers [59], [60], which would reduce salinity in the basin by up to 40% [61]. The entrance to Lake Pontchartrain is <5 km from the mouth of the Pearl River and the system already supports the invasive Rio Grande cichlid (*Herichthys cyanoguttatus*) [57]. Therefore, current salinity patterns are not likely to differ from those in the western Sound. Increasing freshwater input into the basin would increase the probability of Nile tilapia establishment and provide them with a direct connection to adjacent systems, including the Mississippi River. However, restoring natural freshwater flows into Lake Pontchartrain could also aid the recovery of natural predators [62]that could act as a biological control for invasive Nile tilapia [63].

The ecological niche is conceptualized as all the biotic and abiotic factors that affect the expression of a species' distribution and can be further separated into the fundamental and realized niches [64]. Though the focus of this work is to project the fundamental niche of Nile tilapia using salinity and temperature, there are a multitude of biotic and abiotic factors that potentially shape the distribution and spread of this species in the northern GOM. Previous work showed that established Nile tilapia do not compete directly with native sunfishes (Family: Centrarchidae) for food in coastal Mississippi [65] and their aggressive behavior allows them to outcompete other native fishes for space [12]. The spectrum of potential predators of Nile tilapia in the northern GOM is similar to those in their natural range [66] and includes American alligator (*Alligator mississippiensis*), brown pelicans (*Pelecanus occidentalis*), and large piscivorous fishes such as alligator gar (*Atractosteus spatula*) and bull shark (*Carcharhinus leucas*). However, both alligator gar and bull shark abundances have declined in recent decades [62] and, though possible, it is unlikely that Nile tilapia would experience predation rates exceeding those of native fishes. However, the composition of both coastal [48] and freshwater [67] ichthyofaunal assemblages are expected to change with warming temperatures which may result in unpredictable species interactions.

Other abiotic factors that may be important drivers of Nile tilapia spread include dissolved oxygen, pH, depth and substrate type. Though bottom water hypoxia is a normal occurrence in the northern GOM [68], dissolved oxygen concentrations are generally within the optimal range for Nile tilapia [66]. Further, large-scale hypoxic zones associated with the Mississippi River occur in offshore waters outside of the depth preference for Nile tilapia [66]. The size and location of these hypoxic zones are driven principally by increased nutrient loads and the role of future climate change on these zones is unclear [69]. Conversely, while the pH of coastal draining rivers and backwater areas in the northern GOM [70] is within a tolerable range for Nile tilapia [71], the expectation is that pH will decrease due to ocean acidification [72], [73]. Nile tilapia are tolerant of a wide pH range but juvenile mortality is increased in acidic waters (pH<3.0) [71]. Temperature and salinity are the major metabolic modifiers for most tilapiine fishes [71] and, thus, we considered both to be the principal drivers of spread for invasive Nile tilapia in the northern GOM.

Once established, the ecological, evolutionary, and economic impacts of an invasive species can range from negligible to severe [74], [75]. However, there is a growing body of work showing that Nile tilapia can alter the function of aquatic systems through eutrophication [76],[77], altered trophic dynamics [12], and local extinction of native fish populations [66]. Generalizing the ecological impacts of a single invasive species remains a challenging task given the time required for the manifestation of such impacts [44] and lack of funding in this area [78].

Ethics Statement

All animal work conducted in Schofield et al [22] and reproduced in this paper was done in accordance with the 'Guidelines for the Use of Fishes in Research' published by the American Society of Ichthyologists and Herpetologists (http://www.asih.ort/files/fish%20guidelines.doc) and approved by the U.S. Geological Survey, Southeast Ecological Science Center, Institutional Animal Care and Use Committee permit number USGS/FISC 2007-01.

METHODS

We constructed a predictive model based on the derived survival, growth, and reproduction data of Nile tilapia exposed to different combinations of salinity and temperature in an experimental setting (see Chronic salinity-tolerance experiment) [22]. Though a full recount of the experiment is beyond the scope of this paper and can be found in Schofield et al. [22], a brief discussion of the experimental design is warranted. Nile tilapia were housed in individual tanks and gradually acclimated (salinity of 5 per week) to target salinities (0 to 70 in increments of 10) at constant summer (30°C) and winter (14°C) temperatures, reflecting seasonal water conditions in coastal Mississippi. At the time of death or the end of the experiment fish were sacrificed by immersion in an ice water bath, weighed (±0.1 g), measured (±0.1 cm), and frozen. For each individual, survival was estimated with a Kaplan-Meier product limit estimator and both growth and spawning preparedness were quantified.

These data were converted to a binary response for logistic regression. Individuals surviving >65% of the total time at their target salinity were classified as 'survived' and growth was quantified as the change in body mass divided by the total number of experimental days (acclimation plus time at target salinity). Relative growth rates were converted to a binary response such that an individual exhibited either positive (net mass gain) or negative (net mass loss) growth during the experiment. For the reproductive measurements, we chose to restrict our analysis to female Nile tilapia because the maternal contributions to offspring can have a large impact on population dynamics and evolutionary plasticity of future generations [79], [80]. In Schofield et al.'s [22] summer experiment, all female Nile tilapia reared at a salinity ≤20 had GSI values ≥1.8 and produced large, vitellogenic oocytes while individuals reared in higher salinities had GSI values <1.8 and showed a marked reduction in vitellogenic oocyte production. Thus, an individual was deemed "spawning capable" or "spawning incapable" if its GSI value was ≥1.8 or <1.8, respectively. All Nile tilapia in the winter experiment exhibited negative growth and there was no evidence of reproductive development (i.e., GSI<0.75); thus, only survival probabilities were calculated.

We used the logistic regression function (Eq. 1) to model the survival, growth and reproduction probabilities of Nile tilapia from the summer experiment and the survival probabilities for the winter experiment separately:

$$E(y) = \frac{e^{(a+b*X)}}{1+e^{(a+b*X)}} \tag{1}$$

where E(y) is the probability of a positive outcome in survival, growth or reproduction, a and $b$are the intercept and slope in the linear link function, respectively, and X is the salinity covariate predictor. Due to both small sample size (summer survival, n=94; growth, n=76; reproduction, n=38; winter survival, n=70) and 'quasi-complete' separation of the experimental data (i.e., perfect correspondence between survival, positive growth, or reproductive capacity and salinity levels), finite parameter estimates did not exist using a GLM framework for logistic procedures [81]. As a result, we used a Firth-logistic regression (FLR) [82] to estimate survival, growth, and reproduction probabilities as a function of salinity for each season separately. The FLR uses a modified score function to split each original binary observation into "response" and "non-response" components, thus guaranteeing that each level of the predictor has some dichotomy of response variables and eliminating the problems associated with data separation [83]. It also uses penalized maximum likelihood (PML), carried out iteratively until parameter convergence, to estimate the logistic regression parameters, associated standard errors, and 95% confidence intervals [84]. The significance of each model parameter estimate was evaluated with the Wald score (WALD; z-statistic) and the likelihood ratio test (LRT; χ^2) was used to assess goodness of fit by comparing the full model to the intercept only model. Nagelkerkes's R^2 was calculated for each model with the equation (Eq. 2):

$$R_N^2 = \frac{1-e^{\left(-\frac{2}{n}*(LL(Full)-LL(Intercept))\right)}}{1-e^{\left(\frac{2*LL(Full)}{n}\right)}} \tag{2}$$

where e is the base of the natural logarithm, $LL(Full)$ and $LL(Intercept)$ are the penalized log-likelihood's associated with the full and intercept-only models, and n is the sample size [85]. All analyses were done using the logistf package in R [86] at a significance level of 0.05.

Seasonal salinity patterns in the Sound were examined from surface salinity and temperature data acquired from long-term data sets collected by various state and federal agencies in coastal Mississippi since 1973. Though each agency sampled at different temporal and spatial scales, we organized the data into a single relational database for continuously monitored stations. The database was checked for errors (e.g., out of range values) and then queried by temperature to generate salinities corresponding to Schofield et al.›s [22] summer (30±2.5°C) and winter (14±2.5°C) temperature conditions. For each season, we generated 3 data sets: 1) mean salinity across 14 years (1 January1992 through 15 March 2011) for both summer (227 stations) and winter (219 stations) conditions (Figure 1), hereafter normal salinity; 2) mean salinity for three extremely wet years during summer (1997, 2001, and 2003; Figure S1A, 1B; 172 stations) and winter (1998, 2004, 2009; S1C, 1D; 176 stations) conditions; and 3) mean salinity for three extremely dry years during summer (1996, 2000, 2006, 1B; 184 stations) and winter (1999, 2000, 2007; S1C, 1D; 197 stations) conditions. River discharge data was attained from the US Geological Survey (USGS; http://waterdata.usgs.

gov/nwis/rt) and wet and dry years were selected based on abnormally high or low mean discharge, respectively, during the summer and winter months for the Pearl, Wolf, Biloxi, and Pascagoula rivers and maximized spatial coverage for each data set. The latter two data sets were used to examine how climate variability might impact the fundamental niche and potential spread of Nile tilapia.

Universal kriging was used to interpolate salinity between sampling stations in the Sound (4,792 km^2) for each data set. Kriging is a group of geostatistical techniques that uses a set of linear regression routines to construct statistically optimal interpolation of a regionalized variable at unobserved locations (hereafter location) from spatially explicit samples (hereafter stations). Each interpolated value in a given location (0.16 km^2 cell) is a weighted mean of the salinity at each sample (Figure 1) where the weights are based on the fitted semivariogram model derived from sampled values and the spatial configuration of the sampling stations. Due to an expected spatial trend for observed salinity (i.e., salinity decreased toward river mouths) within the Sound, we chose universal kriging. Universal kriging assumes the following model (Eq. 3):

$$Z(s) = \mu(s) + \varepsilon(s) \quad (3)$$

where Z is the interpolated salinity at location s, μ is the deterministic drift or trend modeled as a linear regression from geographic coordinates (WGS84 UTM 16 N projection), and is a stationary random variable that accounts for the spatial autocorrelation among sampling locations. We used the spherical, exponential, and Gaussian semivariogram models [87] to account for spatial autocorrelation and the model with the lowest sum of squares error was selected for salinity interpolation. For each semivariogram, we estimated the nugget, sill, and spatial range. The nugget represents a discontinuity at the origin due to microscale effects or measurement error [88]. The total sill is the sum of nugget effect and partial sill, and represents an estimate of the variance. The range denotes the distance at which the semivariogram reaches the sill, beyond which, there exists minimum spatial autocorrelation in the data of interest (e.g., salinity). For the exponential and Gaussian models, the semivariogram increases asymptotically toward its sill, so we calculated the effective range, defined as the distance at which the semivariance value achieves 95% of the sill. For the exponential and Gaussian models, the effective range is defined as the theorotical range multiplied by 3 and $\sqrt{3}$, respectively. For the spherical model, the effective range is the theoretical range. Kriging was performed in R using gstat package, and the results were imported to ArcGIS 10.0 (ESRI 2009) for projecting the fundamental niche of Nile tilapia.

The derived predictive models (i.e., logistic regression) were applied to the whole Sound using interpolated salinity at each location as the covariate predictor. The simulation results were probabilities of survival, growth and reproduction in summer and probability of survival in winter. The locations with the probability of survival >0.95, probability of positive growth >0.70, and probability of reproduction >0.50 were classified as survival, growth, and reproduction habitats, respectively. Threshold values for each projection were selected at a level that realistically represented the results of Schofield et al. [22]. These were then projected for each season and used to compare changes in spatial coverage of those habitats between normal salinity years and extremely wet or dry years. The areal coverage of each habitat was calculated by multiplying the number of predicted cells by the spatial resolution of the salinity projections (0.16 km^2). Further, we combined the seasonal habitat projections into a series of maps in order to identify 1) areas of the Sound that promote survival, growth, and reproduction in the summer and where overwinter survival was most likely (the highest probability of establishment), 2) areas of the Sound where seasonal pulses in abundance might occur (i.e., survival, growth, and reproduction in the summer but not likely to overwinter), and 3) areas of the Sound that most likely function as a 'salt-bridge' between river systems (i.e., Nile tilapia can survive in the summer, but are not capable of spawning and overwintering is unlikely). A single projection indicating the highest probability of each area was generated for every seasonal combination of normal, wet, and dry years

REFERENCES

1. Kurien J (2005) Responsible fish trade and food security. Rome: FAO.
2. Subasinghe R, Soto D, Jia J (2009) Global aquaculture and its role in sustainable development. Rev Aquac 1: 2–9 Doi: 10. 111/j. 1753-5131. 2008. 01002. x. doi: 10. 1111/j. 1753-5131. 2008. 01002. x
3. Worm B, Hilborn R, Baum JK, Branch TA, Collie JS, et al. (2009) Rebuilding global fisheries. Science 325: 578–585. doi: 10. 1126/science. 1173146
4. De Silva SS, Soto D (2009) Climate change and aquaculture: Potential impacts, adaptation and mitigation. In: Climate Change Implications for Fisheries and Aquaculture: Overview of Current Scientific Knowledge. Cochrane K, De Young C, Soto D, Bahri T, editors. Rome: FAO.
5. Goldburg RJ, Elliott MS, Naylor RL (2002) Marine aquaculture in the United States. Environmental Impacts and Policy Options. Arlington, VA: Pew Oceans Commission.
6. Hubbs C (1968) An opinion of the effects of cichlid release in North America. Trans Am Fish Soc 97: 197–198. doi: 10. 1577/1548-8659(1968)97[197:aooteo]2. 0. co;2
7. Peterson MS, Slack WT, Woodley CM (2005) The occurrence of non-indigenous Nile tilapia, *Oreochromis niloticus*, in coastal Mississippi, USA: ties to aquaculture and thermal effluent. Wetlands 25: 112–121. doi: 10. 1672/0277-5212(2005)025[0112:toonnt]2. 0. co;2
8. Miller RR, Williams JD, Williams JE (1989) Extinction of North American fishes during the past century. Fisheries 13: 22–38. doi: 10. 1577/1548-8446(1989)014<0022:eonafd>2. 0. co;2
9. Lotze HK, Lenihan HS, Bourque BJ, Bradbury RH, Cooke RG, et al. (2006) Depletion, degradation, and recovery potential of estuaries and coastal seas. Science 312: 1806–1809. doi: 10. 1126/science. 1128035
10. Lodge DM, Wiliams S, MacIsaac HJ, Hayes KR, Leung B, et al. (2006) Biological invasions: recommendations for U. S. policy and management. Ecol Appl 16: 2035–2054. doi: 10. 1890/1051-0761(2006)016[2035:birf up]2. 0. co;2

11. Lovell SJ, Stone SF, Fernandez L (2006) The economic impacts of aquatic invasive species: A review of the literature. Agricultural and Resource Economics Review 35: 195–208.

12. Martin CW, Valentine MM, Valentine JF (2010) Competitive interactions between invasive Nile tilapia and native fish: the potential for altered trophic exchange and modification of food webs. PLoS One e14395. 417 doi:10. 1371/journal. pone. 0014395. doi: 10. 1371/ journal. pone. 0014395

13. Canonico-Hyde G (2011) Tilapias in aquaculture: the need for invasion science in decision making to protect and sustain biodiversity. In: Sustainable Fisheries: Multi-Level Approaches to a Global Problem. Taylor WW, Lynch AG, Schechter MG, editors. Bethesda: American Fisheries Society. 113–124.

14. Rahel FJ (2007) Biogeographic barriers, connectivity and homogenization of freshwater faunas: it's a small world after all. Freshw Biol 52: 696–710 doi:10. 1111/J. 1365-2427. 2006. 01708. X. doi: 10. 1111/j. 1365-2427. 2006. 01708. x

15. Hulme PE (2009) Trade, transport and trouble: managing invasive species pathways in an era of globalization. J Appl Ecol 46: 10–18. doi: 10. 1111/j. 1365-2664. 2008. 01600. x

16. Naylor RL, Williams SL, Strong DR (2001) Aquaculture – a gateway for exotic species. Science 294: 1655–1656 doi:10. 1126/SCIENCE. 1064875. doi: 10. 1126/ science. 1064875

17. McAndrew BJ (2000) Evolution, phylogenetic relationships and biogeography. In: Tilapias: Biology and Exploitation. Beveridge MCM, McAndrew BJ, editors. Dordrecht: Kluwer Academic Publishers. 1–32.

18. Casal MVC (2006) Global documentation of fish introductions: the growth in crises and recommendations for action. Biol Invasions 8: 3–11 doi:10. 1007/S10530-005-0231-3. doi: 10. 1007/s10530-005-0231-3

19. Molnar JL, Gamboa RL, Revenga C, Spalding MD (2008) Assessing the global threat of invasive species to marine biodiversity. Front Ecol Environ 6: 485–492. doi: 10. 1890/070064

20. Seehausen O (2006) African cichlid fish: a model system in adaptive radiation research. Proc R Soc B 273: 1987–1998. doi: 10. 1098/rspb. 2006. 3539

21. Blanco JA, Narváez Baradnica JC, Viloria EA (2007) ENSO and the rise and fall of a tilapia fishery in northern Columbia. Fish Res 88: 100–108 doi:10. 1016/J. FISHRES. 2007. 07. 015. doi: 10. 1016/j. fishres. 2007. 07. 015

22. Schofield PJ, Peterson MS, Lowe MR, Brown-Peterson NJ, Slack WT (2011) Survival, growth and reproduction of nonindigenous Nile tilapia (*Oreochromis niloticus* (Linnaeus, 1758)). I. Physiological capabilities to various temperatures and salinities. Mar Freshw Res 62: 439–449. doi: 10. 1071/mf10207

23. Ridha MT (2008) Preliminary observation on salinity tolerance of three sizes of the GIFT and non-improved strains of the Nile tilapia *Oreochromis niloticus*. Eur J Sci Res 24: 373–377.

24. Brummett RE, Ponzoni RW (2009) Concepts, alternatives, and environmental considerations in the development and use of improved strains of tilapia in African aquaculture. Rev Fish Sci 17: 70–77 doi:10. 1080/10641250802083174. doi: 10. 1080/10641260802083174

25. Grammer GL, Slack WT, Peterson MS, Dugo MA (2012) Nile tilapia *Oreochromis niloticus* (Linnaeus, 1758) establishment in temperate Mississippi, USA: multi-year survival confirmed by otolith ages. Aquat Invasions. 7 (in press). .

26. Genovesi P (2007) Limits and potentialities of eradication as a tool for addressing biological invasions. In: Nentwig W, editor. Berlin: Springer Verlag. 385–403. doi: 10. 1007/978-3-540-36920-2_22

27. Walther GR, Post E, Convey P, Menzel A, Parmesan C, et al. (2009) Ecological responses to recent climate change. Nature 416: 389–395. doi: 10. 1038/416389a

28. Brown JA, Scott DM, Wilson RW (2007) Do estuaries act as saline bridges to allow invasion of new freshwater systems by nonindigenous fish species? In: Biological Invaders in Inland Waters: Profiles, Distribution and Threats. Gherardi F, editor. Dordrecht: Springer. 401–414. doi: 10. 1007/978-1-4020-6029-8_21

29. Hellmann JJ, Byers JE, Bierwagen BG, Dukes JS (2008) Five potential consequences of climate change for invasive species. Cons Biol 22: 534–543 doi:10. 1111/j. 1523-1739. 2008. 00951. x. doi: 10. 1111/j. 1523-1739. 2008. 00951. x

30. Blackburn TM, Pyšek P, Bacher S, Carlton JT, Cuncan DP, et al. (2011) A proposed unified framework for biological invasions. Trends in Ecology and Evolution 26: 333–339. doi: 10. 1016/j. tree. 2011. 03. 023

31. Reeves SA, Usher MB (1989) Application of a diffusion model to the spread of an invasive species: The coypu in Great Britain. Ecol Modell 47: 217–232. doi: 10. 1016/0304-3800(89)90002-1

32. Hastings A (1996) Models of spatial spread: A synthesis. Biological Conservation 78: 143–148. doi: 10. 1016/0006-3207(96)00023-7

33. Holway DA (1998) Factors governing rate of invasion: A natural experiment using Argentine ants. Oecologia 115: 206–212. doi: 10. 1007/s004420050509

34. DeVaney SC, McNyset KM, Williams JB, Peterson AT, Wiley EO (2009) A tale of four "Carp": Invasion potential and ecological niche modeling. PLoS One 4: e5451 doi: 10. 1371/journal. pone. 0005451. doi: 10. 1371/journal. pone. 0005451

35. Hobbs RJ, Mooney HA (1998) Broadening the extinction debate: Population deletions and additions in California and western Australia. Conserv Biol 12: 271–283. doi: 10. 1111/j. 1523-1739. 1998. 96233. x

36. Kearney M, Porter WP (2009) Mechanistic niche modeling: combining physiological and spatial data to predict species› ranges. Ecol Lett 12: 334–350 doi: 10. 1111/j. 1461-0248. 2008. 01277. x. doi: 10. 1111/j. 1461-0248. 2008. 01277. x

37. Guisan A, Graham CH, Elith J, Huettmann F (2007) NCEAS Species Distribution Modelling Group (2007) Sensitivity of predictive species distribution models to change in grain size. Divers Distrib 13: 332–340. doi: 10. 1111/j. 1472-4642. 2007. 00342. x

38. Hanberry BB, He HS, Dey DC (2012) Sample sizes and model comparison metrics for species distribution models. Ecol Modell 227: 29–33. doi: 10. 1016/j. ecolmodel. 2011. 12. 001

39. Cheung WWL, Lam VWY, Sarmiento JL, Kearney K, Watson R, et al. (2010) Large-scale redistribution of maximum fisheries catch potential in the global ocean under climate change. Glob Chang Biol 16: 24–35. doi: 10. 1111/j. 1365-2486. 2009. 01995. x

40. Girvetz EH, Zganjar C, Raber GT, Maurer EP, Kareiva P, et al. (2009) Applied Climate-Change Analysis; The Climate Wizard Tool. PLoS One 4 (12) e8320. doi: 10. 1371/journal. pone. 0008320

41. Dukes JS, Mooney HA (1999) Does global change increase the success of biological invaders? Trends in Ecology and Evolution 14: 135–139. doi: 10. 1016/ s0169-5347(98)01554-7

42. Thuiller W, Richardson DM, Midgley GF (2007) Will climate change promote alien invasions? In: Biological Invasions. Nentwig W, editor. Berlin: Springer Verlag. 197–211. doi: 10. 1007/978-3-540-36920-2_12

43. Genovesi P, Scalera R, Brunel S, Roy D, Solarz W (2010) Towards an early warning and information system for invasive alien species (IAS) threatening biodiversity in Europe. Copenhagen: European Environment Agency.

44. Leprieur F, Brosse S, García-Berthou E, Oberdorff T, Olden JD, et al. (2009) Scientific uncertainty and the assessment of risks posed by non-native freshwater fishes. Fish and Fisheries 10: 88–97. doi: 10. 1111/j. 1467-2979. 2008. 00314. x

45. Genovesi P (2005) A strategy to prevent and mitigate the impacts posed by invasive alien species in Europe. Biol Invasions 6: 145–147.

46. Intergovernmental Panel on Climate Change (2007) Climate change 2007: the physical science basis. Contribution of Working Group I to the fourth assessment report of the IPCC. Cambridge: Cambridge University Press.

47. Gritti ESB, Smith B, Sykes MT (2006) Vulnerability of Mediterranean Basin ecosystems to climate change and invasion by exotic plant species. J Biogeog 33: 145–157. doi: 10. 1111/j. 1365-2699. 2005. 01377. x

48. Fodrie FJ, Heck KL, Powers SP, Graham W, Robinson KL (2009) Climate-related, decadal-scale assemblage changes of seagrass-associated fishes in the northern Gulf of Mexico. Glob Chang Biol 16: 48–59. doi: 10. 1111/j. 1365-2486. 2009. 01889. x

49. Roberts ME (Personal Communication) Mississippi Museum of Natural Science, Jackson, MS. doi: 10. 1016/b978-0-240-81475-9. 50013-x

50. Crooks JA (2005) Lag times and exotic species: the ecology and management of biological invasions in slow motion. Ecoscience 12: 316–329. doi: 10. 2980/i1195-6860-12-3-316. 1

51. Turner GF, Robinson RL (2000) Reproductive biology, mating systems and parental care. In: Beveridge MCM, McAndrews BJ, editors. Tilapias: biology and exploitation. 33–58. doi: 10. 1007/978-94-011-4008-9_2

52. Peterson MS, Slack WT, Brown-Peterson NJ, McDonald JL (2004) Reproduction in non-native environments: establishment of the Nile Tilapia *Oreochromis niloticus* in coastal Mississippi watersheds. Copeia 4: 842–849. doi: 10. 1643/ce-04-134r1

53. Whitfield AK, Blaber JM (1979) The distribution of the freshwater chichlid *Sarotherodon mossambicus* in estuarine systems. Environ Biol Fishes 4: 77–81. doi: 10. 1007/bf00005931

54. Burger J, Cooper K, Gochfeld DJ, Saliva JE, Safina C, et al. (1992) Dominance of*Tilapia mossambic*, an introduced fish species, in three Puerto Rican estuaries. Estuaries 15: 239–245. doi: 10. 2307/1352698

55. Adams AJ, Wolfe RK (2007) Occurrence and persistence of non-native *Cichlasoma urophthalmus* (Family Cichlidae) in estuarine habitats of south-west Florida (USA): environmental controls and movement patterns. Mar Freshw Res 58: 921–930. doi: 10. 1071/mf07086

56. Idelberger CF, Stafford CJ, Erickson SE (2011) Distribution and abundance of introduced fishes in Florida›s Charlotte Harbor estuary. Gulf Carrib Res 23: 13–22.

57. Lorenz OT, O›Connell MT (2011) Establishment and post-hurricane survival of the non-native Rio Grande cichlid (*Herichthys cyanoguttatus*) in the greater New Orleans metropolitan area. Southeastern Naturalist 10: 673–686. doi: 10. 1656/058. 010. 0407

58. Lopez JA (2009) The environmental history of human-induced impacts to the Lake Pontchartrain Basin in southeastern Louisiana since European settlement. J Coast Res 54: 1–11. doi: 10. 2112/si54-021. 1

59. Lane RR, Day JW, Kemp JP, Demcheck DK (2001) The 1994 experimental opening of the Bonnet Carré Spillway to divert Mississippi River water into Lake Pontchartrain, Louisiana. Ecol Eng 17: 411–422. doi: 10. 1016/s0925-8574(00)00170-1

60. Gramling C (2012) Rebuilding wetlands by managing the muddy Mississippi. Science 335: 521–521. doi: 10. 1126/science. 335. 6068. 520

61. Georgiou IY, McCorquodale JA, Shindler J, Retana AG, FitzGerald DM, et al. (2009) Impact of multiple freshwater diversions on the salinity distribution in the Ponchartrain estuary under tidal forcing. J Coast Res 54: 59–70. doi: 10. 2112/si54-008. 1

62. O'Connell MT, Shepherd TD, O'Connell MU, Myers RA (2007) Long-term declines in two apex predators, bull sharks (*Carcharhinus leucas*) and alligator gar (*Atractosteus spatula*), in Lake Pontchartrain, an oligohaline estuary in southeastern Louisiana. Estuaries Coast 30: 567–574.

63. Ianni RC (2011) Monitoring diets and growth rates of native predatory fish stocked to suppress non-native tilapia. MS Thesis, Nicholls State University, Thibodaux, Louisiana.

64. Hutchinson GE (1957) Concluding remarks. Cold Spring Harbor Symposium on Quantitative Biology 22: 414–427. doi: 10. 1101/sqb. 1957. 022. 01. 039

65. Peterson MS, Slack WT, Waggy GL, Finley J, Woodley CM, et al. (2006) Foraging in non-native environments: comparison of Nile tilapia and three co-occuring native centrarchids in invaded coastal Mississippi watersheds. Environ Biol Fish 76: 283–301. doi: 10. 1007/s10641-006-9033-4

66. Lowe-McConnell RH (2000) The role of tilapias in ecosystems. In: Tilapias: Biology and Exploitation. Beveridge MCM, McAndrew BJ, editors. Dordrecht: Kluwer Academic Publishers. 129–162.

67. Mulholland PJ, Best GR, Coutant CC, Horngerger GM, Meyer JL, et al. (1997) Effects of climate change on freshwater ecosystems of the south-eastern United States and the Gulf coast of Mexico. Hydrol Process 2: 949–970. doi: 10. 1002/(sici)1099-1085(19970630)11:8<949::aid-hyp513>3. 0. co;2-g

68. Engle VD, Summers JK, Macauley JM (1999) Dissolved oxygen conditions in northern Gulf of Mexico estuaries. Environ Monit Assess 57: 1–20. doi: 10. 1007/bf00545979

69. Bianchi TS, DiMarco SF, Cowan JH, Hetland RD, Chapman P, et al. (2010) The science of hypoxia in the northern Gulf of Mexico: A review. Sci Total Environ 408: 1471–1484. doi: 10. 1016/j. scitotenv. 2009. 11. 047

70. Rakocinski CF, Peterson MS, VanderKooy SJ, Crego GJ (1997) Biodiversity patterns of littoral tidal river fishes in the Gulf coastal plain region of Mississippi. Gulf Mex Sci 1: 2–16.

71. Ross LG (2000) Environmental physiology and energetics. In: Tilapias: Biology and Exploitation. Beveridge MCM, McAndrew BJ, editors. Dordrecht: Kluwer Academic Publishers. 89–128. doi: 10. 1007/978-94-011-4008-9_4

72. Caldeira K, Wickett ME (2003) Anthropogenic carbon and ocean pH. Science 424: 365. doi: 10. 1038/425365a

73. Feely RA, Sabine CL, Lee K, Berelson W, Kleypas J, et al. (2004) Impact of anthropogenic CO2 on the CaCO3 system in the oceans. Science 305: 363–366. doi: 10. 1126/science. 1097329

74. Ricciardi A, Kipp R (2008) Predicting the number of ecologically harmful species in an aquatic system. Divers Distrib 14: 374–380. doi: 10. 1111/j. 1472-4642. 2007. 00451. x

75. Cucherousset J, Olden JD (2011) Ecological impacts of non-native freshwater fishes. Fisheries 36: 215–230. doi: 10. 1080/03632415. 2011. 574578

76. Starling FLRM, Lazzaro X, Cavalcanti C, Moreira R (2002) Contribution of omnivorous tilapia to eutrophication of a shallow tropical reservoir: evidence from a fish kill. Freshw Biol 47: 2443–2452. doi: 10. 1046/j. 1365-2427. 2002. 01013. x

77. Figueredo CC, Giani A (2005) Ecological interactions between Nile tilapia (*Oreochromis niloticus*) and the phytoplanktonic community of the Furnas Reservoir (Brazil). Freshw Biol 50: 1391–1403. doi: 10. 1111/j. 1365-2427. 2005. 01407. x

78. Coblentz BE (1990) Exotic organisms: a dilemma for conservation biology. Cons Biol 4: 261–265. doi: 10. 1111/j. 1523-1739. 1990. tb00287. x

79. McCormick MI (2006) Mothers matter: crowding leads to stressed mothers and smaller offspring in marine fish. Ecology 87: 1104–1109. doi: 10. 1890/0012-9658(2006)87[1104:mmclts]2. 0. co;2

80. Gagliano M, McCormick MI (2007) Maternal condition influences phenotypic selection on offspring. J Anim Ecol 76: 174–182 doi:10. 1111/j. 1365-2656. 2006. 01187. x. doi: 10. 1111/j. 1365-2656. 2006. 01187. x

81. Albert A, Anderson JA (1984) On the existence of maximum likelihood estimates in logistic regression models. Biometrika 71: 1–10. doi: 10. 1093/biomet/71. 1. 1

82. Firth D (1993) Bias reduction of maximum likelihood estimates. Biometrika 80: 27–38. doi: 10. 1093/biomet/80. 1. 27

83. Heinze G, Schemper M (2002) A solution to the problem of separation in logistic regression. Stat Med 21: 2409–2419 Doi: 10. 10002/sim. 1047. doi: 10. 1002/sim. 1047

84. Bull SB, Lewinger JB, Lee SSF (2007) Confidence intervals for multinomial logistic regression in sparse data. Stat Med 26: 903–918. doi: 10. 1002/sim. 2518

85. Tabachnick BG, Fidell LS (2001) Using Multivariate Statistics. Boston: Allyn and Bacon.

86. R Development Core Team (2008) R: A language and environment for statistical computing. Vienna, Austria: R Foundation for Statistical Computing.

87. Griffith DA, Layne LJ(1999) A casebook for spatial statistical analyses: a compliation of analyses of different thematic data sets. Oxford, UK: Oxford University Press.

88. Wester R, Oliver MA (2007) Geostatistics for Environmental Scientists. Second Edition. John Wiley & Sons, LTD. West Sussex, England.

CHAPTER 11 REPRODUCTIVE TRAITS IN BLACK- CHINNED TILAPIA FEMALES

Moussa Guèye, Mbaye Tine, Justin Kantoussan, Papa Ndiaye, Omar Thiom Thiaw, JeanJacques Albaret

INTRODUCTION

Hypersaline conditions are increasingly observed in estuarine ecosystems as a consequence of global climate changes. This is especially the case for some Sahelian estuaries in West Africa, where the reduced freshwater influx and water evaporation have resulted in an overall increase of salinity. This region has experienced a succession of drought periods since the 1960's, the impact of which was higher in the estuaries with no or small river connection. In such estuaries, the freshwater inputs which essentially come from groundwater discharges and rainfall are largely excided by the loss via evaporation [1], [2]. This has resulted in an inversion of the salinity gradient in some estuaries such as the Sine Saloum River (Senegal), with salinities increasing from downstream to upstream where they can exceed 130 psu [3](Figure 1). The salinity levels in these estuaries also change very significantly between the dry season and the rainy season [3], [4] with amplitudes that can reach 70 psu. Such spatiotemporal variations of salinity constitute serious abiotic obstacles that could deeply impact normal biological function of a species such as growth and reproduction.

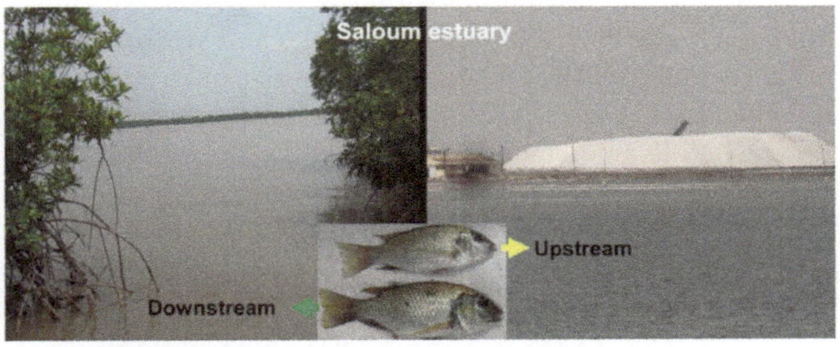

Figure 1. Dissimilarities between downstream and upstream parts of the Saloum estuary (Senegal).

The salinity gradient in the estuary is inverted, with lower salinity at the estuary mouth (downstream) than in the upper reaches. The estuary mouth is characterised by a discontinuous mangrove whereas in the upstream part, the mangrove has disappeared because of very high salinities and destructive human activities. Fish inhabiting the upstream part (hypersaline zones) of the estuary have impaired growth performance compared to those living in the downstream zone. Fish analysed in this study were collection in upstream of the Saloum estuary where the salinity is higher.

Some fish populations have developed adaptive strategies including the regulation of growth and reproduction to cope with these unusually stressful salinity conditions [5], [6], [7], [8]. The reproductive strategy of a fish species is the overall

Citation: Guèye M, Tine M, Kantoussan J, Ndiaye P, Thiaw OT, Albaret J-J (2012) Reproductive Traits in Black- Chinned Tilapia Females doi:10.1371/journal.pone.0029464.

pattern of reproduction common to all individuals and covers a range of life-history traits including the age and size at first sexual maturity, gonadal development, fecundity and gamete size. Individual fish can, however, develop alternative reproductive tactics which are variations with respect to the normal reproduction pattern of the species, to respond to fluctuation in the environment. Both, the overall strategy and tactical variations are adaptive and aim at ensuring the survival of the species in specific environmental conditions [9]. The reproductive tactics or compensations imply that, depending on the environmental conditions, a female can choose between allocating energy to either survival or somatic growth and reproduction. The compensation most commonly involves modifications in fecundity and life span [10]. Roff [11] proposed two models for energy allocation during reproductive compensation: (i) maintenance of body weight and adjustment of gamete production, and (ii) maintenance of a constant number of eggs at the expense of somatic tissues. The first type of compensation occurs when the stress compromises the development of the ovary during vitellogenesis, which can lead to atresia of the eggs and subsequent resorption.

The tilapias are widely distributed in tropical regions where they have colonised a wide range of water bodies as an introduced or native species. Previous studies have demonstrated that the fecundity and egg size in tilapia can be influenced by the food quantity [12], the fishing pressure exerted on the fish [13], [14], but also by environmental conditions [13], [15], [16],[17], [18]. Tilapias are capable of allocating most of their energy reserves to reproduction to the detriment of growth if environmental conditions change [19], [20]. In natural environments a tilapia species can, therefore, show a great variability in its reproductive traits [20], [21]. Among the tilapias, the genus *Sarotherodon* including the species *Sarotherodon melanotheron*and *Sarotherodon galileus* is widely distributed in West-African coastal, estuarine and lagoon ecosystems. The black-chinned tilapia *S. melanotheron* is particularly notable for its ability to tolerate a wide range of environmental salinities [22], [23], [24], [25]. The species is also known to be widely tolerant to temperature variations [3], [4], [26], [27], and dissolved oxygen conditions at the scales of both daily and seasonal fluctuations. Although *S. melanotheron* can tolerate a broad range of salinities, fish inhabiting extremely hypersaline waters of the Sine Saloum estuary exhibited changes in fitness-related traits. It has been demonstrated that black-chinned tilapias living in hypersaline zones of the Saloum (Figure 1) had impaired growth performance and precocious reproduction [4], which have been also reported in another estuarine species, *Ethmalosa fimbriata* [27], [28]. These phenotypic differences have been interpreted as indicative of hypersaline stress.

This study attempted to better understand the mechanisms of salinity adaptation in *S. melanotheron*, in particular the reproductive strategies and/or tactics which may underlie its exceptional euryhalinity. Our study focuses on variation of some reproduction parameters (reproductive period, fecundity and egg size) in natural populations of *S. melanotheron* from three ecosystems with different salinities: Guiers lake (0 psu), Hann bay (37 psu) and upper part of the Saloum estuary (66–127 psu). Reproduction parameters were quantified over two consecutive sampling years. The condition factor of the fish was measured in parallel and taken as a proxy of physiological status. Condition factor is a morphometric index frequently used to evaluate the wellbeing or physiological status of fish, based on the principle that heavier individuals of a particular length which have a higher body weight, are in better "condition".

RESULTS

Correlations between fecundity and body weight

Table 1 shows the relationship between fecundity and the fish body weight. There was a significant and positive correlation between absolute fecundity and fish body weight in the all ecosystems ($P<0.01$). The correlation was significantly stronger ($P<0.05$) in Guiers lake ($R^2=0.68$; $P<0.001$) compared to Hann bay ($R^2=0.33$; $P<0.001$) and Saloum estuary ($R^2=0.49$;$P<0.001$).

Table 1. Relationship between fecundity and total body weight

Relation	Locations	N	a	b	r^2	F-statistic	*df*	*p*
Fecundity-Body weight	Guiers lake	20	1.519	212.174	0.68	37.8	18	***
	Hann bay	53	1.393	673.046	0.33	25.39	51	***
	Saloum	84	4.639	40.069	0.49	77.89	82	***

N = number of observations; *a* and *b* = parameters of the linear regression y = ax+b; r^2 = coefficient of regression; *df* = number of freedom degree; *p* = probability level; significant levels: * = 0.05; ** = 0.01; *** = 0.001.
doi:10.1371/journal.pone.0029464.t001

Fish condition factor

The average condition factor considering both years together was significantly different between locations or salinities (Figure 2). The best condition factor was observed in fish collected at the freshwater (Guiers lake) and seawater (Hann bay) locations. Fish caught in the Saloum estuary, the most saline location, had the lowest condition, significantly lower than the other sites (ANOVA, $P<0.05$). No significant difference in condition factor was found between Guiers lake and Hann bay (Figure 2). The average condition factor considering the years separately (Table 2) showed similar values in freshwater and seawater and lower values in hypersaline water. There was no significant difference in average condition factor between 2003 and 2004 in each location (Table 2).

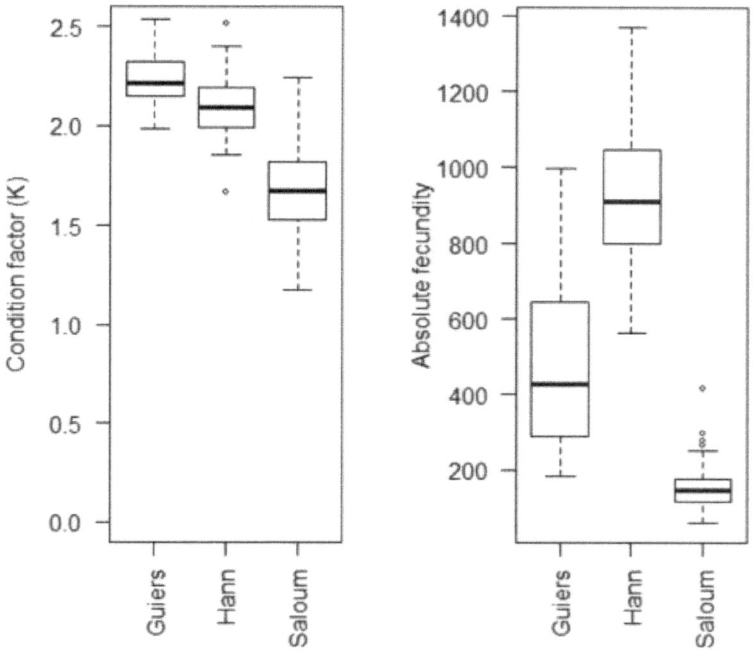

Figure 2. Condition factor and absolute fecundity of the black-chinned tilapia *S. melanotheron* **from three wild populations.**

The condition factor and absolute fecundity were calculated considering samples collected in both years (2003 and 2004) together. Data are illustrated in box plots that contained the median (horizontal line) as well as the 25th and 75th percentiles (bottom and top edges of the boxes).

Table 2. Mean ± (SD) of condition factor and reproductive parameters of the black-chinned tilapia *S. melanotheron* **from three wild populations adapted to dif**ferent environmental salinities

Ecosystem	Guiers lake		Hann bay		Saloum	
Year	2003	2004	2003	2004	2003	2004
Condition factor	2.34±0.15[a]	2.28±0.19[a]	2.08±0.14[a]	2.13±0.12[a]	1.60±0.14[a]	1.86±0.14[a]
Absolute fecundity	482±202[a]	430±221[a]	868±238[a]	936±276[a]	130±52[a]	168±58[a]
Relative fecundity	2706±215[a]	3241±1235[a]	5843±1520[a]	5308±1824[a]	6899±1884[a]	6111±1524[a]
Oocyte weight (g)	1.58±0.03[a]	1.01±0.31[a]	0.75±0.26[a]	0.80±0.30[a]	0.55±0.23[a]	0.66±0.20[a]
Oocyte diameter (mm)	2.79±0.73[a]	3.14±0.48[a]	2.92±0.45[a]	3.07±0.45[a]	2.67±0.40[a]	2.75±0.32[a]
Egg weight (g)	6.23±4.80[a]	4.89±3.09[a]	7.39±2.87[a]	8.16±3.51[a]	0.84±0.45[a]	1.24±0.39[a]
L_{50}±SD (cm)	13.83±0.28[a]	14.66±0.30[a]	11.08±0.37[a]	11.71±0.33[a]	7.84±0.24[a]	7.43±0.44[a]

For each variable, same superscript in the column indicates that there is no significant difference ($P\geq0.05$) among years (2003 and 2004) in the same ecosystem.
doi:10.1371/journal.pone.0029464.t002

Reproductive cycle and size at first maturity

The reproduction period of *S. melanotheron* in Guiers lake extended from January to August, followed by a period of sexual rest that occurred from October to December (Figure 3A). The peak of sexual activity was recorded in June in 2003 and in March in 2004, suggesting variations in the reproductive cycle over the years. In Hann bay, the reproduction period of the species extended from January to October with a period of sexual rest in November and December (Figure 3B). There were also some differences in the peak of sexual activity, which was observed in August in 2003 and in May in 2004. In each of these ecosystems, the breeding season covered both dry and rainy seasons but the largest part of the reproductive cycle occurred during the dry season. Upstream of the Saloum estuary, the reproductive activity was intense in rainy season (from June to August) when the salinity in the estuary was lower (Figure 3C). Fish sampled in dry season (October to May) was characterized by low gonadosomatic index (*GSI*) and an absence of sexual stages 4, 5 and 6.

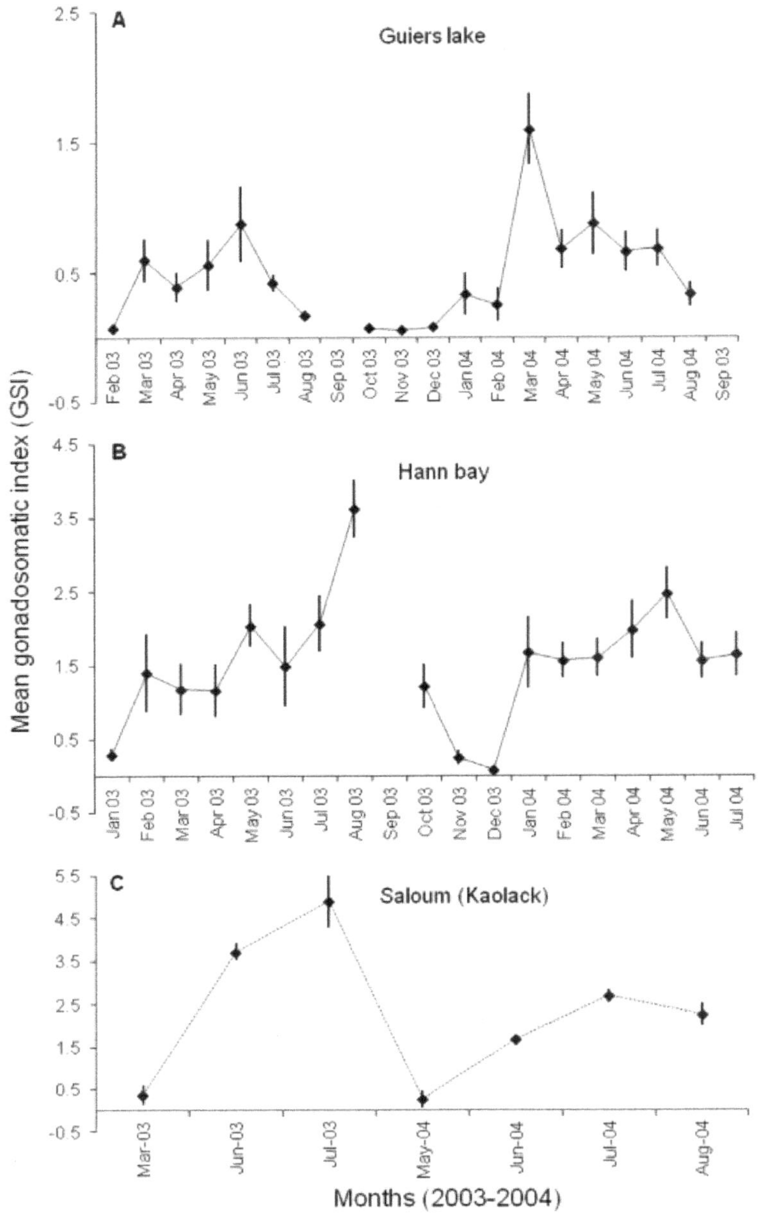

Figure 3. Reproductive cycle of *S. melanotheron* **in** Guiers lake, Hann bay and Saloum estuary.

Data are expressed as the mean GSI ± SD. The space in the curves indicates the lack of samples in September. The reproductive cycle of the species in the Saloum estuary was previously determined by Panfili *et al.* [4]. Therefore, fish were sampled only during the spawning season to confirm the breeding periods determined by these authors.

Table 2 indicates the size at first sexual maturity of black-chinned tilapia females (L_{50}) during the reproductive periods in 2003 and 2004. The L_{50} were higher at Guiers lake location and lower upstream of the Saloum estuary. The L_{50} at Hann bay location was intermediate between those observed in the two other locations. In each ecosystem, the L_{50} did not differ significantly between 2003 and 2004.

Fecundity in black-chinned Tilapia Populations

The average absolute fecundity varied significantly between sites and salinities. Fish sampled in Hann bay had significantly higher absolute fecundity (Kruskal Wallis test; $P<0.001$) than those sampled in the upstream of the Saloum estuary or in Guiers lake (Figure 2). The lowest absolute fecundity was observed in fish sampled in the Saloum estuary. The comparison of absolute fecundity between Guiers lake and the Saloum estuary revealed significantly higher values at Guiers lake. The relative fecundity exhibited a different pattern between locations, being highest at the most saline station of the Saloum estuary (Kruskal Wallis test; $P<0.001$), and lowest at the least saline location, Guiers lake (Figure 4). It did not, however, reveal any significant difference between the Saloum estuary and Hann bay. The average absolute and relative fecundities did not reveal significant differences between 2003 and 2004 in all the ecosystems (Table 2). The relative fecundity was higher in rainy season by comparison to dry season in the Saloum estuary whereas this relationship was not found for the other two locations (Figure 5).

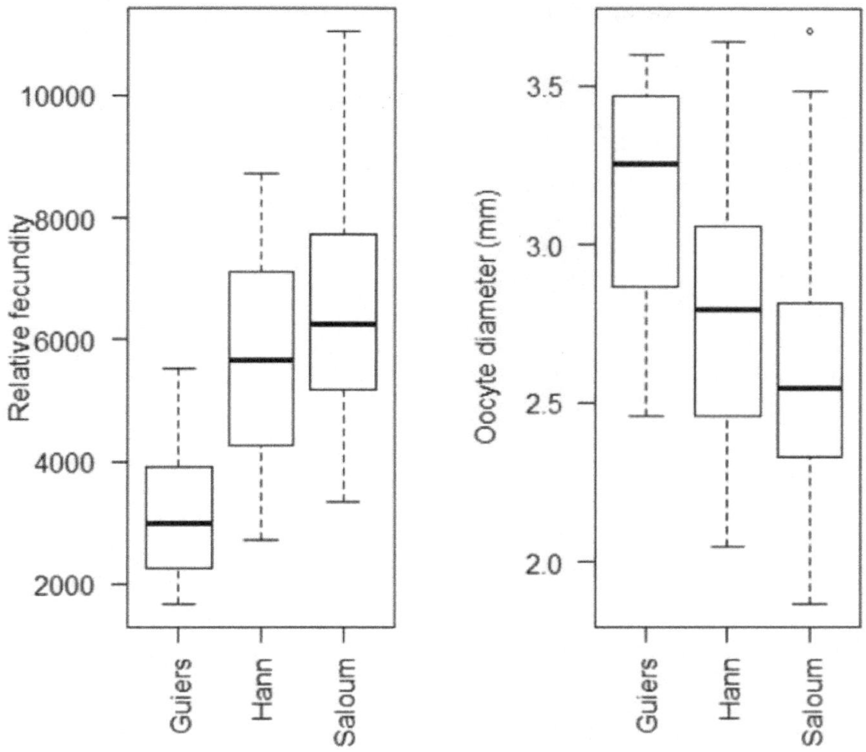

Figure 4. Relative fecundity and oocyte diameter of the black-chinned tilapia *S. melanotheron* **from three wild populations.**

These reproductive parameters were calculated considering samples collected in both years (2003 and 2004) together. Data are illustrated in box plots that contained the median (horizontal line) as well as the 25th and 75th percentiles (bottom and top edges of the boxes).

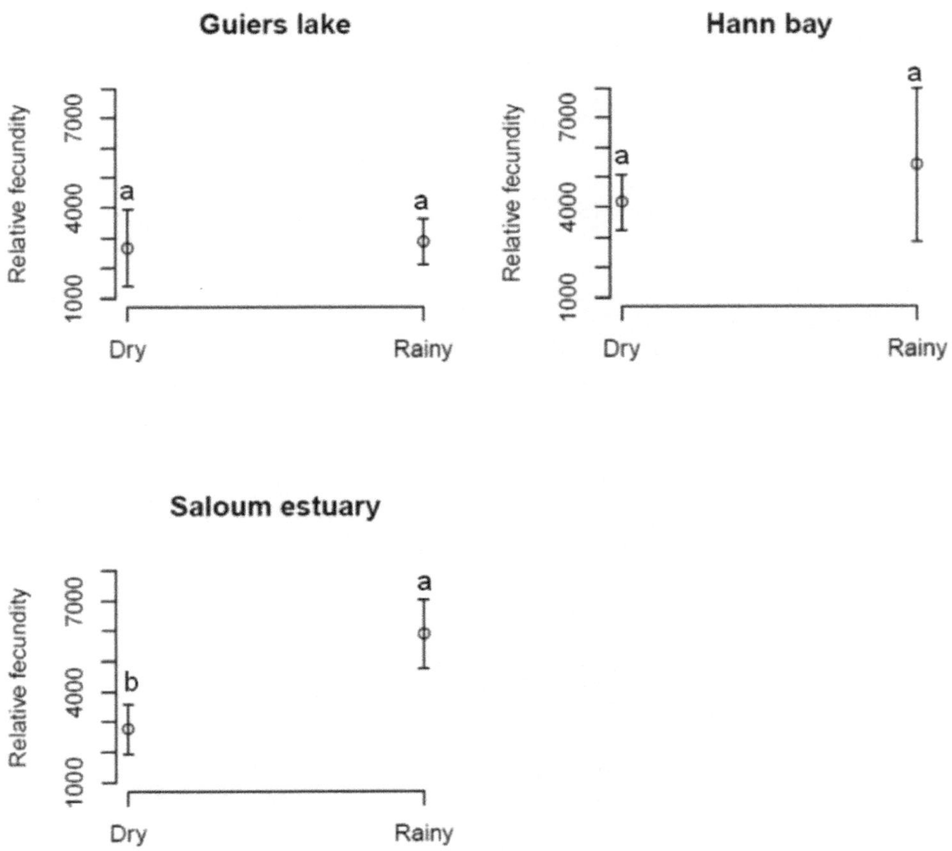

Figure 5. Seasonal variation in relative fecundity in the black-chinned tilapia *S. melanotheron*.

Relative fecundity was calculated on females from three wild populations sampled in both the rainy and the dry season. Data are expressed as the mean ± SD.

Oocyte size and total weight of eggs

Fish sampled in Guiers lake had significantly higher oocyte diameter than those sampled in the Saloum estuary (Tukey test; $P<0.05$) (Figure 4). At Hann bay where the salinity was intermediate, there were intermediate oocyte diameters. The oocyte weight (Figure 6) exhibited a similar pattern between locations, being highest at the Guiers lake, but lowest at the Saloum estuary (Kruskal Wallis test; $P<0.05$). The average oocyte weight at Hann bay location was intermediate to those observed at Guiers lake and the Saloum estuary. By contrast, the total weight of eggs exhibited a different pattern between locations (Figure 6). It is significantly higher at Hann bay than in Guiers lake and Saloum estuary. The average total egg weight was significantly lower at the most saline location. The fish sampled in Guiers lake have total egg weight intermediate to those from the two other locations. The oocyte size (oocyte diameter and weight) and total weight of eggs did not significantly differ between 2003 and 2004 in each ecosystem (Table 2). There was also no significant difference of these parameters between the rainy and the dry season in all locations.

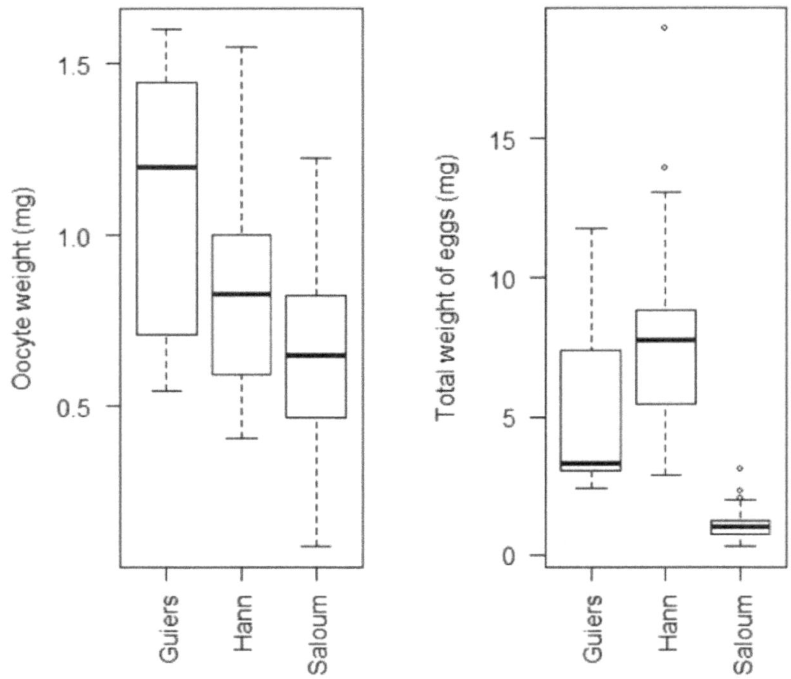

Figure 6. Oocyte weight and total weight of eggs of the black-chinned tilapia *S. melanotheron*.

Oocyte weight and total weight of eggs were calculated considering samples collected in both years (2003 and 2004) together. Data are illustrated in box plots that contained the median (horizontal line) as well as the 25th and 75th percentiles (bottom and top edges of the boxes).

DISCUSSION

The results demonstrate that the marine *S. melanotheron* population (Hann bay) has significant high absolute fecundity and spawning weight (total weight of eggs) by comparison to those from freshwater or hypersaline water environments. Fish relative fecundity correlates significantly with their total body weight in each environment. The relative fecundity in the Saloum estuary was higher in rainy season compared to the dry season whereas there were no significant seasonal variations in the two other locations. The data also indicate an inverse relationship between relative fecundity and oocyte size in fish living in freshwater and hypersaline water, suggesting different reproductive strategies at these two extreme salinities.

The differences in reproductive traits among populations may reflect environmentally induced phenotypic plasticity and/or genetic variations. In this study we do not have measures of genetic differentiation, but previous studies have revealed high genetic differentiation levels in*S. melanotheron*, even at microgeographical scales [29], [30], suggesting no or very limited gene flow among populations. This high population genetic structure has been assumed to be directly linked to the low larval dispersion ability of the species, due to a mouth-brooding reproductive behaviour. Analyses of strontium/calcium ratios in otoliths also support the sedentary character of *S. melanotheron* [26], which can significantly contribute to population genetic isolation. It can be assumed, therefore, that the *S. melanotheron* populations analysed in this study are genetically different. However, the within-population (upstream Saloum) variability in relative fecundity between the dry and rainy seasons strongly suggests the variations in reproductive traits among *S. melanotheron* populations reflect environmental differences. This conclusion is consistent with those of previous studies [4], [27], [31] that differences in life-history traits of *S. melanotheron* in the Gambia and Senegalese estuaries are related to the environmental salinity rather than genetic differences among populations.

Salinity, temperature and dissolved oxygen are among the abiotic factors which are likely to impact fish fecundity and egg size [13], [15], [16], [17], [18]. The water temperature measured in our study area varied slightly (24.9 to 26.7°C; Table 3) between locations and seasons as previously reported by Simier *et al.* [32] in the Saloum estuary. The amount of dissolved oxygen (DO) is expected to be limiting in the extremely saline environments, because the presence of salt limits the amounts of oxygen that can be dissolved in water. However, previous studies have demonstrated that the level of saturation in DO (81%) of the Saloum water is satisfying [33]. This observation is consistent with the conclusion of other studies that the DO is not a limiting resource in Senegalese estuaries [34]. Therefore, the water salinity is the most constraining environmental factor for *S. melanotheron* in our study area, and seems to be predominant over the other environmental factors.

Table 3. Sample characteristics of the black-chinned tilapia *S. melanotheron* **and some historical environmental data** [48], [49], [50] **of the study area.**

Variable	Guiers lake	Hann bay	Saloum
Sample size	901	574	862
Min fork length (mm)	48	85	49
Max fork length (mm)	283	292	164
Mean fork length (mm)	191.50±43.93	205.85±30.05	110.47±12.76
Salinity (psu)	0	37	66–127
Water temperature (°C)	24.9	25	26.7
Oxygen (mg l^{-1})	6.70	4.84	-
Transparency (m)	0.79	-	2.5
Conductivity (µs cm^{-1})	174	-	71
Chrorophyll-*a* (µg l^{-1})	23.8	16.83	2.3

Reproductive cycle of *S. melanotheron*

The black-chinned tilapia *S. melanotheron* exhibited reproductive cycles with an extended intense reproduction period (between 9 and 10 months) in Hann bay and Guiers lake, and a short intense reproduction period (3 months) in the upstream part of the Saloum estuary. These results are congruent with those reported by Panfili *et al.* [4] in the Saloum and the Gambia estuaries. However, in contrast to our results, Legendre and Ecoutin [17] have observed that the reproductive cycle of *S. melanotheron* in Ebrie lagoon (Ivory Coast) is continuous throughout the year with an intense reproductive activity in the dry season. Moreover, Faunce[35] reported that the breeding season of same species in mangrove ecosystems of Florida State (USA) characterised by a semi-tropical climate, extends from April to October. All together, these results indicate that the breeding in *S. melanotheron* is dependent on local environmental conditions. Our results also showed smaller size at first maturity in the upstream part of the Saloum estuary, which can be attributed to the extremely high salinities in this area.

Among-Population Variation in Fecundity and Condition

The absolute fecundity of the black-chinned tilapia was higher in fish adapted to seawater (37 psu) compared to those living in freshwater (0 psu) or hypersaline water (66–127 psu). Likewise, the spawn weight (i.e. total weight of eggs), indicative of reproductive effort, was higher in seawater. Taken together, these results suggest higher reproductive activity in seawater, the optimal salinity for this species [4]. *S. melanotheron* is a marine species of tilapia, which may therefore have evolved mechanisms adapted to have their minimal energetic requirements for maintenance of hydromineral balance in seawater. Such lower energetic cost for osmoregulation would allow supplying more energy to other biological functions, which is evidenced by the best growth performances of *S. melanotheron* in seawater [4], [36]. This interpretation is in accordance with the low expression levels of the gene coding for the Na^+-K^+-ATPase 1α pump in fish inhabiting Hann bay compared to those living in Guiers lake and Saloum estuary [37]. Indeed, the Na^+-K^+-ATPase is a membrane protein which maintains ion gradients required for cell homeostasis. Therefore, its activity in the gills is related to active ion secretion or absorption in hyper- and hypo-osmotic conditions, respectively. These findings are supported by the poorer condition factors recorded in hypersaline water in this study, although the absence of differences between freshwater and seawater suggest that there may be other factors acting on the fish condition. The specific reasons why fish condition is higher at Guiers lake compared to the Saloum estuary, and did not differ with that of Hann bay are unclear. One possible explanation is that fish in Guiers lake do not need to respond to sudden fluctuations in salinity, which are energetically costly contrary to fish inhabiting the Saloum estuary where the salinity is very unstable. A more likely explanation is that the pollutant loads that characterise the Hann bay negatively impact the fish condition. They may thus interfere with the effects of the salinity and lead to undetectable differences in condition factor between this bay and Guiers lake.

Differences in food resources at sites and/or efficiency at foraging and prey availability can also influence reproductive parameters and fish condition. Food resources could be limited in the environments with highest salinities. Individuals in these areas will invest more in maintenance of hydromineral balance than their counterparts living in seawater with more energy available for other biological functions. While an impact of food resources on fish reproduction and condition cannot be excluded, it is clear that food availability is indirectly related to the salinity which is the main environmental constraint in the Saloum estuary. In fact, the mangrove and its associated fauna are considered as major food resource exploited by fishes in the estuaries [38], [39], [40], [41]. Upstream of the Saloum estuary, the mangrove has disappeared (Figure 1) as a result of very high salinities and destructive human exploitation [42], which may significantly impact the availability of food resources exploited by fish species in these areas. Furthermore, concomitant decreases of growth rate and condition factor with increase of environmental salinity were observed in the Saloum estuary [4], [31],[36], suggesting an indirect impact of hypersaline conditions on food availability.

The highest fecundities recorded in rainy season, when the salinity in the Saloum estuary was lower by comparison to dry season (higher salinity), are an evidence for an impact of environmental salinity on the reproduction in *S. melanotheron*. The reduced salinity in rainy season would lower energy requirement for osmoregulation and, therefore, allow better energy supply to the reproduction. This reproductive tactic of producing large amounts of eggs in the rainy season when the salinity conditions in the estuary are more favourable would ensure a higher survival chance to the larvae and, therefore, lead to a higher larval recruitment.

Reproductive Strategies in *S. Melanotheron*

Two reproductive patterns (Figure 4) of *S. melanotheron* can be distinguished from this study: (*i*) low fecundity and a large egg size observed at Guiers lake, and (*ii*) high fecundity with small egg size in Saloum population. Fish collected at Hann bay have relative fecundity and oocyte sizes intermediate to those from the other two sites. It has been demonstrated in fishes that, depending on environmental conditions, females can produce either large number of small eggs or small number of large eggs, due to their flexibility in allocating the energetic reserves between these two modes of reproduction. Many studies have thus reported the existence of trade-off between fecundity and egg size in fishes [43], [44] including tilapia species [13], [15],[16], [17], [18]. In the current study, the high fecundity and small egg size observed in the *S. melanotheron* populations inhabiting the upper Saloum estuary could be an adaptive strategy to extremely high salinities that prevailed in this area. Such a high fecundity would increase the rate of larvae survival and, therefore, enable the species to thrive despite the extremely constraining salinity conditions. Furthermore, small eggs in hypersaline water could decrease the surface exchange between intracellular and ambient mediums and, therefore, reduce the energy expenditure for the maintenance of hydromineral balance (S. Gilles, IRD unpublished data). The interpretation can be, however, different at Guiers lake. Indeed, although freshwater represents an extreme salinity for *S. melanotheron*, the large egg size is in discordance with decrease of hydromineral surface exchange. Low fecundity, large egg and size at first maturity have been suggested to be characteristic of population facing stable environmental conditions. This is because in these habitats, fishes are subjected to high inter-specific competition, which can limit the amount of food available for consumption and lead to high juvenile mortality. In such situation, fishes will maintain egg quality at the expense of the number, and the egg production can therefore vary with individual size. It has been also reported that eggs of large size with a relatively small number is an adaptation to poor food supply for the juveniles [45]. Therefore, in addition to the extreme salinity conditions, the larger egg and lower relative fecundity associated with the larger size at first maturity in Guiers lake where the environmental conditions are relatively stable, could reflect limited food resources.

This study provides new insights into the influence of environmental salinity on life history-traits in *S. melanotheron*, and the findings are somewhat similar to those previously reported by Panfili *et al.* [4]. These latter authors did not, however, analyse fish living permanently in seawater or freshwater, and consequently there is no evidence for difference in life history-traits with fish from hypersaline waters. By including Hann bay and Guiers lake populations to our study, we do not only cover the maximum salinity gradient where *S. melanotheron* is encountered, but we also provide the opportunity of comparing reproductive parameters among populations inhabiting lower and higher salinities and those living at salinities close to iso-osmotic conditions. This has enabled to better highlight the relationship between life history-traits and osmoregulation and assume that these parameters might be in trade-off in *S. melanotheron*. Such a trade-off would occur only if both traits (reproductive parameters and metabolic efforts needed to maintain osmotic homeostasis) are phenotypically plastic in response to salinity variations. Within-population variations in relative fecundity between the dry and rainy seasons in upstream of the Saloum estuary suggest that the latter are environmentally induced. However, further investigations by common garden experiments on variations of life-history traits and osmotic stress indicators in relation to the salinity are needed to establish beyond doubt whether these parameters are plastic.

MATERIALS AND METHODS

Study Area

With a catchment area of 300 km^2 at high water, 60 km long and 77 km wide, Guiers lake is located in north Senegal at 18°30W longitude between 15°30 and 16°N latitude (Figure 7). At low water, its catchment area decreases considerably to become 200 km^2 with a depth from 2 to 3.5 m north and 1 to 1.5 m south. The flat surrounding land, the porous nature of the soils and the low local rainfall greatly reduce the amount of runoff into the lake from its own catchment (not more than 10%). For its water supply, Guiers lake depends mainly on the Senegal River. Before the construction of the dam in Senegal River and the bridge at Richard-Toll in 1947, the Taouey channel brought water from the river to the lake during the flooding and partially emptied it into the river at low water [46].

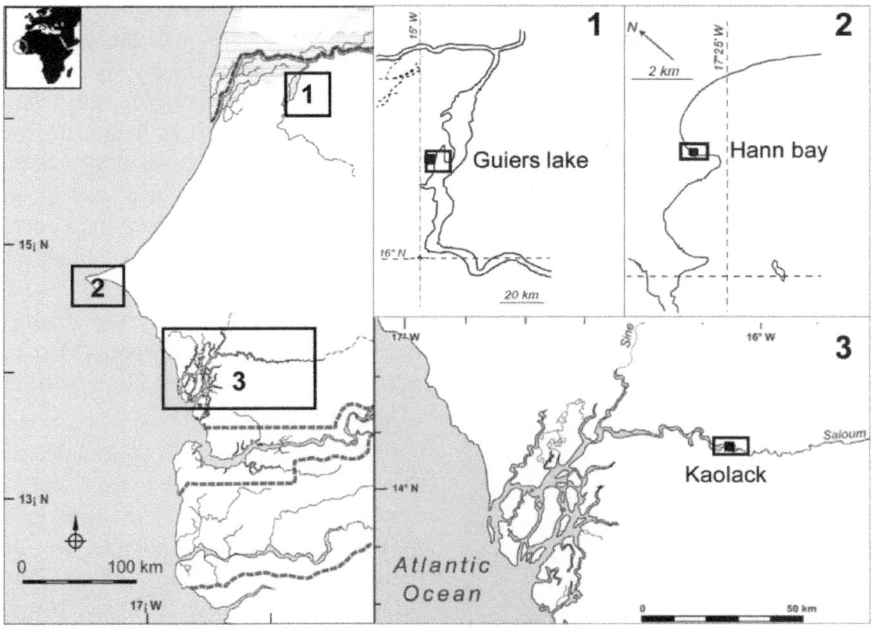

Figure 7. Sampling locations of the black-chinned tilapia *S. melanotheron*.

Fish were collected in 2003 and 2004 from three locations in Senegal with different salinities: Guiers lake (0 psu), Hann bay (37 psu) and upstream of the Saloum estuary (Kaolack; 66–127 psu).

Hann bay (Figure 7) is about 20 km long and is situated on the Atlantic coast near the Cap Vert peninsula (Dakar, Senegal). It is a site of an important small-scale fishery and is one of the main centers for fishery landings in Senegal. For several years, Hann bay has been affected by eutrophication with a pronounced increase in its nutrient concentrations. Periodic upwellings of cold and nutrient-rich waters occur quite frequently. The combination of this natural nutrient-enrichment phenomenon and the current pollution explains the episodes of blooms of macroscopic algae (Ulvales) and natural phytoplankton [47]. Despite this pollution within the bay, a large quantity of tilapia (mainly *S. m. heudelotii*) is landed every day by the purse seine and especially beach seine fishery.

The Sine Saloum estuary is located between 13°55′ and 14°10′ North and 16°03′ and 16°50′ west and comprises three main branches that are from north to south, the Saloum, the Diomboss and the Bandiala (Figure 7). The estuary drains a catchment of 29,720 km^2 with a very low slope. The Sine Saloum region is characterized by an extended dry season from November to June, and a short rainy season from July to October. The estuary does not receive freshwater inputs except by precipitation. Therefore, salinity levels change considerably between the rainy season and the dry season [3], [4]. In rainy season, the salinity in the estuary is very unstable and decreases considerable due to the input of freshwater by precipitation whereas it increases during the dry season because of intense evaporation. The combined effect of limited freshwater inputs and intense evaporation has resulted in an overall increase of salinity levels and an inversion of salinity gradient (Figure 1). No specific permits were required to perform experiments in these areas.

Sampling Design

The samples used in this study were collected in accordance with good animal practice as outlined by French Research Institute for Exploitation of the Sea (IFREMER) in a training course on how to handle fish and promote their welfare under experimental conditions. IFREMER do not approve or give a permit for studies of wild populations of fish but only provide a code of conduct to follow to minimize the suffering in experiments involving fish. Study approval by another academic ethic committee (permit number or approval ID) was not necessary as all procedures carried out with the black-chinned tilapia in this study fish are conformed to IFREMER recommendations.

Three natural populations of *S. melanotheron* were sampled in Senegal between January 2003 and August 2004. Two sampling locations, Guiers lake (freshwater) and Hann bay (seawater) do not particularly undergo salinity variations throughout the year (Figure 7). The other sampling location, Kaolack (hypersaline water) located upstream of the Saloum estuary experiences considerable seasonal salinity variations (Figure 7). The reproductive cycle of *S. melanotheron* has never been studied at Guiers lake and Hann bay, therefore fish were monthly sampled in these ecosystems. By contrast, in the Saloum estuary where the reproductive cycle of the species was previously determined by Panfili *et al.* [4], fish were sampled only during the spawning season to confirm the breeding periods determined by these authors. For each location, the salinity and temperature (Table 3) were measured *in situ* with a refractometer (ATAGO) and a thermometer, respectively. Fish sampling was carried out by local fishermen using beach seine net or castnet with small mesh size. All fish were killed by anaesthetization with a lethal dose of 2-phenoxyethanol, and then preserved in 95% ethanol until processing by dissection. In the laboratory, fish were measured (fork length, FL, in mm) and weighed (total mass, W, in g). They were then sexed and the stage of gonad maturity was recorded according to Legendre and Ecoutin [17]. Briefly, the stage 1 corresponds to immature individuals, the stage 2 to beginning of maturation and stage 3 to mature individuals. Stage 4 corresponds to females ready to reproduce, stage 5 to ripe females and stage 6 to post-spawning individuals. The gonads were extracted, weighted and then preserved in 95% ethanol.

Fish Condition Factor

The condition factor (K) is a morphometric index frequently used as proxy to evaluate physiological status of fish based on the principle that individuals of a given length, exhibiting higher weight are in better condition. Assuming this relationship, inter population variation of this index has been used to investigate salinity impacts. The condition factor can be influenced by differences in sexual stage. For this reason, we have subtracted the gonad weight from the total body weight to avoid possible bias related to sexual maturity stage. The condition factor was calculated using the remaining weight of samples collected in both years together, but also by considering the two years separately. The condition factor was calculated using the standard formula:

$$K = \frac{W}{FL^3} * 10^5$$

where W=body mass without gonads in g and LF=fork length in mm

Gonadosomatic index and size at First Maturity

The gonads used for the analysis of fecundity and, the oocyte diameter and weight should be in similar stages of vitellogenic development to avoid any bias in the comparison of egg sizes, which means they should have completed their growth [21]. This implies that the gonadosomatic index (GSI) values should not be positively correlated with the oocyte diameter or weight. Therefore, we classified the gonads into GSI classes of 0.5% for each ecosystem and then determined the GSI threshold above which the oocyte weight and diameter are no longer increasing. This threshold was reached at GSI≥4% in Hann bay and Saloum and at 3% in Guiers lake. The GSI was calculated using the standard formula:

$$GSI = \frac{W_g}{W_b} * 100$$

where W_g=gonad weight; W_b=total body weight.

Average size at first maturity (L_{50}) was defined as the size (the fork Length) at which 50% of individuals in the population reached sexual maturity during the reproduction period. The L_{50} was determined by modelling the proportion of mature individuals according to their length class for different populations using the R software (http://www.r-project.org). The L_{50} was estimated using the logistic function expressed by the standard formula:

$$M = \frac{1}{1 + e^{[-A(L - L_{50})]}}$$

where M=percentage of mature females by length class, L=central value of the length class and A is the constant of the model.

Fecundity and the egg size

The absolute fecundity is the number of oocytes likely to be released at the next spawning whereas the relative fecundity is the fecundity per kilogram of fish. In *S. melanotheron*, the oocyte distribution in the ovary is multimodal and stage 4 females are more likely to lay oocytes with the highest modal diameter. Therefore, only the stage 4 gonads from females were used to estimate the fecundity. All oocytes in samples were counted manually because the oocytes of *S. melanotheron* are large and easy to manipulate. In upstream of the Saloum estuary, the gonad contained a low number of oocytes, therefore, the whole content was counted. By contrast, at Hann bay and Guiers lake where female gonads contained a large number of oocytes, only a sample of about 50% of the ovary was weighed and counted. The total oocyte count of the gonad was then determined by reporting this weight to the total weight of the gonad.

To determine the oocyte diameter and weight, only those oocytes belonging to the largest size mode in the gonads and whose growth was completed (GSI greater than the threshold) were used. The oocyte diameter was determined using Image J software developed the National Institute of Health (http://rsb.info.nih.gov/nih-image). The software enables processing of digital photos to determine the oocyte number in the sample and the mean oocyte diameter (each being measured along the largest Feret diameter). It also allows determining the minimum and maximum diameter of the oocytes in the sample as well as the oocyte size distribution. As the oocytes are ellipsoidal in *S. melanotheron*, the largest and smallest diameters were determined. This was possible because the software fits an ellipse to each egg, giving its major and minor diameters. The major and minor diameters were measured in parallel on an oocyte sample from 15 females using a binocular microscope. Comparison of the results revealed great similarity between the two methods of measurements, which enabled us to validate the method of determining the oocyte diameter by Image J software. The mean weight of an oocyte was determined by weighing 100 oocytes using a mini scales with hundredth gram resolution.

Statistical Analyses

For the fecundity, oocyte diameter, oocyte weight, we performed preliminary analyses of the variance homogeneity and normality of the data using Bartlett and Kolmogorov-Smirnov tests, respectively. One-way ANOVA followed by multiple comparison Tukey test was performed on data of oocyte diameter with normal distribution and uniform variance to test differences between locations. When the data was not normally distributed and did not have uniform variance as for the fecundity and oocyte weight, a Kruskal-Wallis non-parametric analysis of variance (ANOVA) was performed to reveal significant differences in means between locations. Taking all the individual data from the locations, the strength of the correlations between fecundity, oocyte diameter, oocyte weight, egg weight and total body weight were assessed by covariance analysis. These tests were performed with the R software (http://www.r-project.org). For all tests, a probability of less than 5% and a confidence of 95% are considered as fiducial level of significance.

ACKNOWLEDGMENTS

The authors are especially grateful to GAMET team at IRD Montpellier, particularly Marc Legendre and Etienne Barras for their helpful comments and suggestions on sample treatment and analysis. They are also grateful to the two anonymous reviewers for their comments that have helped improve this manuscript.

AUTHOR CONTRIBUTIONS

Conceived and designed the experiments: MG JJA. Performed the experiments: MG JJA. Analyzed the data: MT JK MG. Contributed reagents/materials/analysis tools: PN OTT. Wrote the paper: MT JK MG. Read and approved the final manuscript: All authors.

REFERENCES

1. Pagès J, Citeau J (1990) Rainfall and salinity of sahalian estuary between 1927 and 1987. Journal of Hydrology 113: 325–341.
2. Savenije HHG, Pagès J (1992) Hypersalinity: a dramatic change in the hydrology of Sahalian estuaries. Journal of Hydrology 135: 157–174.
3. Panfili J, Thior D, Ecoutin JM, Ndiaye P, Albaret JJ (2006) Influence of salinity on the size at maturity for fish species reproducing in contrasting West African estuaries. Journal of Fish Biology 69: 95–113.
4. Panfili J, Mbow A, Durand J-D, Diop K, Diouf K, et al. (2004) Influence of salinity on the life-history traits of the West African black-chinned tilapia (*Sarotherodon melanotheron*): comparison between the Gambia and Saloum estuaries. Aquatic Living Resourses 17: 65–74.
5. Duponchelle F, Cecchi P, Corbin D, Nunez J, Legendre M (1999) Spawning season variations of female Nile tilapia, *Oreochromis niloticus*, from man-made lakes of Ivory Coast. Environmental Biology of Fishes 56: 377–389.
6. Duponchelle F, Panfili J (1998) Variations in age and size at maturity of female Nile tilapia, *Oreochromis niloticus*, populations from man-made lakes of Ivory Coast. Environmental Biology of Fishes 52: 453–465.
7. Stearns SC, Crandall RE (1984) Plasticity for age and size at sexual maturity: a life history response to unavoidable stress. In: Potts GW, Wootton RJ, editors. Fish Reproduction: Strategies and Tactics. pp. 13–33. Academic Press, London.
8. Stewart KM (1988) Change in condition and maturity of the *Oreochromis niloticus*population of Freguson's Gulf, Lake Turkana, Kenya. Journal of Fish Biology 33: 181–188.
9. Lévêque C, Paugy D (1999) Les poissons des eaux continentales africaines. Éditions de l›IRD, Paris, France.
10. Schreck CB, Contreras-Sanchez W, Fitzpatrick MS (2000) Effects of stress on fish reproduction, gamete quality, and progeny. Aquaculture 197: 3–24.
11. Roff DA (1982) Reproductive uncertainty and the evolution of iteroparity: why don›t flatfish put all their eggs in one basket? Canadian Journal of Fisheries and Aquatic Sciences 38: 968–977.
12. Bagenal TB (1969) Relationship between egg size and fry survival in Brown Trout*Salmo trutta* L. Journal of Fish Biology 1: 349–353.
13. De Silva SS (1986) Reproductive biology of *Oreochromis mossabicus* populations of man-made lakes in Sri Lanka: a comparative study. Aquaculture and Fisheries Management 17: 31–38.
14. Laë R, Fagianelli DJ, Fagianelli E (1984) La pêche artisanale individuelle sur le système lagunaire togolais: description des pêcheries et estimation de la production au cours d›un cycle annuel 1983–1984. Documents ORSTOM, série Hydrobiologie, Lomé, Togo.
15. Duponchelle F (1997) Reproduction du tilapia (Pisces, Cichlidae) *Oreochromis niloticus*(Linnaeus, 1758) dans les retenues artificielles de Côte d'Ivoire: analyse comparative des modalités de production et approche expérimentale de leur déterminisme. PhD Thesis Université de Bretagne Occidentale, Rennes.
16. Legendre M, Ecoutin JM (1989) Suitability of brackish water tilapia species from the Ivory Coast for lagoon aquaculture. 1 - Reproduction. Aquatic Living Resourse 2: 71–79.
17. Legendre M, Ecoutin JM (1996) Aspects of the reproductive strategy of *Sarotherodon melanotheron*: comparison between a natural population (Ébrié lagoon, Côte d›Ivoire) and different cultured populations. In: Pullin RSV, Lazard J, Legendre M, Amon Kothias J;-B, Pauly D, editors. pp. 320–325. Proceeding of the 3rd international symposium on tilapia in aquaculture ICLARM, Manila.
18. Peters HM (1963) Fecundity, egg weight and oocyte development in tilapias (Cichlidae, Teleostei). In: Pauly D, editor. ICLARM Translation No 2 ICLARM, Manila.
19. Fryers G, Iles TD (1972) The cichlid fishes of the great lakes of Africa. 641 p. Oliver and Boyed, London.
20. Lowe-McConnell RH (1982) Tilapias in fish communities. In: Pullin RSV, Lowe-McConnell RH, editors. The biology and culture of tilapias. pp. 83–113. ICLARM, Manila.
21. Duponchelle F, Cecchi P, Corbin D, Nunez J, Legendre M (2000) Variations in fecundity and egg size of female Nile tilapia, *Oreochromis niloticus*, from man-made lakes of Ivory. Coast Environmental Biology of Fishes 57: 155–170.
22. Philippart J, Ruwet JC (1982) Ecology and distribution of Tilapias. In: Pullin RSV, Lowe-McConnell RH, editors. The biology and culture of Tilapias. pp. 16–60. ICLARM Conf. Proc. Manilla, Philippines.
23. Ouattara NI, Teugels GG, N'Douba V, Philippart J-C (2003) Aquaculture potential of the black-chinned tilapia, *Sarotherodon melanotheron* (Cichlidae). Comparative study of the effect of stocking density on growth performance of landlocked and natural populations under cage culture conditions in Lake Ayame (Côte d›Ivoire). Aquaculture Research 34: 1223–1229.
24. Lemarié G, Baroiller JF, Clota F, Lazard J, D A (2004) A simple test to estimate the salinity resistance of fish with specific application to *O. niloticus and S. melanotheron*. Aquaculture 240: 575–587.

25. Campbell D, Mahatane A, Aleem SO (1986) Mass synchronized spawning of Tilapia guineensis. ARAC Working Paper No1 ARAC/WP 1/86 African Regional Aquaculture Centre, Port Harcourt, Nigeria.

26. Diouf K, Panfili J, Labonne M, Aliaume C, Tomás J, et al. (2006) Effects of salinity on strontium:calcium ratios in the otoliths of the West African black-chinned tilapia*Sarotherodon melanotheron* in a hypersaline estuary. Environmental Biology of Fishes 77: 9–20.

27. Labonne M, Morize E, Scolan P, Lae R, Dabas E, et al. (2009) Impact of salinity on early life history traits of three estuarine fish species in Senegal. Estuarine, Coastal Shelf Science 82: 673–681.

28. Panfili J, Durand J-D, Mbow A, Guinand B, Diop K, et al. (2004) Influence of salinity on life history traits of the bonga shad *Ethmalosa fimbriata* (Pisces, Clupeidae): comparison between the Gambia and Saloum estuaries. Marine Ecology Progress Series 270: 241–257.

29. Gourène B, Pouyaud L, Agnèse JF (1993) Importance de certaines caractéristiques biologiques dans la structuration génétique des espèces de poissons: le cas de*Ethmalosa fimbriata* et *Sarotherodon melanotheron*. Journal Ivoirien Océano Limnol Abidjan 2: 55–69.

30. Pouyaud L, Desmarais E, Chenuil A, Agnèse JF, Bonhomme F (1999) Kin cohesiveness and possible inbreeding in the mouth brooding tilapia *Sarotherodon melanotheron* (Pisces Cichlidae). Molecular Ecology 8: 803–812.

31. Diouf K, Guilhaumon F, Aliaume C, Ndiaye P, Do Chi T, et al. (2009) Effects of the environment on fish juvenile growth in West African stressful estuaries. Estuarine, Coastal Shelf Science 83: 115–125.

32. Simier M, Blanc L, Alioume C, Diouf PS, Albaret JJ (2004) Spatial and temporal structure of fish assemblages in an "inverse estuary", the Sine Saloum system (Senegal). Estuary Coastal and Shelf Sciences 59: 69–86.

33. Ecoutin JM, Simier M, Albaret JJ, Lae R, Tito de Morais L (2010) Changes over a decade in fish assemblages exposed to both environmental and fishing constraints in the Sine Saloum estuary (Senegal). Estuarine, Coastal and Shelf Science *in press* 1–9.

34. Albaret J-J, Simier M, Darboe FS, Ecoutin J-M, Rafray J, et al. (2004) Fish diversity and distribution in the Gambia Estuary, West Africa, in relation to environmental variables. Aquatic Living Resourses 17: 35–46.

35. Faunce CH (2000) Reproduction of blackchin tilapia, *Sarotherodon melanotheron*, within an impounded mangrove ecosystem in east central Florida. Environmental Biology of Fishes 57: 353–361.

36. Tine M, de Lorgeril J, Dio p K, Bonhomme F, Panfili J, et al. (2007) Growth hormone and Prolactin-1 gene transcription in natural populations of the black-chinned tilapia*Sarotherodon melanotheron* acclimatised to different salinities. Comparative Biochemistry and Physiology Part B 147: 541–549.

37. Tine M, McKenzie DJ, Bonhomme F, Durand J-D (2011) Salinity-related variation in gene expression in wild populations of the black-chinned tilapia from various West African coastal marine, estuarine and freshwater habitats. Estuarine, Coastal and Shelf Science 91: 102–109.

38. Blaber SJM, Cyrus DP, Albaret J-J, Chong VC, Day JW, et al. (2000) Effects of fishing on the structure and functioning of estuarine and nearshore ecosystems. ICES Journal of Marine Science 57: 590–602.

39. Nanjo K, Kohno H, Sano M (2008) Food habits of fishes in the mangrove estuary of Urauchi River, Iriomote Island, southern Japan. Fisheries Science 74: 1024–1033.

40. Thayer GW, Colby DR, Hettler WF (1987) Utilization of the red mangrove prop root habitat by fishes in south Florida. Marine Ecology Progress Series 35: 25–38.

41. Vidy G (2000) Estuarine and mangrove systems and the nursery concept: which is which? The case of the Sine Saloum system (Senegal). Wetlands Ecology and Management 8: 37–51.

42. Gning N, Le Loc›h F, Thiaw OT, Aliaume C, Vidy G (2010) Estuarine resources use by juvenile Flagfin mojarra (*Eucinostomus melanopterus*) in an inverse tropical estuary (Sine Saloum, Senegal). Estuarine, Coastal and Shelf Science 86: 683–691.

43. Albaret JJ (1982) Reproduction et fécondité des poissons d'eau douce de Côte d'Ivoire. Revue Hydrobiologie Tropicale 15: 347–371.

44. Duarte CM, Alcaraz M (1989) To produce many small or few large eggs: a size-independent tactic of fish. Oceologia 80: 401–404.

45. Lahti E, Muje P (1991) Egg quality and female condition in vendace (*Coregonus albda L.*) before and during spawning. Hydrobiologia 209: 175–182.

46. Reizer C (1974) Définition d'une politique d'aménagement des ressources halieutiques d'un écosystème complexe par l›étude de son environnement abiotique, biotique et anthropique. Le fleuve Sénégal Moyen et Inférieur: Dissertation Arlon, Fondation universitaire luxembourgeoise.

47. Bouvy M, Briand E, Boup MM, Got P, Leboulanger C, et al. (2008) Effects of sewage discharges on microbial components in tropical coastal waters (Senegal, West Africa). Marine & freshwater research 59: 614–626.

48. Marius C (1985) Mangroves du Sénégal et de la Gambie: écologie pédologie géochimie mise en valeur et aménagement. 335 p. Thèse de Doctorat l›Université Louis Pasteur, ORSTOM: Collection Travaux et Documents, France.

49. Kâ S, Pagano M, Bâ N, Bouvy M, Leboulanger C, et al. (2006) Zooplankton distribution related to environmental factors and phytoplankton in a shallow tropical lake (Lake Guiers, Senegal, West Africa). Internat Rev Hydrobiol 91: 389–405.

50. Bettarel Y, Arfi R, Bouvier T, Bouvy M, Briand E, et al. (2008) Virioplankton distribution and activity in a tropical eutrophicated bay. Estuarine, Coastal and Shelf Science 80: 425–429.

SECTION IV - FISH FARMING INTENSIFICATION, THREATS TO FISH HEALTH AND ALTERNATIVE FISH HEALTH MANAGEMENT.

CHAPTER 12 BIOLOGICAL CONTROL OF CATFISH PATHOGENS

Chao Ran, Abel Carrias, Malachi A. Williams, Nancy Capps, Bui C. T. Dan, Joseph C. Newton, Joseph W. Kloepper, Ei L. Ooi, Craig L. Browdy, Jeffery S. Terhune, Mark R. Liles

INTRODUCTION

Aquaculture farming of the channel catfish, *Ictalurus punctatus*, has been one of the most successful animal production industries in North America in the past 30 years and currently represents the largest aquaculture industry in the United States. Over 90% of all catfish produced in the U.S. are raised in Alabama, Arkansas, Louisiana, and Mississippi and are primarily grown in earthen ponds ranging in size from 2 to 10 ha [1], [2].

Feed inputs associated with high-density fish culture stimulate the proliferation of opportunistic bacteria [3]. The combination of high-density fish culture with rapidly changing water temperature and chemical composition of aquaculture ponds places stress on fish, thereby resulting in favorable conditions for the onset and spread of disease. Enteric Septicemia of Catfish (ESC), caused by the Gram negative bacterium *Edwardsiella ictaluri* [4], is the most important endemic infectious disease in the channel catfish aquaculture industry [5]. Losses resulting from ESC were reported in over 78% of all operations with outbreaks being reported in 42% of catfish production ponds, with an economic loss between $20 and $30 million yearly[1], [2], [6].

Another important pathogen in channel catfish is *Aeromonas hydrophila*, which is the primary causative agent of motile aeromonad septicaemia (MAS) [7] and can infect multiple fish species including tilapia, catfish, goldfish, common carp, and eel [8]. In 2009 and 2010, *A. hydrophila* was identified as the etiologic agent of a disease epidemic in farmed channel catfish, resulting in higher mortality rates than typical for MAS with over five million pounds of catfish lost in the Alabama commercial catfish industry. The *A. hydrophila* strains (e.g., strain AL09-119) isolated from diseased fish during this epidemic are highly virulent in aquaria disease challenge trials compared to *A. hydrophila* reference strains [8].

Pangasianodon hypophthalmus Sauvage, commonly known as the striped catfish, is the native catfish in the Mekong Delta of Vietnam. The farming sector of *P. hypophthalmus* has recorded the highest growth rate in volume compared to any other aquaculture commodity globally over the last decade [9], [10]. The sector accounted for 687,000 and 1,094,879 t production in 2007 and 2008, respectively, the latter amounting to 34% of the total aquaculture production in Vietnam, the fifth-ranked nation in global aquaculture production [11]. Furthermore, over 90% of the farmed catfish is processed and exported to more than 100 countries globally [10]. Bacillary necrosis of *Pangasius* spp. (BNP), also

Citation: Ran C, Carrias A, Williams MA, Capps N, Dan BCT, Newton JC, et al. (2012) Biological Control of Catfish Pathogens. doi:10.1371/journal.pone.0045793

caused by *E. ictaluri*, is an economically significant disease for the striped catfish aquaculture industry in the Mekong Delta, which can cause 50–90% mortality and occurs in 98% of farms [9].

Chemotherapy by oral administration of antibiotics in fish feeds is the most common treatment for bacterial diseases. However, the indiscriminant application of antibiotics may result in many problems including the spread of drug-resistant pathogens, environmental hazards and food safety problems. This has fostered an increased interest in alternatives to antibiotics. Probiotics, which have various health-promoting properties and minor adverse side effects, are gaining an increasing scientific and commercial interest in aquaculture practice. The beneficial effects of probiotics involve improvement of feed utilization, modulation of intestinal microflora, enhancement of immune responses and antagonism to pathogens. The most commonly used probiotics in aquaculture are lactic acid bacteria and *Bacillus* spp. [12]. *Bacillus* species have advantages as probiotics in that their spore-forming ability allows greater viability after pelleting and high survival rates after exposure to gastric acid [13], [14], [15]. *Bacillus* spp. has been reported to have various beneficial attributes when applied to fish [16], [17], [18], [19], [20],[21], [22].

Few studies have been conducted to investigate probiotic bacteria for mitigating infectious diseases in channel catfish, and no studies have been reported using direct administration in feed. Queiroz and Boyd [23] applied a commercial probiotic product, Biostart, which contained a few species of *Bacillus* spp., to channel catfish pond water and demonstrated that survival and net production of fish treated with *Bacillus spp.* were significantly greater than the control. However, the bacteria used in this previous study were not isolated specifically for use in channel catfish nor were their antimicrobial activities against important pathogens of channel catfish characterized.

In this research an extensive collection of *Bacillus* strains (n=160) isolated from soil and from the intestine of channel catfish (n=17) was tested for *in vitro* antimicrobial activity against *E. ictaluri, A. hydrophila*, and other bacterial and fungal pathogens of channel catfish. *Bacillus*strains that showed effective antibiosis were evaluated for their respective survival in the intestine of channel catfish. The biological control activity of the best performing *Bacillus*strains when amended onto feed was investigated using channel and striped catfish disease challenge studies in an aquarium system. The safety of selected *Bacillus* strains was also assessed in terms of the presence of plasmids and resistance to antibiotics.

MATERIALS AND METHODS

Ethics Statement

All experiments conducted with vertebrate animals (catfish) were approved by the Institutional Animal Care and Use Committee (IACUC) review boards at Auburn University or Novus International Inc. in accordance with the respective animal welfare guidelines specified in the United States or Vietnam.

Bacterial Strains

E. ictaluri strain S97-773 was used for the primary screening for *Bacillus* antibiosis and for ESC challenge experiments since this strain is highly pathogenic for channel catfish, has previously been used in challenge studies and was obtained from the Southeastern Cooperative Fish Disease Laboratory (SCFDL), Auburn University. *E. ictaluri* strain R-4383, *E. ictaluri* strain Alg-08-200, *E. tarda* AL09-82, *S. iniae* USDA 2009, *Y. ruckeri* ATCC 29473, and*F. columnare* AL09-10 were from the collection of pathogenic isolates at the SCFDL, and *S. ferax* was obtained from Carolina Biological (catalog # 156271, Burlington, NC). *E. ictaluri*NLF33 were isolated from diseased striped catfish in Vietnam. *A. hydrophila* ML09-119 was isolated from a diseased channel catfish with MAS in 2009. The collection of soil-derived*Bacillus* strains (n=160) was provided by the laboratory of Dr. Joseph Kloepper (Department of Entomology and Plant Pathology, Auburn University). *Bacillus subtilis* 1E17 was obtained from the *Bacillus* Genetic Stock Center (http://www.bgsc.org/).

Isolation of *Bacillus* spp. Strains from the Intestine of Channel Catfish and Evaluation of Antimicrobial Activity

Healthy catfish (7–10 cm) were killed by administration of an overdose of MS-222 (tricaine mesylate, FDA approved for anesthesia in fin fish), and the digestive tracts were removed in their entirety. Approximately 1.0 g of the intestinal tissue with gut content was homogenized in 9.0 ml of sterile saline (0.9% w/v) using a sterile mortar and pestle. Ten-fold serial dilutions were prepared to 10^{-6} in fresh 1x phosphate-buffered saline (PBS), and 0.1 ml was spread over the surface of triplicate plates of tryptone soy agar (TSA) with incubation at 28°C for 48 h[24]. *Bacillus*-like colonies were picked at random, purified by streaking for isolated colonies on fresh media, and examined for inhibition against the growth of *E. ictaluri* using the double-layer soft agar method [25]. For the soft agar overlay, the bacterial isolates were grown in 5 ml of tryptone soy broth (TSB) for 24 h at 30°C. A volume of 5 µl was then spotted onto triplicate plates of TSA and incubated for a further 24 h. Soft agar (0.7% w/v agar) prepared with TSB was melted, cooled and seeded with an inoculum of log-phase *E. ictaluri* strain S97-773 to achieve slight turbidity (i.e., ~10^7 cells/ml). The bacterial cell suspension in soft agar was immediately poured over the TSA plates and incubated for 24 h at 30°C whereupon the presence of zones of clearing in the growth of of *E. ictaluri* were recorded (in mm) as evidence of growth inhibition. Cultures that were regarded as inhibitory to *E. ictaluri* were characterized by Gram staining and 16S rRNA gene sequencing using the 'universal bacteria' primer set 27F and 1492R [26]. A consensus 16S rRNA sequence was produced using ChromasPro (Technelysium Pty Ltd., Queensland, Australia), and each sequence was compared to the GenBank non-redundant nucleotide database by BLASTn. *Bacillus* spp. strains were cryopreserved at −80°C. The collection of soil-derived *Bacillus* strains (n=160) was tested for antimicrobial activity against *E. ictaluri* using the same method.

Fifty *Bacillus* strains with antagonistic activity against *E. ictaluri* S97-773 were tested for their inhibitory activity against other *E. ictaluri* strains (*E. ictaluri* R-483, *E. ictaluri* Alg-08-

200).*Bacillus* strains that showed antimicrobial activity against all three *E. ictaluri* strains were evaluated further for their activity to inhibit the growth of *A. hydrophila* strain AL09-119. Twenty-one *Bacillus* strains that showed significant antimicrobial activity against both *E. ictaluri* and *A. hydrophila* were tested for their activity against several other channel catfish pathogens including *E. tarda, S. iniae, Y. ruckeri, S. ferax* with the soft agar overlay method described above.

The antimicrobial activity against *F. columnare* was tested by an agar well diffusion method. For the well diffusion assay, the *Bacillus* strains were grown in 5 ml of TSB for 48 h at 30°C. After centrifugation at 3,600×g for 10 min, the culture supernatant was filtered through a 0.2 µm filter. Then 200 µl of the filter-sterilized supernatant was added to a round well (approx. 10 mm in diameter) made in a *F. columnare* growth medium (FCGM) agar plate [27]. After the supernatant was absorbed into the agar medium, a log-phase *F. columnare* culture grown in FCGM broth was spread thoroughly over the plate using a sterile cotton swab. The plates were incubated for 48 h at 30°C. The zones of clearing in the growth of the lawn of *F. columnare*were measured by the same method as in double-layer soft agar protocol.

Bacillus strains AB01, AP79, AP143, AP193, and AP254 were sent to Vietnam and their *in vitro* antimicrobial activity was tested against *E. ictaluri* NLF33, the causative agent of BNP in striped catfish. A broth culture of *E. ictaluri* was adjusted to 10^6 CFU/mL and evenly swabbed onto TSA plates. Three wells were punched from the agar plate and 50 µL of a 10^8 CFU/mL of a *Bacillus* cell-free supernatant (48 h culture in TSB) was added into each well. Zones of inhibition were measured after 24 hours incubation at 30°C.

Preparation of *Bacillus* Spores and Spore-amended Feed

Bacillus spores were prepared by the method described by Kenny and Couch [28] with some modifications. *Bacillus* strains were grown in TSB at 30°C overnight. The cell suspension was spread onto spore preparation agar (peptone 3.3 g/l, beef extract powder 1.0 g/l, NaCl 5.0 g/l, K_2HPO_4 2.0 g/l, KCl 1.0 g/l, $MgSO_4$. $7H_2O$ 0.25 g/l, $MnSO_4$ 0.01 g/l, lactose 5 g/l, agar 15 g/l) using a sterile cotton swab and incubated at 28°C for 5 to 7 days. To collect the spores, 5 ml of sterile distilled water was added to the plate and the spores were suspended in water using an inoculation loop. The spore suspension was then incubated at 85°C for 15 min to kill the vegetative cells. The concentration of the spore suspension was determined by 10-fold serial dilution in 1 x PBS and spreading onto TSA. The final concentration of the spore suspension was manipulated with sterile water to 1.25×10^{10} CFU/ml for the intestinal survival assay and 10^9 CFU/ml for the challenge study. To prepare spore-amended feed, 80 ml of the spore suspension was sprayed onto 1000 g commercially available slow-sinking pelleted fish feed (2 mm, 40% protein, Zeigler, Gardners, PA) using a bleach- and ethanol-sterilized pump sprayer to achieve approximately 8% (v/w) spore suspension application. The feed was then mixed thoroughly with 30 ml fish oil. The control feed was ammended solely with fish oil.

Inoculation and Quantification of *Bacillus* spp. in the Intestine of Channel Catfish

Fingerling channel catfish (7–10 cm) were distributed into twenty-three 60 L tanks each containing 15 L water and three fish. Fish were starved for one week prior to the experiment. Catfish feed was amended in separate batches with the 21 *Bacillus* strains that showed good antimicrobial activity against both *E. ictaluri* and *A. hydrophila* using the spore application method described above. Each unique *Bacillus* strain-amended feed (~10^9 CFU/g feed) was given to one aquarium tank. The fish were fed once daily with spore-amended feed or control feed for one week, and thereafter all fish received the control feed for three days. One tank was used as the control and received untreated fish feed for the duration of the experiment. Daily feeding rate was 3% of total body weight.

At the end of the experiment, all of the fish were killed by administration of an overdose of MS-222. The intestine was removed, weighed, and then homogenized in sterile saline (0.9% w/v). The final volume of the homogenized intestine sample was adjusted to 2 ml by sterile saline. Homogenized samples were then serially diluted in sterile saline and spread on TSA and incubated at 28°C for 48 h. Three representative colonies with the same morphology as the applied *Bacillus* strain were randomly picked from the plate, purified on new plates and identified by 16S rRNA gene sequencing as described previously and compared with the known 16S rRNA gene sequence from each respective *Bacillus* strain. For the control and treatment groups, only the unique colony morphology corresponding to that of the amended*Bacillus* strains was recorded. Cultured counts for each *Bacillus* strain recovered from the intestine were determined as CFU/g of intestine sample.

Aquarium Challenge Studies

In the first challenge, five *Bacillus* strains (AB01, AP143, AP193, AP254, and AP79) were selected for evaluation of their biological control of ESC in an aquarium challenge. Five*Bacillus* treatments and one control, each with four replicate aquaria, were included. Each replicate aquarium was stocked with 25 fingerling channel catfish weighing about 13 g. Fish were acclimated to commercial dry feed for one week. Fish from each treatment group were then fed with an experimental diet supplemented with spores of one *Bacillus* strain (8×10^7CFU/g) at a daily feeding rate of 2.5% fw/bw (feed weight/body weight) for two weeks. Fish in the control group received normal feed only.

Fish were challenged by immersion for 45 minutes in 10 L of water containing 4.5×10^6 CFU/ml*E. ictaluri* S97-773. All fish from the same group were immersed in a single container. The challenge condition for the control group was the same as other treatments except that Brain Heart Infusion (BHI) medium was added instead of *E. ictaluri* culture. Mortalities were monitored over a 21-day period, and dead fish were dissected and the presence of *E. ictaluri*confirmed by microbiological examination of kidney and liver swabs on TSA. The final mortality was calculated when there was no more mortality for five consecutive days after challenge. The identity of the recovered *E. ictaluri* was confirmed by biochemical analyses that included indole and hydrogen sulfide production, catalase and oxidase activities, hydrolysis of esculin and gelatin, nitrate reduction, and acid formation from carbohydrates.

Fish were reared in a recirculating system during the acclimation period. Upon initiation of*Bacillus* feeding and during the challenge phase, a static system was incorporated with a 20–30 min daily water exchange that resulted in turnover of approximately half of the aquarium water. Sponge biofilters and daily removal of uneaten/waste materials were incorporated to improve water quality. Water temperature was kept at 26±2°C. During the static phase, the central room heating system in conjunction with submersible aquarium water heaters was used to control the required water temperature, and a water heater system was used to control the temperature of the incoming water during water exchange.

In a second challenge trial using channel catfish, flow-through conditions were used to reduce catfish mortality. In this challenge experiment, four *Bacillus* treatments (AP79, AP143, AP193, AB01) and one control each with four replicate aquaria were included. Each aquarium was stocked with 20 fingerling channel catfish (~12 g). A lower dose of *E. ictaluri* S97-773 (8×10^5CFU/ml) was used to challenge fish and starting immediately after challenge the aquaria were flushed for 5–8 hours a day. All other conditions in this challenge were the same as described above. Mortalities were monitored over a 21-day period after challenge, and presence of *E. ictaluri* in the dead fish was confirmed as previously described.

A third challenge trial was conducted to evaluate the protective effect of four *Bacillus* strains (AP79, AP193, AP254 and AB01) against *E. ictaluri* for striped catfish. Four *Bacillus*treatments and one control each with four replicate tanks were included in this study. Each tank was stocked with 18 striped catfish (~14 g). Striped catfish were administered feed amended with *Bacillus* spores (~10^7 CFU/g feed) or control feed for two weeks and the fish were transferred to 80 L tanks for a bath challenge with *E. ictaluri* NLF33. Fish were immersed for 30 min in static, aerated aquaria at a dose of ~10^6 CFU/mL to target about 70% mortality in the control group. The control and test diets were offered throughout the trial. The recording of mortality and confirmation of *E. ictaluri* in dead fish were conducted as above.

Plasmid Analysis

Plasmid DNA was extracted from *Bacillus* strains AP79, AP143, AP193, and AB01 by the alkaline lysis method [29]. *Bacillus subtilis* 1E17 containing plasmid pC194 was used as a positive control. The extracted DNA was analyzed by a Chef-DR II pulsed field electrophoresis system (Bio-Rad, Hercules, CA). Pulse time ranged from 1 to 15 seconds for 15 h at 6 V/cm. The gel was stained with ethidium bromide and visualized using an AlphaImager HP gel documentation system (ProteinSimple, Santa Clara, CA).

Antibiotic Resistance Analysis

The susceptibility of *Bacillus* strains AP79, AP143, AP193, and AB01 to carbenicillin, ampicillin, spectinomycin, oxacillin, vancomycin, cephalothin, novobiocin, sulfadiazine, amikacin, erythromycin, neomycin, penicillin, chloramphenicol, sulfamethoxazole, norfloxacin, gentamicin and ciprofloxacin was determined by disc diffusion test following procedures outlined by National Committee for Clinical Laboratory Standards [30]. A log-phase culture of each strain was diluted to a concentration of approximately 1×10^8 to 2×10^8 CFU/ml (McFarland standard 0.5). The inoculum was then seeded onto a Mueller-Hinton agar plate using a cotton swab. Antibiotic-impregnated discs (BD Biosciences) were placed on seeded plates, and the diameter of the zone of growth inhibition was measured after 18 h of incubation at 37°C. The experiments were repeated three times and the average diameter of inhibition zones was calculated.

Statistics

A completely randomized design was used in this research. Data were presented as mean ± standard error (SE). Challenge data were subjected to analysis of variance in SAS 9.2. Differences between means were tested by Tukey's range test and were considered significant when probability (P) values < 0.05 were obtained.

RESULTS

Characterization of *Bacillus* Isolates

Each of the *Bacillus* strains isolated from soil or catfish intestine that exhibited inhibitory activity against both *E. ictaluri* and *A. hydrophila* was capable of endospore formation. Each pure *Bacillus* culture was ribotyped, indicating that most of the *Bacillus* strains were within the*B. subtilis* group (inclusive of *B. amyloliquefaciens*) (Table 1). The 16S rRNA gene sequences for the 21 strains were submitted to GenBank (Accession Numbers JX094283 to JX094303). Two strains of *B. pumilus* (AP18 and AP280) and one strain of *B. methylotrophicus* (AP191) were also within the collection. Strain AP76 was identified as *Bacillus cereus* and thus eliminated for further evaluation due to the potential for foodborne illness. All of the *Bacillus*strains that were subsequently determined to have *in vivo* biological control activity were within the *B. subtilis* group. This phylogenetic affiliation was also confirmed by comparison of genome sequences to the GenBank nr/nt database for strains AB01, AP79, AP143, and AP193, indicating >80% average nucleotide identity to a previously sequenced genome within the *B. subtilis* group (data not shown).

Table 1. Antimicrobial activity of 22 *Bacillus* **strains against**

multiple aquatic pathogens.

Phylogeny	Strain	*Aeromonas hydrophila*	*Edwardsiella ictaluri*	*Edwardsiella tarda*	*Flavobacterium columnare*	*Saprolegnia ferax*	*Streptococcus iniae*	*Yersinia ruckeri*
B. subtilis group	AB01	+	+	+	+++	+	++	++
B. pumilus	AP18	+	+	-	−	+	+	−
B. subtilis group	AP71	+	++	+	−	−	++	+
B. cereus	AP76	+	++	++	−	+	++	++
B. subtilis group	AP77	+	+++	++	−	+	+	++
B. subtilis group	AP79	++	++	+	−	−	++	++
B. subtilis group	AP102	+	++	++	++	+	−	++
B. subtilis group, *B. amyloliquefaciens*‡	AP143	++	++	++	−	+	+	++
B. subtilis group	AP183	+	++	++	−	−	++	+
B. subtilis group	AP189	++	+++	++	−	−	++	+
Bacillus methylotrophicus	AP191	++	+++	+	+	+	++	−
B. subtilis group, *B. amyloliquefaciens*‡	AP193	++	++	++	++	+	+	−
B. subtilis group	AP215	+	+++	+	−	+	+	++
B. subtilis group	AP218	+	++	+	−	+	+	+
B. subtilis group	AP219	++	++	+	+	+	++	+
B. subtilis	AP254	+	++	+	++	−	−	−
B. pumilus	AP280	+	++	+	−	+	+	+
B. subtilis group	AP295	+	++	+	−	+	+	++
B. subtilis group	AP301	+	++	+	+++	+	++	+
B. subtilis group	AP303	++	++	++	−	+	+	++
B. subtilis group	AP305	++	++	++	−	−	+	++

‡Draft genome sequences are available for these *Bacillus* strains, so the phylogenetic affiliation is inferred from a comparison of these *Bacillus* strain genome sequences with previously sequenced *Bacillus* genomes.
Note that (+) indicates a zone of inhibition up to 5 mm, (++) indicates a zone of inhibition from 5 mm to 1 cm, (+++) indicates a zone of inhibition greater than 1 cm, and (−) indicates no observable zone of inhibition.
doi:10.1371/journal.pone.0045793.t001

Antimicrobial Activity of *Bacillus* Strains

The *B. subtilis* strain AB01 isolated from the catfish intestine showed significant antimicrobial activity against *E. ictaluri.* From the collection of soil-derived *Bacillus* strains, 49 strains showed significant antagonism against *E. ictaluri.* All of the 50 *Bacillus* strains also showed inhibitory activity against *E. ictaluri* R-4383 and *E. ictaluri* Alg-08-200. A total of 21 *Bacilllus*strains showed potent antibiotic activity against both *E. ictaluri* and *A. hydrophila* (Table 1). The 21 *Bacillus* strains selected were tested for their activity against multiple pathogens in aquaculture. All of the strains were antagonistic against multiple catfish pathogens, including Gram-negative and -positive bacteria, and the oomycete *Saprolegnia. Bacillus* strains AB01, AP219, and AP301 showed antimicrobial activity against all of the tested pathogens (Table 1). Also, all five of the *Bacillus* strains tested for biological control of BNP (AP79, AP143, AP193, AP254, AB01) showed significant antagonistic activity against *E. ictaluri* NLF33 (data not shown).

Survival and Persistence of *Bacillus* Strains in the Intestine of Channel Catfish

Bacillus spores administered to channel catfish via feed for one week followed by three days of control feed were recovered from the catfish intestine. Over 10^7 CFU/g of introduced*Bacillus* was observed in the gut for strains AB01, AP76, AP77, AP79, AP143, and AP254 (Fig. 1). For strains AP18, AP280, and AP303, the counts of recovered bacteria were relatively low, and they were eliminated from further investigation. None of the 21 *Bacillus* strains were recovered from the control group. In all cases the 16S rRNA gene sequence determined from representative colonies matched the 16S rRNA gene sequence from the respective *Bacillus*strain that was added to catfish feed (data not shown).

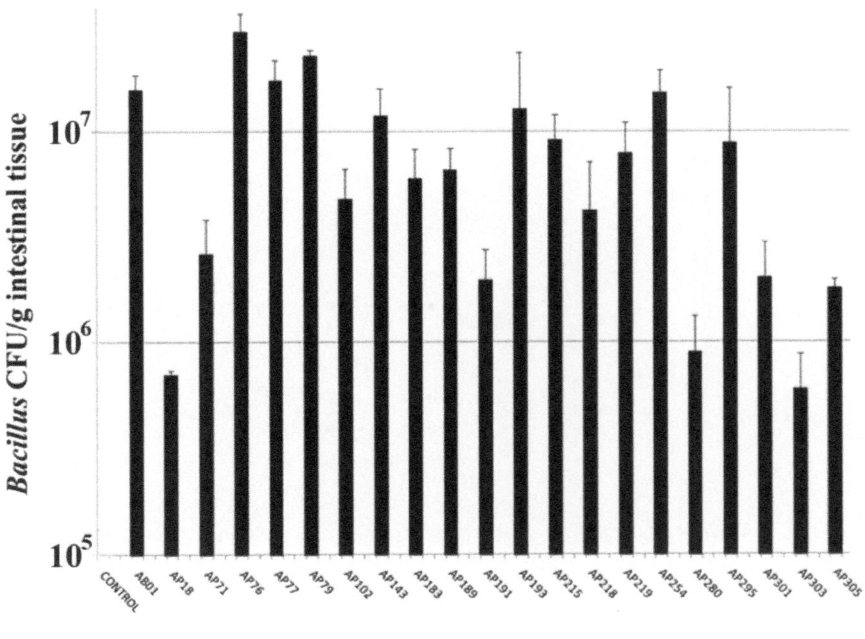

Figure 1. ***Bacillus*** **strain CFUs/g of catfish intestine, after feeding with** *Bacillus*-amended or non-amended feed (n=3 animals per ***Bacillus*** **strain).**

Challenge Study

In the first immersion challenge, the mean mortality of control group was 98.0%, a very high level of mortality that was likely a consequence of the persistence of *E. ictaluri* within the aquaria water under static conditions. Treatment groups that were fed with feed amended with spores of *Bacillus* strains AP143 or AB01 showed significantly reduced mortality compared with the control (P<0.05), with 83.1% and 84.8% mortality, respectively. There was no significant difference in the mortality observed between the two strains. The treatment groups fed with *Bacillus* strains AP79, AP193, or AP254 (with mortality 89.0%, 95.0%, and 93.7%, respectively) did not show significant differences compared with the control (Fig. 2A, Table 2). In every challenge experiment a control group was included that was not challenged with *E. ictaluri*, and in every case this control group had no observed mortalities, demonstrating that mortalities were due to infection with *E. ictaluri*.

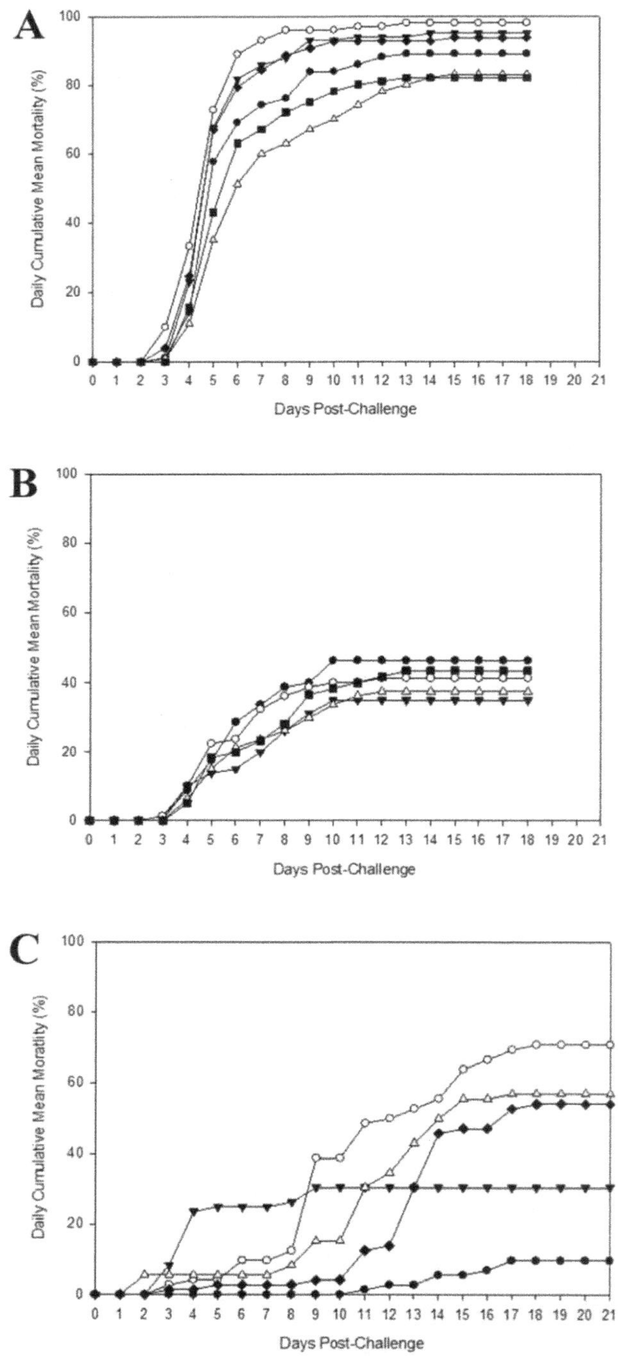

Figure 2. Daily mean cumulative mortality of (A) channel catfish in static system with 20–30 min daily water exchange and (B) channel catfish with 5–7 h flow through water daily, or (C) striped catfish in static system with 20–30 min daily water exchange, fed with and without addition of *Bacillus* **strains and challenged with** *E. ictaluri*.

All values are means of four replicates per treatment. Treatments: (open circle) Control, (closed circle) AP79, (closed triangle) AP193, (open triangle) AB01, (closed square) AP143, and (closed diamond) AP254.

Table 2. Mortality (%) (± SE) of groups of fish that received feed amended with different *Bacillus* **strains or control feed and were challenged with** *E. ictaluri* **(n=4).**

Treatment	Channel catfish challenge (Fig. 2A)	Channel catfish challenge (Fig. 2B)	Striped catfish challenge (Fig. 2C)
Control	98.0 ± 1.2^{a}	41.3 ± 5.9^{a}	70.8 ± 7.3^{a}
AB01	84.8 ± 2.0^{bc}	37.5 ± 9.5^{a}	56.9 ± 6.6^{ab}
AP143	83.1 ± 2.9^{bc}	43.3 ± 14.8^{a}	Not determined
AP193	95.0 ± 3.0^{ab}	35.0 ± 5.4^{a}	30.6 ± 23.7^{ab}
AP254	93.7 ± 2.8^{ab}	Not determined	54.2 ± 11.4^{ab}
AP79	89.0 ± 2.7^{abc}	46.3 ± 5.2^{a}	9.7 ± 6.6^{b}

Means in the same column sharing a common superscript letter were not significantly different. (P>0.05) as determined by Tukey's test.

For the second immersion challenge, flow-through conditions were adopted post-challenge to reduce catfish mortality, and this was successful as only 41.3% mortality was observed in the control group. However, under the flow-through conditions of this challenge the mortality in the treatment groups ranged from 35.0% to 46.3% with no significant differences observed between any of the treatment groups and the control (Fig. 2B, Table 2).

In the striped catfish challenge experiment, the catfish were again challenged by immersion and then maintained under static conditions similar to those used in the first channel catfish challenge. Under these conditions the control group had 70.8% mortality which is consistent with the higher mortality observed under static conditions when re-infection can occur (Fig. 2C, Table 2). The treatment group fed with feed amended with spores of strain AP79 had the lowest (9.7%) cumulative mortality and this was significantly different from the control (P<0.05). Catfish fed with spores of strains AP193, AP254 or AB01 had 30.6%, 54.2% and 56.9% mortality, respectively.

Plasmid and Antibiotic Resistance Study

An analysis of plasmid DNA extracted from *Bacillus* strains AP79, AP143, AP193, and AB01 was conducted by PFGE. We did not observe the presence of any plasmid within AP79, AP193 and AB01 but the positive control did show the presence of plasmid pC194 (data not shown). A plasmid was evident from strain AP143. High coverage contigs from the assembled genome sequence of AP143 (data not shown) were screened by BLASTn and one of the contigs showed high identity (99%) with plasmid pBSG3 from *Bacillus amyloliquefaciens* B3[31]. BLASTx analysis of the predicted open reading frames from this plasmid did not show any similarity to genetic loci involved in antibiotic resistance or virulence. Evaluation of antibiotic susceptibility determined that all of these strains were susceptible to all of the tested antibiotics to varying degrees. They were all highly susceptible to carbenicillin, cephalothin, sulfamethoxazone and ciprofloxacin (>25 mm diameter inhibition zone). Ampicillin, penicillin, vancomycin, novobiocin, amikacin, erythromycin, neomycin, chloramphenicol, norfloxacin and gentamicin also inhibited their growth effectively (20–25 mm zones of inhibition), whereas spectinomycin, oxacillin, sulfadiazine showed moderate inhibition (15–20 mm inhibition zones). These strains showed very similar antibiograms, with the variation in the diameter of inhibition zones less than 10% of the average diameter for each of the antibiotics tested.

DISCUSSION

The results of this study indicate that specific strains within the *Bacillus subtilis* group showed promise for biological control of disease in catfish aquaculture. It provided another evidence of the efficacy of the *Bacillus* strains in boosting disease resistance of fish [16], [17], [19], [20]. This study is the first to select probiotic bacteria for control of ESC and other pathogens in catfish and to evaluate their biocontrol efficacy via feed administration.

Gatesoupe [32] concluded that probiotics for aquaculture should be antagonistic to pathogens, colonize intestines, and increase resistance of the host to pathogens. Ideally probiotic bacteria should be selected by considering all three criteria. However, it is difficult to evaluate potential probiotic bacterial strains for the second and third criterion on a large number of candidate bacteria. Therefore, *in vitro* antimicrobial activity was the primary criteria by which a large number of strains were evaluated, with candidate bacterial strains that did not show antagonistic activity eliminated from further study. The primary objective of this research project was to identify bacterial strains that can be applied for the control of *E. ictaluri, A. hydrophila*, as well as other bacterial and omycete pathogens of catfish. Since the bacterial pathogens *E. ictaluri* and *A. hydrophila* are responsible for the majority of the mortality observed currently in catfish aquaculture, the ability of a *Bacillus* strain to inhibit the growth of these two pathogens was of paramount importance and only the strains capable of inhibiting both pathogens were selected for testing in aquarium disease challenges.

The ability of a probiotic bacterial strain to colonize and survive within or on its host is also an important criterion for strain selection. However, in many cases the probiotic bacteria may not permanently colonize the gastrointestinal tract but instead achieve a sustained transient state [24], [33]. Even transient bacteria may be efficient at mediating biological control of disease if the cells are introduced artificially via food either continuously or semi-continuously [32], [34]. High population levels of several *Bacillus* strains were recovered from catfish intestines three days post-feeding with *Bacillus*-spore amended feed. For *Bacillus* strains with high counts in the intestine, colonies with the same morphology as the applied *Bacillus* strain dominated the TSA plates, and the ribotype of the representative colonies confirmed their identity as the applied *Bacillus* strain. Considering that bacterial population levels in the intestine should decline after cessation of feeding with the spore-containing diets, the maximal level of *Bacillus*strain CFU/g of intestinal tissue reached during the feeding regime may be higher. The bacterial population levels here (10^6–10^7 CFU/g for most of the strains) are in general agreement with previous studies involving fish [24], [33], [35], [36]. These results demonstrate that some of the *Bacillus* strains evaluated in this study can persist within the catfish gastrointestinal tract for at least three days. However, at this point the degree of persistence and ability to colonize the intestinal mucosa are unknown for each strain. A more detailed experiment evaluating the colonization and/or persistence of specific *Bacillus* strains within the catfish intestine will be conducted to help understand the biocontrol mechanism(s) of *Bacillus*strains and guide the duration and timing of *Bacillus* feeding. Future studies will also examine the impact of each *Bacillus* strain on the intestinal microbiota and the health and growth of the fish in the absence of aquaculture pathogens.

Three aquarium disease challenges were conducted in this study, two of which evaluated biological control of ESC in channel catfish. In the first channel catfish disease challenge, a very high mortality (98.0%) due to ESC was observed in the control group. Ideally, an aquarium disease challenge would result in a mortality of 60%–70%, which more accurately simulates the natural development of ESC in an aquaculture pond. The high mortality was probably the consequence of the incorporation of a static system after the immersion challenge, wherein the *E. ictaluri* persisted in the aquaria for an extended period of time and bacterial cells shed in feces could potentially infect other fish within the same aquarium. Despite the higher mortality observed in this first challenge, two *Bacillus* strains (AP143, AB01) provided significant protection to channel catfish. However, the degree of mortality reduction for the *Bacillus* treatment groups compared with control was lower in the first challenge compared to a later challenge with striped catfish. This discrepancy might be due to the much higher mortality in the first challenge, which could have reduced the biological control capacity of the *Bacillus* strains. Presumably the degree of biological control would be greater in an ESC challenge if lower mortality (~70%) was obtained in the control group. In addition, at the lower doses of *E. ictaluri* that catfish are typically exposed to in an aquaculture pond the degree of biocontrol provided by *Bacillus* strains would presumably be of an even greater magnitude.

One solution to reduce channel catfish mortality during an aquarium disease challenge is to use a flow-through system, but we discovered that the use of a flow-through system also resulted in the loss of *Bacillus* biological control activity. The lack of a protective effect for*Bacillus* strains when catfish were maintained in flow-through conditions after immersion challenge is likely a consequence of the removal of the *Bacillus* cells from the aquarium water, thereby preventing the persistence of the *Bacillus* cells within the aquarium that would naturally occur within an aquaculture pond. This suggests that a more pond-like environment wherein the probiotic is maintained within the water, and potentially the skin and gills of the fish, may be more conducive for effective biological control of disease. Since a static aquarium system is a better model for evaluating the biological control of disease that would occur in an aquaculture pond, we considered the results of the first disease challenge more relevant to the evaluation of *Bacillus* strains for their future adoption in pond-scale aquaculture.

Considering the importance of striped catfish for the Vietnamese aquaculture industry, the best performing *Bacillus* strains were sent to collaborators in Vietnam to evaluate their biological control activity for striped catfish against *E. ictaluri*. Significant reductions in mortality were obtained in this experiment, with an especially large reduction in mortality observed for fish fed with feed amended with *Bacillus* strain AP79. The biological control activity observed for *Bacillus* strains in this striped catfish disease challenge was comparable with previous studies in other fish species [17], [20]. However, in this study fish were fed with *Bacillus*-amended feed for two weeks before challenge, which is a relatively short period of time compared with the feeding duration time of one to two months in previous studies [16], [20]. The duration of time for probiotic feeding is a factor that can influence the degree of protection activity against infection in challenge [17]. Thus, the degree of protection afforded by the *Bacillus* strains in this study might be even better if a longer feeding duration was adopted.

It is interesting that the *Bacillus* strains that showed a significant protective effect in the disease challenges were different for the two catfish species. This could reflect a biologically meaningful difference in the interactions between *Bacillus* strains and their respective host. Also, there could be unique tripartite interactions between host, pathogen, and probiotic bacteria that could be influenced by environmental factors. Clearly more research is needed to understand the complex interplay between host, pathogen and probiotic *Bacillus* strains, and how to manipulate the environment to achieve the optimal biological control of disease. Further studies using an aquarium disease model with static conditions need to be conducted to optimize important parameters for challenge such as dosage and timing with the best performing *Bacillus* strains, with subsequent studies at a pond-scale to evaluate biological control efficacy within an aquaculture pond ecosystem.

One of the safety requirements for live bacteria directly consumed by humans is the absence of any acquired resistance to clinically important antibiotics [37]. Although the *Bacillus* strains used in this research were not for direct consumption by humans, they might be consumed inadvertently, as their hosts were cultured for food. Thus, it is important to analyze antibiotic resistance in probiotic strains and to distinguish the natural resistance, which is one of the phenotypic characteristics of a species, and acquired (i.e., transferable) resistance, which is associated with occurrence of plasmids. Also, pathogenicity and enterotoxin production are closely associated with plasmids [38]. Each of the four selected strains was susceptible to a broad spectrum of antibiotics tested with very similar susceptibility profile. Strains AP79, AP193 and AB01 did not contain any plasmids, ensuring their inability to conjugally transfer any plasmid that might confer antibiotic resistance or other traits. Although a plasmid was extracted from AP143, draft genome sequences from this strain were used to identify plasmid sequences and there were no predicted genes involved in antibiotic resistance or virulence.

Diffusible antimicrobial compounds were clearly involved in the *in vitro* antagonistic activity observed in soft agar overlay and in diffusion tests. The relative importance of secondary metabolites for *in vivo* biological control is unknown compared to other mechanisms such as enhancement of immune response, competitive exclusion of pathogens and modulation of intestinal microbiota. Future studies will investigate the relative contribution of antibiotic compound(s) production to the biological control activity of the *Bacillus* strains by comparing the mortality reduction activity of transposon mutagenized *Bacillus* strains with no antagonism activity with the wild type strains in aquarium challenge experiments.

In conclusion, a collection of *Bacillus* strains was identified that are antagonistic to the primary pathogens of catfish and are beneficial to both channel catfish and striped catfish when administered on feed for the control of ESC and BNP, respectively. These bacteria have potential application in aquaculture as a cost-effective alternative to the current use of antimicrobial compounds.

AUTHOR CONTRIBUTIONS

Conceived and designed the experiments: CR AC ELO JCN JWK CLB JST MRL. Performed the experiments: CR AC MAW NC BCTD ELO. Analyzed the data: CR AC ELO MRL. Contributed reagents/materials/analysis tools: JWK CLB MRL. Wrote the paper: CR JWK JST MRL.

REFERENCES

1. US Department of Agriculture, USDA (2003a) Part I: Reference of Fingerling Catfish Health and Production Practices in the United States. Fort Collins, CO, USA: National Health Monitoring System.
2. US Department of Agriculture, USDA (2003b) Part II: Reference of Foodsize Catfish Health and Production Practices in the United States. Fort Collins, CO, USA: National Health Monitoring System.
3. Austin B, Stuckey LF, Robertson P, Effendi I, Griffithc D (1995) A probiotic strain of*Vibrio alginolyticus* effective in reducing diseases caused by *Aeromonas salmonicida, Vibrio anguillarum and Vibrio ordalii.* J Fish Dis 18: 93–96. doi: 10.1111/j.1365-2761.1995.tb01271.x
4. Hawke JP (1979) A bacterium associated with disease of pond cultured channel catfish, *Ictalurus punctatus*. J Fish Res Board Can 36: 1508–1512. doi: 10.1139/f79-219
5. Hawke JP, Khoo LH (2004) Infectious diseases. In: Tucker CS, Hargreaves JA, editors. Biology and Culture of Channel Catfish. Amsterdam, The Netherlands: Elsevier. 387–443.
6. Wagner BA, Wise DJ, Khoo LH, Terhune JS (2002) The epidemiology of bacterial diseases in food-size channel catfish. J Aquat Anim Health 14: 263–272. doi: 10.1577/1548-8667(2002)014<0263:teobdi>2.0.co;2
7. Harikrishnan R, Nisha Rani M, Balasundaram C (2003) Hematological and biochemical parameters in common carp, *Cyprinus carpio*, following herbal treatment for *Aeromonas hydrophila* infection. Aquaculture 221: 41–50. doi: 10.1016/s0044-8486(03)00023-1
8. Pridgeon JW, Klesius PH, Mu X, Song L (2011) An *in vitro* screening method to evaluate chemicals as potential chemotherapeutants to control *Aeromonas hydrophila*infection in channel catfish. J Appl Microbiol 111: 114–124. doi: 10.1111/j.1365-2672.2011.05030.x
9. Phan LT, Bui TM, Nguyen TTT, Gooley GJ, Ingram BA, et al. (2009) Current status of farming practices of striped catifish, *Pangasianodon hypophthalmus* in the Mekong Delta, Vietnam. Aquaculture 296: 227–236. doi: 10.1016/j.aquaculture.2009.08.017
10. Phuong NT, Oanh DTH (2009) Striped catfish (*Pangasianodon hypophthalmus*) aquaculture in Viet Nam: an unprecedented development within a decade. In: De Silva SS, Davy FB, editors. Success Stories in Asian Aquaculture. Dordrecht, Bangkok and Ottawa: Springer, NACA and IDRC. 133–149.
11. De Silva SS, Ingram BA, Nguyen PT, Bui TM, Gooley GJ, et al. (2010) Estimation of nitrogen and phosphorus in effluent from the striped catfish farming sector in the Mekong Delta, Vietnam. Ambio 39: 504–514. doi: 10.1007/s13280-010-0072-x
12. .Wang YB, Li JR, Lin J (2008) Probiotics in aquaculture: challenges and outlook. Aquaculture 281: 1–4. doi: 10.1016/j.aquaculture.2008.06.002
13. .Casula G, Cutting SM (2002) *Bacillus* probiotics: spore germination in the gastrointestinal tract. Appl Environ Microbiol 68: 2344–2352. doi: 10.1128/aem.68.5.2344-2352.2002
14. Hong HA, Duc LH, Cutting SM (2005) The use of bacterial spore formers as probiotics. FEMS Microbiol Rev 29: 813–835. doi: 10.1016/j.femsre.2004.12.001
15. Hyronimus B, Le Marrec C, Hadji Sassi A, Deschamps A (2000) Acid and bile tolerance of spore-forming lactic acid bacteria. Int J Food Microbiol 61: 193–197. doi: 10.1016/s0168-1605(00)00366-4
16. Ai QH, Xu HG, Mai KS, Xu W, Wang J, et al. (2011) Effects of dietary supplementation of *Bacillus subtilis* and fructooligosaccharide on growth performance, survival, non-specific immune response and disease resistance of juvenile large yellow croaker,*Larimichthys crocea*.. Aquaculture 317: 155–161. doi: 10.1016/j.aquaculture.2011.04.036
17. Aly SM, Ahmed YA, Ghareeb AA, Mohamed MF (2008) Studies on *Bacillus subtilis*and *Lactobacillus acidophilus*, as potential probiotics, on the immune response and resistance of Tilapia nilotica (*Oreochromis niloticus*) to challenge infections. Fish Shellfish Immunol 25: 128–136. doi: 10.1016/j.fsi.2008.03.013
18. Kumar R, Mukherjee SC, Ranjan R, Nayak SK (2008) Enhanced innate immune parameters in *Labeo rohita* (Ham.) following oral administration of *Bacillus subtilis*. Fish Shellfish Immunol 24: 168–172. doi: 10.1016/j.fsi.2007.10.008
19. Nayak SK, Swain P, Mukherjee SC (2007) Effect of dietary supplementation of probiotic and vitamin C on the immune response of Indian major carp, *Labeo rohita*(Ham.). Fish Shellfish Immunol 23: 892–896. doi: 10.1016/j.fsi.2007.02.008
20. Newaj-Fyzul A, Adesiyun AA, Mutani A, Ramsubhag A, Brunt J, et al.. (2007) *Bacillus subtilis* AB1 controls *Aeromonas* infection in rainbow trout

(*Oncorhynchus mykiss*, Walbaum). J Appl Microbiol 103 (5), 1699–1706.

21. Salinas I, Cuesta A, Esteban MÁ, Meseguer J (2005) Dietary administration of*Lactobacillus delbrüeckii* and *Bacillus subtilis*, single or combined, on gilthead seabream cellular innate immune responses. Fish Shellfish Immunol 19: 67–77. doi: 10.1016/j.fsi.2004.11.007
22. Salinas I, Abelli L, Bertoni F, Picchietti S, Roque A, et al. (2008) Monospecies and multispecies probiotic formulations produce different systemic and local immunostimulatory effects in the gilthead seabream (*Sparus aurata* L.). Fish Shellfish Immunol 25: 114–123. doi: 10.1016/j.fsi.2008.03.011
23. Queiroz J, Boyd CE (1998) Effects of a Bacterial Inoculum in Channel Catfish Ponds. J World Aquacult Soc 29: 67–73. doi: 10.1111/j.1749-7345.1998.tb00300.x
24. Irianto A, Austin B (2002) Use of probiotics to control furunculosis in rainbow trout,*Oncorhynchus mykiss* (Walbaum). J Fish Dis 25: 333–342. doi: 10.1046/j.1365-2761.2002.00375.x
25. Jack RW, Wan J, Gordon J, Harmark K, Davidson BE, et al. (1996) Characterization of the chemical and antimicrobial properties of piscicolin 126, a bacteriocin produced by*Carnobacterium piscicola* JG126. Appl Environ Microbiol 62: 2897–2903.
26. Weisburg WG, Barns SM, Pelletier DA, Lane DJ (1991) 16S Ribosomal DNA Amplification for Phylogenetic Study. J Bacteriol 173: 697–703.
27. Farmer B (2004) Improved methods for the isolation and characterization of*Flavobacterium columnare*. Thesis. Baton Rouge, LA: Department of Pathobiological Sciences, Louisiana State University.
28. Kenney DS, Couch TL (1981) Mass production of biological agents for plant disease, weed and insect control. In: Papavizas GC, editors. Biological Control in Crop Production BARC Symposium No. 5. Totowa, NJ: Allenheld and Osmum. 143–150.
29. Birnboim HC, Doly J (1979) A rapid alkaline extraction procedure for screening recombinant plasmid DNA. Nucleic Acids Res 7: 1513–1523. doi: 10.1093/nar/7.6.1513
30. National Committee for Clinical Laboratory Standards (2012) Performance Standards for Antimicrobial Disk Susceptibility Test; Approved Standard-Ninth Edition. Wayne, PA: Clinical and Laboratory Standards Institute.
31. Qiao JQ, Tian DW, Huo R, Wu HJ, Gao XW (2011) Functional analysis and application of the cryptic plasmid pBSG3 harboring the RapQ-PhrQ system in *Bacillus amyloliquefaciens* B3. Plamid 65: 141–149. doi: 10.1016/j.plasmid.2010.11.008
32. Gatesoupe FJ (1999) The use of probiotics in aquaculture. Aquaculture 180: 147–165. doi: 10.1016/s0044-8486(99)00187-8
33. Robertson PAW, O'Dowd C, Burrells C, Williams P, Austin B (2000) Use of*Carnobacterium* sp. as a probiotic for Atlantic salmon (*Salmo salar* L.) and rainbow trout (*Oncorhynchus mykiss*, Walbaum). Aquaculture 185: 235–243. doi: 10.1016/s0044-8486(99)00349-x
34. Gournier-Chateau N, Larpent JP, Castellanos I, Larpent JL (1994) Les Probiotiques en Alimentation Animale et Humaine. Paris: Technique et Documentation Lavoisier. 192.
35. Jöborn A, Olsson JC, Westerdah MA, Conway PL, Kjelleberg S (1997) Colonization in the fish intestinal tract and production of inhibitory substances in intestinal mucus and faecal extracts by *Carnobacterium* sp. strain K1. J Fish Dis 20: 383–392. doi: 10.1046/j.1365-2761.1997.00316.x
36. Gildberg A, Mikkelsen H (1998) Effects of supplementing the feed to Atlantic cod (*Gadus morhua*) fry with lactic acid bacteria and immuno-stimulating peptides during a challenge trial with *Vibrio anguillarum*.. Aquaculture 167: 103–113. doi: 10.1016/s0044-8486(98)00296-8
37. Sorokulova I (2008) Preclinical testing in the development of probiotics: a regulatory perspective with *Bacillus* strains as an example. Clin Infect Dis 46: SS92–95. doi: 10.1086/523334
38. .Pannucci J, Okinaka RT, Sabin R, Kuske CR (2002) *Bacillus anthracis* pXO1 plasmid sequence conservation among closely related bacterial species. J Bacteriol 184: 134–141. doi: 10.1128/jb.184.1.134-141.2002

CHAPTER 13

GEOGRAPHICAL DISTRIBUTION OF PROTOZOAN AND METAZOAN PARASITES OF FARMED NILE TILAPIA OREOCHROMIS NILOTICUS (L) (PERCIFORMES: CICHLIDAE) IN YUCATÁN, MÉXICO

Amelia Paredes-Trujillo, Iván Velázquez-Abunader, Edgar Torres-Irineo, David Romero and Víctor Manuel Vidal-Martínez

Keywords: Nile tilapia farms, Parasites, Geographical distribution, Gyrodactylus cichlidarum, Management practices, Yucatán México

BACKGROUND

In the last two decades, tilapia has become one of the most commercially important freshwater fish species in global aquaculture [1, 2]. In the Americas, the increased production of farmed tilapia has been due to its adaptability to a diverse array of production systems, including extensive pond culture, semi-intensive cage culture, intensive flow-through tank culture, raceway culture and various highly intensive indoor methods [3]. In 2009, México produced 71,358 tons of tilapia and was the sixth largest producer of tilapia in the Americas after Brazil, Honduras, Colombia, Ecuador, and Costa Rica [1]. The production methods ranged from stocking fingerlings by release into reservoirs to intensive methods in ponds, lake cages, tanks, and shrimp ponds [3]. However, the intensification of crop densities in tilapia aquaculture, deficient management practices and a lack of biosecurity plans as defined by the World Organization for Animal Health (OIE) have led to the spread and establishment of non-native parasitic diseases [4,5]. Some studies have identified poor management conditions such as stocking density, fish source, high concentrations of nitrates and low frequency of water exchange in cultured tilapia as being the factors associated with reports of parasitic (Cichlidogyrus spp., Coccidia, Trichodina sp., and Gyrodactylus sp.) and bacterial diseases [6, 7, 8]. Apart from deadly bacteria such as Streptococcus iniae Pier, 1976 [5, 8], the most common health problems in tilapia aquaculture are due to helminth parasites, especially monogeneans, which have produced economic losses attributed to slow growth, reduced fertility rates, and high mortality rates [9]. In Yucatán (a tropical state in southeastern México), the Nile tilapia [Oreochromis niloticus (L.)] culture is a rapidly growing commercial activity in rural areas. The monogeneans Cichlidogyrus spp. are the most frequent and abundant parasites in farmed Nile tilapia in Yucatán, yet their geographical distribution and the associated risk factors have not been previously reported [10].

Mapping techniques have been useful as descriptive analytical tools in numerous epidemiological studies, mainly regarding diseases that represent public health problems (e.g., leishmaniasis, schistosomiasis, and trypanosomiasis) [11]. These studies have described disease spread throughout different geographical areas, identified high-risk regions, and observed the natural historical variation of disease. These

Citation: Amelia Paredes-Trujillo, Email author, Iván Velázquez-Abunader, Edgar Torres-Irineo, David Romero and Víctor Manuel Vidal-Martínez, Contributed equally, Geographical distribution of protozoan and metazoan parasites of farmed Nile tilapia Oreochromis niloticus (L) (Perciformes: Cichlidae) in Yucatán, México, DOI: 10.1186/s13071-016-1332-9.

epidemiological maps have been useful as powerful monitoring tools in aquatic animal health for preventing geographical translocation of potential pathogens and determining the occurrence or distribution of exotic or endemic diseases, including changes in prevalence during different time periods [12]. Additionally, this approach could assist risk-based surveillance and help to monitor and predict the impact of environmental changes on the prevalence and severity of emerging endemic diseases (i.e. increasing in prevalence or range) [13, 14]. We hypothesise that the use of geographical tools such as maps, in combination with multivariate and nonlinear statistical analyses for determining meaningful environmental and management variables, will be valuable for adequate sanitation management in the Nile tilapia farms of Yucatán. Therefore, the aims of this study were to describe the geographical distribution of protozoan and helminth parasites of farmed Nile tilapia in Yucatán and to analyse the potential statistical associations between the prevalence and mean abundance of these parasites with management and environmental variables, with particular emphasis on Gyrodactylus cichlidarum Paperna, 1968, a well-known monogenean pathogen of the Nile tilapia.

METHODS

We conducted a census of all 29 Nile tilapia farms currently operating in Yucatán and registered with the Yucatán Aquatic Animal Health Committee (CESAY), which, in turn, graciously provided us with transportation and contact with the farm owners. To describe the parasite populations, we used measures of infection such as prevalence, mean abundance and mean intensity [15]. Prevalence was defined as the number of individuals of a host species infected with one or more particular parasite species, divided by the total number of hosts examined for that parasite species (expressed as a percentage) [15]. The abundance was defined as the total number of individuals of a particular parasite species in an individual from a particular host species [15]. The mean abundance was defined as the total number of individuals of a particular parasite species in a sample taken from a particular host species, divided by the total number of hosts of that species examined (including both infected and uninfected hosts) [15]. The mean intensity was defined as the average intensity of a particular species of parasite among the infected members of a particular host species [15].

We constructed a database with the information of the measures of infection of the ectoparasites (prevalence and mean abundance) of the 29 extant Nile tilapia farms in Yucatán. The information for the database was obtained from a census program undertaken by the senior author who visited the Nile tilapia farms during 2013 and from a historical database containing information regarding the measures of infection by parasites from all 29 Nile tilapia farms (2011 to 2012). This database was produced by the Aquatic Pathology Laboratory at CINVESTAV - Mérida, based on the parasitological analysis of Nile tilapia requested by CESAY. In both cases, the same measures of infection were used: prevalence and mean abundance of all the helminth and protozoan species infecting Nile tilapia at each farm. The sample size was 20 fish per farm, and the choice of this sample size was based on three considerations. Firstly, we assumed that the prevalence was 20 %, which was sufficient for us to collect species of ectoparasites with relatively high prevalence values, that is, the ones relevant in aquaculture [16]. Secondly, we accepted a probability α (alpha) that 5 % of all random samples drawn had no infected fish when the population was indeed infected. Thirdly, we assumed that the sensitivity of the sampling technique used (the stereomicroscope as a method for parasite detection) was 75 % due to the possibility of human error. We based this assumption on the fact that, even when the technicians at CINVESTAV are well trained in the parasitological search with the stereomicroscope, new personnel may be at risk of missing the parasites if occurring in very low numbers or in the very small larval stages. Because the values of sensitivity of microscopy as a diagnostic test for parasites of Nile tilapia have not been published, we assumed the level of sensitivity to be 75 %, which is similar to the one reported for the sensitivity of microscopy as a diagnostic test for malaria (77 %) [17].

With respect to specificity, we assumed a confidence level of 95 % due to the low probability for the misclassification of the parasites by our group, which has extensive experience in the taxonomy and identification of the monogeneans of cichlids [18, 19]. Thus, assuming a Poisson distribution for the probability of detecting species with at least 20 % of prevalence, the monthly sample size was obtained using the formula $n=4/$ prev, where n is the fish sample size; the number 4 originated from the formula derived from a Poisson distribution – Ln ($\alpha=0\cdot05$(the accepted probability))/ 0.75 (sensitivity of the diagnostic method); and prev is the prevalence in the fish population (in this case, the assumed 20 %) [16]. Therefore, the formula $n=4/$ prev, as $n=4/0.2=20$ fish per farm was applied. At each farm, the fish were collected using a 2-m diameter cast net with a 2.5-cm mesh from a randomly selected production tank. We obtained the temperature (C), conductivity (μS/cm), dissolved oxygen (mg/l), salinity (ppt), nitrites (mg/l), nitrates (mg/l) and ammonium (mg/l) levels using a multiparameter meter (YSI-85) [20]. The live fish were transported in containers with artificial aeration to the Aquatic Pathology Laboratory at CINVESTAV - Mérida. Once in the laboratory, each individual Nile tilapia was kept in a container with freshwater and an oxygen supply and was euthanized with a 100 mg/L of benzocaine until opercular movements ceased; the brain was severed posteriorly by spiking. Immediately afterward, each fish was measured to obtain the total length (TL, cm), standard length (SL, cm) and total weight (W, g).

Skin, gills, scales from the lateral line and fins were examined under a stereomicroscope for ectoparasites. For endoparasites, the organs were separated individually into Petri dishes with 0.07 % saline solution. Liver, kidney, brain, spleen, heart, muscle, mesentery and intestine were reed by compression between two 10 cm squared pieces of glass. Once the parasites were found, they were counted, preliminarily identified and fixed depending on the taxonomic group [18]. The monogeneans were removed with paintbrushes, stained with ammonium picrate and identified according to suitable literature (e.g. [18, 21]). The protozoans were stained with silver nitrate and identified [22].

To display the geographical location of each Nile tilapia farm and that of the parasites infecting Nile tilapia at each farm in Yucatán, maps were created with the geographical coordinates of the farms by using the mapplots package [23] in the statistical software R, version 2.9.1. To determine whether the spatial distribution of the mean abundance of each parasite species was spatially random or clustered, Ripley's K (r) function was used under a homogeneous Poisson assumption [24]. This function is flexible because it takes into account all possible pairs of points and not merely the nearest pairs [25]. We plotted the Ripley's K function vs the radius of complete spatial randomness to assess if the parasites species presented spatial clusters [25]. If the observed Ripley's K function was above the theoretical Ripley's K function, then the point pattern was not random (i.e. clustered). If a point fulfilled the assumption of non-randomness, then mean density (= mean abundance) was estimated using an isotropic Gaussian kernel smoothing. For n points, the shape of a kernel smoothing estimator may be expressed as follows [24, 25]:

$$\widehat{\lambda}(x) = \frac{1}{h^2}\sum_{t=1}^{n} \kappa\left(\frac{\|x - x_i\|}{h}\right)/q(\|x\|)$$

where κ (u) is a bivariate symmetrical kernel function, q (||x||) represents a border correction to compensate for missing observations that occur when x is close to the border region, and the bandwidth h measures the level of smoothing. In this step, the analysis was made using an R implementation for analysing spatial point patterns in the package spatstat [26].

A total of 39 management and environmental variables related to the Nile tilapia production cycle, including the education level of the farmers, were collected from 21 of the 29 farms (Table 3). To obtain information on the management procedures at each production unit, a standardised questionnaire was designed and applied following the method described [27] (Additional file 1). The management variables were then semi-quantitatively categorised according to [27]. From the 29 farms, only 21 owners were willing to answer the questionnaire. The farms were then classified based on (i) the mean annual production, number of workers, and nearby communities; (ii) the biosecurity measures including disinfection of equipment, tank cleaning, and vector presence; (iii) the presence of facilities/management such as quarantine area, bathroom, footbaths, warehouse, guardhouse; and (iv) an established monitoring procedure to collect information on environmental variables such as nitrates, nitrites, salinity and temperature (Table 3).

We performed a redundancy analysis (RDA) using the Canonical Correspondence Analysis software CANOCO to determine whether there were statistical associations between management and environmental variables and the abundance values of each parasite species per individual fish. RDA is a direct ordination analysis that identifies compositional gradients in a biological data set (e.g. a parasite population or community) as a response to measured environmental factors (in this case, environmental factors obtained from each Nile tilapia farm) [28].

Redundancy analysis was used because the lengths of the ordination axes were less than two standard deviations [29]. In this case, it is assumed that most of the response curves for the number of individuals of the metazoan and protozoan parasite species, with respect to the environmental variables, will be linear. If the lengths of the ordination axes are longer than two standard deviations, then unimodal models, such as canonical correspondence analysis (CCA), should be used [29, 30].

RDA is a direct extension of regression analysis to model multivariate response for dependent variables (in this case, the abundance values of each parasite species) regressed on several explanatory independent variables with linear constraints on the regression coefficients [28, 29]. Since RDA is similar to a multivariate linear regression, linearity is an important requirement. Variables with large deviations from linearity were log-transformed, and the improvement in linearity re-checked. The significance of all statistical analyses was established at an $\alpha = 0.05$ unless otherwise stated. To determine the statistical significance ($p < 0.05$) of the canonical axes for mean abundance of parasites with respect to the environmental and management variables, a Monte Carlo simulation based on 4,999 permutations was used [28, 30].

We used a generalized additive model of location, shape and scale (GAMLSS), which includes nonparametric and nonlinear smoothing functions [31], to determine the potential effect of the independent management and environmental variables on the mean abundance of Gyrodactylus cichlidarum (the dependent variable). The GAMLSS allows flexibility when specifying the frequency distribution of the response variable such as Gaussian, negative binomial or Poisson distributions, among 39 other available specific distributions (31). This algorithm also allows us to model all distribution parameters as functions of independent variables. The model allows modelling not only of the mean (or location) but also of other parameters of the distribution of the dependent variable such as the linear and/or nonlinear, parametric and/or additive nonparametric and functions of independent variables and/or random effects. Hence, GAMLSS is especially suited for modelling a dependent variable that does not follow an exponential family distribution or when the scale or shape of the frequency distribution of this dependent variable changes with the independent variables [31]. In this case, we chose the Poisson distribution because the value of the Akaike information criterion (AIC) was the lowest when compared with different frequency distributions (e.g. negative binomial or Gaussian distributions) [32]. The GAMLSS was fitted assuming a Poisson error distribution because the dependent variable (mean abundance of Gyrodactylus cichlidarum) consisted of counts and the $\eta = \ln(\bar{y})$ function was used, where $\bar{y}$ is the dependent variable and ηη is the link function [31, 32]. The GAMLSS model also included data for sites where G. cichlidarum was absent, and the best model was selected based on the lowest AIC, lowest global deviance value and highest percentage of explained deviance [32]. A stepwise model selection was applied to select the independent variables. The GAM was fitted using regression cubic splines (cs) for the smooth terms regarding only these continuous variables. The GAM was fitted using the GAMLSS package of the R software [33].

RESULTS

A total of 580 Nile tilapia from all 29 farms in Yucatán were examined, and 11 species were recorded. The values of prevalence and mean abundance of the monogenean and protozoan species for each farm are presented in Table 1. The monogeneans were represented by nine species, followed by the protozoans with two species (Table 1). Cichlidogyrus sclerosus Paperna & Thurston, 1969 and Cichlidogyrus tilapiae Paperna, 1960 were the most frequent and abundant monogeneans at every farm (Table 1). Fish in all farms were infected with C. sclerosus, with an overall prevalence of 74 % (confidence intervals (CI) of 70–77 %) and a mean abundance of 73.83 ± 134.29 parasites per fish (Table 1). The pathogenic monogenean Gyrodactylus cichlidarum was found at 26 of 29 farms visited, with an overall prevalence of 31 %. Table 2 presents the values for prevalence, mean abundance, and mean intensity of each monogenean and protozoan species for the number of fish examined in each of the 29 extant farms in Yucatán.

Table 1: Species composition and measures of infection of the ectoparasites (protozoans and helminths) of Oreochromis niloticus (L.) for the 29 extant farms in Yucatán, México. The calculation of these measures of infection was based on the 580 fish examined across all 29 tilapia farms

Parasite species	Infection site	Estimated prevalence[a] (%)	Abundance	Intensity	Intensity	Intensity
		(CI-95 %)	Mean ± SD	Mean ± SD	Median	Range
PROTOZOA						
Trichodina sp.	Skin	41 (36–45)	12.04 ± 49.95	29.3 ± 74.68	10	1–980
Vorticella sp.	Skin	8 (5–8)	1.22 ± 7.20	14.88 ± 20.89	6	1–82
MONOGENEA						
Cichlidogyrus sclerosus Paperna & Thurston, 1969	Gills	74 (70–77)	73.83 ± 134.29	99.25 ± 147.51	46	1–11178
Cichlidogyrus tilapiae Paperna, 1960	Gills	65 (61–70)	32.35 ± 74.24	51.83 ± 91.52	28	1–885
Cichlidogyrus dossoui Paperna, 1960	Gills	44 (39–48)	8.64 ± 22.82	19.51 ± 31.06	12	1–356
Cichlidogyrus longicornis Paperna & Thurston, 1969	Gills	39 (33–42)	6.36 ± 20.71	16.37 ± 30.69	8	1–246
Cichlidogyrus quaestio Douëllou, 1993	Gills	2 (1–1)	1 ± 0.26	1.7 ± 0.94	2	1–4
Cichlidogyrus sp.	Gills	42 (37–46)	15.98 ± 47.99	38.01 ± 68.19	14	1–746
Cichlidogyrus halli (Price & Kirk, 1967)	Gills	4 (1–4)	1 ± 3.64	14.6 ± 12.57	7	1–44
Gyrodactylus cichlidarum Paperna, 1968	Fins, skin	31 (26–33)	2.00 ± 8.67	6.49 ± 14.86	9	1–173
Enterogyrus malmbergi Bilong Bilong, 1988	Intestine	13 (10–15)	1.06 ± 9.42	8.02 ± 24.93	3	1–169

[a]Expected apparent prevalence following [16]

Abbreviations: CI Confidence intervals, SD Standard deviation

Table 2. Measures of infections with protozoan and helminth ectoparasites of Oreochromis niloticus (L.) (n=20) for the 29 extant tilapia farms in Yucatán, México

Farms (n = 29)	Measures of infection	Cs	Ct	Cd	Cl	Cq	Ci	Ch	Gy	Tr	Vo	En
	P	70 (65-74)	95 (90-99)	65 (59-69)	40 (35-47)	0	55 (51-54)	0	40 (35-47)	40 (35-47)	0	10 (6-15)
1	MA	14.1 ± 22.03	34.8 ± 38.92	9.45 ± 14.73	14.8 ± 54.65	0	16.95 ± 43.28	0	1.95 ± 3.17	12.5 ± 24.84	0	0.15 ± 0.48
	MI	20.14 ± 24.05	49.71 ± 39.09	13.5 ± 16.24	21.14 ± 76.28	0	24.21 ± 55.58	0	2.78 ± 3.31	17.85 ± 31.72	0	0.21 ± 0.70
	P	70 (65-74)	50 (45-56)	55 (50-60)	35 (32-40)	0	20 (16-26)	0	15 (11-20)	35 (32-40)	5 (1-8)	20 (16-26)
2	MA	41.5 ± 54.26	4.3 ± 6.49	9.4 ± 10.44	2.1 ± 3.21	0	1.3 ± 2.77	0	0.45 ± 1.23	3.3 ± 5.69	0.1 ± 0.44	0.25 ± 0.55
	MI	59.28 ± 56.28	6.14 ± 6.93	13.42 ± 7.91	3 ± 2.30	0	1.85 ± 1.91	0	0.64 ± 1.73	4.71 ± 5.94	0.14 ± 0.44	0.35 ± 0.5
	P	85 (79-89)	85 (79-89)	20 (16-26)	20 (16-26)	20 (16-26)	50 (45-56)	0	20 (16-26)	65 (59-69)	10 (6-15)	15 (11-20)
3	MA	153.3 ± 174	115.9 ± 182	5.6 ± 16.57	1.65 ± 5.18	0.25 ± 0.55	24.7 ± 43.66	0	0.65 ± 1.59	6.95 ± 8.84	0.15 ± 0.48	1.05 ± 3.80
	MI	180.35 ± 176	136.35 ± 190	28 ± 30.06	8.25 ± 9.87	1.25 ± 0.54	49.4 ± 51.66	0	3.25 ± 2.21	10.69 ± 9.07	1.5 ± 1	7 ± 8.66
	P	80 (76-88)	40 (35-47)	50 (45-56)	50 (45-56)	10 (6-15)	35 (32-40)	5 (1-8)	15 (11-20)	50 (45-56)	5 (1-8)	10 (6-15)
4	MA	52.4 ± 97.80	10.5 ± 19.33	5.95 ± 9.09	5.9 ± 8.42	0.1 ± 0.30	8.4 ± 18.30	0.2 ± 0.89	0.3 ± 0.80	4.9 ± 8.69	0.75 ± 3.35	0.1 ± 0.30
	MI	65.5 ± 105	26.25 ± 23.28	11.9 ± 9.80	11.8 ± 8.50	1 ± 0	24 ± 24.97	4 ± 0.89	2 ± 1	9.8 ± 10.31	15 ± 3.35	1 ± 0
	P	100 -	55 (51-54)	30 (26-36)	75 (70-80)	0	50 (45-56)	0	64 (59-68)	85 (79-89)	35 (32-40)	35 (32-40)
5	MA	151 ± 216.46	47.25 ± 113.89	4.2 ± 7.89	8.65 ± 8.58	0	28.2 ± 45.87	0	3.5 ± 3.69	35.3 ± 60.38	2.55 ± 5.22	0.75 ± 1.64
	MI	161 ± 216.46	85.90 ± 144.88	14 ± 8.39	11.53 ± 8.02	0	56.4 ± 51.72	0	5.38 ± 3.24	41.52 ± 63.68	7.28 ± 6.79	2.14 ± 0.89
	P	95 (90-99)	55 (50-60)	50 (45-56)	35 (32-40)	0	40 (35-47)	15 (11-20)	5 (1-8)	55 (50-60)	5 (1-8)	0
6	MA	38.8 ± 34.30	8.4 ± 12.14	9.65 ± 19.50	3.8 ± 6.51	0	24.25 ± 76.05	2.8 ± 8.21	0.05 ± 0.22	20.35 ± 53.98	0.1 ± 0.44	0
	MI	40.84 ± 33.97	15.27 ± 12.84	19.3 ± 24.42	10.85 ± 6.71	0	60.62 ± 114	18.66 ± 14.04	1 ± 0	37 ± 69.70	2 ± 0	0
	P	80 (76-88)	55 (50-60)	55 (50-60)	45 (40-51)	0	45 (40-51)	5 (1-8)	35 (32-40)	55 (50-60)	5 (1-8)	0
7	MA	66.05 ± 94.31	11.7 ± 13.72	3.8 ± 7.19	3.05 ± 4.71	0	11 ± 20.19	0.4 ± 1.78	1.05 ± 2.23	68.7 ± 217	0.05 ± 0.22	0
	MI	82.56 ± 99.06	21.27 ± 11.56	6.90 ± 8.64	6.77 ± 4.94	0	24.44 ± 24.46	8 ± 8	3 ± 3	124.90 ± 286	1 ± 1	0
	P	80 (76-88)	80 (76-88)	40 (35-47)	55 (50-60)	0	55 (50-60)	10 (6-15)	35 (32-40)	65 (59-69)	0	25 (20-29)
8	MA	87.5 ± 221	20.35 ± 34.05	4.75 ± 8.28	4.8 ± 6.56	0	15.9 ± 24.31	1.9 ± 6.85	1.75 ± 3.05	14.2 ± 26.03	0	0.45 ± 0.88
	MI	109.37 ± 244	25.43 ± 36.48	11.87 ± 9.47	8.72 ± 6.64	0	28.90 ± 26.64	19 ± 15.55	5 ± 3.26	21.84 ± 29.87	0	1.8 ± 0.83
	P	95 (90-99)	45 (40-51)	65 (59-69)	55 (50-60)	0	60 (54-66)	0	60 (54-66)	85 (79-89)	25 (21-29)	25 (20-29)
9	MA	81.05 ± 143	10.95 ± 22.34	12.8 ± 17.46	22.35 ± 34.10	0	6.9 ± 8.51	0	1.7 ± 1.86	36.45 ± 35.23	13 ± 25.34	1.3 ± 2.73
	MI	85.31 ± 146	24.33 ± 28.61	19.69 ± 18.32	40.63 ± 37.31	0	11.5 ± 8.21	0	2.83 ± 1.58	42.88 ± 34.36	52 ± 22.71	5.2 ± 3.19
	P	100 -	50 (45-56)	75 (69-80)	75 (69-80)	0	65 (59-69)	15 (11-20)	40 (35-47)	10 (6-15)	10 (6-15)	15 (11-20)
10	MA	189.1 ± 187	63.6 ± 152	19.7 ± 32.34	10.8 ± 10.92	0	54.15 ± 166	2.2 ± 7.64	4.1 ± 8.29	0.65 ± 2.30	0.55 ± 2.03	0.4 ± 1.09
	MI	190.1 ± 187	127.21 ± 199	26.26 ± 35.14	14.4 ± 10.31	0	83.30 ± 203	14.66 ± 16.77	10.25 ± 10.71	6.5 ± 4.94	5.5 ± 4.94	2.66 ± 1.52
	P	95 (90-99)	75 (69-80)	70 (65-74)	65 (59-69)	0	65 (59-69)	0	30 (26-36)	5 (1-8)	10 (6-15)	5 (1-8)
11	MA	111.6 ± 121	52.5 ± 68.55	27.7 ± 46.99	9.8 ± 16.29	0	19.9 ± 38.04	0	2.1 ± 4.21	0.1 ± 0.44	1.4 ± 6.02	0.1 ± 0.44
	MI	117.47 ± 122	70 ± 71.16	39.57 ± 52.17	15.07 ± 18.28	0	30.61 ± 44.01	0	7 ± 5.13	2 ± 2	14 ± 18.38	2 ± 2
	P	60 (54-66)	50 (45-56)	35 (32-40)	35 (32-40)	0	20 (16-26)	0	10 (6-15)	20 (16-26)	20 (16-26)	0
12	MA	12.55 ± 23.26	17.25 ± 39.95	3.6 ± 7.80	4.45 ± 10.59	0	4.2 ± 10.02	0	0.3 ± 0.97	1.25 ± 2.84	1.4 ± 3.66	0
	MI	20.91 ± 27.28	34.5 ± 52.05	10.28 ± 10.60	12.71 ± 15.26	0	21 ± 12.90	0	3 ± 1.41	6.25 ± 3.09	7 ± 5.71	0
	P	20 (16-26)	45 (40-51)	5 (1-8)	10 (6-15)	0	10 (6-15)	0	15 (11-20)	5 (1-8)	45 (40-51)	0
13	MA	3.2 ± 9.34	4.8 ± 11.83	0.1 ± 0.44	0.3 ± 0.97	0	0.4 ± 1.23	0	0.7 ± 2.10	0.1 ± 0.44	4.4 ± 8.18	0
	MI	16 ± 16.73	10.66 ± 16.18	2 ± 2	3 ± 1.41	0	4 ± 0	0	4.66 ± 3.78	2 ± 2	9.77 ± 9.99	0
	P	100 -	70 (65-74)	85 (79-89)	55 (50-60)	0	80 (76-88)	0	70 (65-74)	70 (65-74)	5 (1-8)	15 (11-20)
14	MA	195.2 ± 247	79.5 ± 130	42.3 ± 76.69	20.2 ± 51.85	0	59 ± 76.07	0	3.3 ± 4.25	7.65 ± 11.54	0.4 ± 1.78	1.6 ± 4.76
	MI	198.20 ± 246	113.57 ± 144	49.76 ± 81.18	36.72 ± 66.64	0	73.75 ± 78.54	0	4.71 ± 4.39	10.92 ± 12.49	8 ± 8	10.66 ± 8.38
	P	95 (90-99)	90 (86-96)	60 (54-66)	50 (45-56)	0	70 (65-74)	0	40 (35-47)	45 (40-51)	0	10 (6-15)
15	MA	112.4 ± 51.73	82.35 ± 84.85	13.4 ± 20.11	10.9 ± 17.05	0	48.35 ± 60.81	0	1.35 ± 1.87	4.05 ± 6.55	0	0.25 ± 0.71
	MI	118.31 ± 45.67	91.5 ± 84.62	22.33 ± 21.93	21.8 ± 18.70	0	69.07 ± 62.15	0	3.37 ± 1.30	9 ± 7.21	0	2.5 ± 1.15
	P	85 (54-66)	35 (32-40)	30 (26-36)	50 (45-56)	0	25 (21-29)	15 (11-20)	35 (32-40)	70 (65-74)	5 (1-8)	5 (1-8)
16	MA	93.4 ± 145	18.55 ± 40.59	3.2 ± 8.46	7.3 ± 10.75	0	8 ± 22.05	0.85 ± 2.80	3 ± 8.73	5.45 ± 7.10	0.15 ± 0.68	0.05 ± 0.22
	MI	109.88 ± 104	53 ± 53.60	10.66 ± 12.81	14.6 ± 10.91	0	32 ± 35.29	5.66 ± 5.50	8.57 ± 13.22	7.78 ± 7.25	3 ± 3	1 ± 1
	P	100 -	60 (54-66)	35 (32-40)	30 (26-36)	0	55 (50-60)	5 (1-8)	10 (6-15)	0	0	0
17	MA	153.25 ± 118	47.9 ± 62.67	7.1 ± 11.58	3.65 ± 8.48	0	28.95 ± 36.67	0.05 ± 0.22	0.15 ± 0.48	0	0	0
	MI	158.25 ± 120	79.83 ± 63.26	20.28 ± 10.62	12.16 ± 12.20	0	52.63 ± 36.15	1 ± 1	1.5 ± 0.70	0	0	0
	P	70 (65-74)	85 (54-66)	80 (76-88)	40 (35-47)	0	55 (50-60)	0	30 (26-36)	50 (45-56)	0	15 (11-20)
18	MA	12.15 ± 11.24	30.4 ± 52.74	11.4 ± 11.21	2.2 ± 3.98	0	3.6 ± 5.28	0	1 ± 2.15	4.7 ± 8.20	0	0.25 ± 0.71
	MI	17.35 ± 9.34	35.76 ± 55.67	14.25 ± 10.76	5.5 ± 4.72	0	6.54 ± 5.64	0	3.33 ± 2.87	9.4 ± 9.64	0	1.66 ± 1.15
	P	90 (86-96)	85 (54-66)	65 (59-69)	40 (35-47)	0	75 (69-80)	0	10 (6-15)	15 (11-20)	0	15 (11-20)
19	MA	34.35 ± 22.52	18.75 ± 14.71	8.7 ± 14.18	10.05 ± 28.94	0	6.9 ± 8.96	0	0.2 ± 0.61	0.25 ± 0.63	0	0.25 ± 0.71
	MI	38.16 ± 20.31	22.05 ± 13.39	13.38 ± 15.62	25.12 ± 40.99	0	9.2 ± 9.29	0	2 ± 0	1.66 ± 0.57	0	1.66 ± 1.15
	P	95 (90-99)	80 (76-88)	70 (65-74)	70 (65-74)	0	65 (59-69)	20 (16-26)	55 (50-60)	35 (32-40)	10 (6-15)	60 (55-66)
20	MA	101.1 ± 144	18.5 ± 19.37	19 ± 19.54	8.4 ± 9.38	0	28.45 ± 47.53	5.3 ± 11.62	2.9 ± 6.17	14.55 ± 29.46	0.1 ± 0.30	20.35 ± 44.17
	MI	106.42 ± 146	23.12 ± 19.32	27.14 ± 18.69	12 ± 9.34	0	43.76 ± 54.54	26.5 ± 11.93	5.27 ± 7.86	41.57 ± 39.35	1 ± 0	33.91 ± 53.37
	P	5 (1-8)	80 (76-88)	15 (11-20)	20 (16-26)	15 (11-20)	0	0	50 (45-56)	55 (50-60)	0	5 (1-9)
21	MA	0.1 ± 0.43	9.4 ± 8.45	1.4 ± 4.01	2.2 ± 5.77	0.4 ± 1.02	0	0	3.95 ± 6.51	21.9 ± 36.76	0	0.05 ± 0.21
	MI	2 ± 2	11.75 ± 7.75	9.33 ± 7.02	11 ± 9.59	2.66 ± 1.15	0	0	7.9 ± 7.60	39.81 ± 43.27	0	1 ± 1
	P	40 (35-47)	60 (54-66)	0	7 (4-11)	0	13 (9-18)	7 (4-11)	5 (1-8)	33 (29-39)	20 (16-26)	0
22	MA	68.93 ± 165	19.26 ± 44.91	0	13.33 ± 51.63	0	0.93 ± 2.71	0.8 ± 3.09	2.4 ± 4.25	4.93 ± 9.83	6.53 ± 20.09	0
	MI	73.85 ± 171	32.11 ± 55.36	0	200 ± 200	0	7 ± 4.24	12 ± 12	4.5 ± 5.04	14.8 ± 12.47	32.66 ± 39.3	0

	P	64 (61-69)	43 (39-49)	0	0	0	0	0	0	14 (11-19)	0	0
23	MA	76 ± 112	4.71 ± 7.25	0	0	0	0	0	0.07 ± 0.26	0.28 ± 0.82	0	0.07 ± 0.26
	MI	118.22 ± 122	11 ± 7.34	0	0	0	0	0	0	2 ± 1.41	0	1 ± 1
	P	60 (54-66)	20 (16-26)	13 (9-18)	20 (16-26)	0	0	0	30 (26-36)	60 (54-66)	0	0
24	MA	3.33 ± 4.13	0.8 ± 1.69	0.26 ± 0.70	0.46 ± 0.99	0	0	0	20.6 ± 44.96	47.73 ± 55.84	0	0
	MI	5.55 ± 4.00	4 ± 1	2 ± 0	2.33 ± 0.57	0	0	0	38.62 ± 56.99	79.55 ± 51.09	0	0
	P	20 (16-26)	30 (26-36)	15 (11-20)	25 (21-29)	0	20 (16-26)	0	20 (16-26)	10 (6-15)	5 (1-8)	35 (31-40)
25	MA	0.65 ± 1.53	0.5 ± 1	0.15 ± 0.36	0.35 ± 0.67	0	3.6 ± 12.96	0	0.65 ± 2.03	0.5 ± 1.67	2.7 ± 12.07	0.65 ± 1.38
	MI	3.25 ± 1.89	1.66 ± 1.21	1 ± 0	1.4 ± 0.54	0	18 ± 26.82	0	3.25 ± 3.86	5 ± 2.82	54 ± 0	1.85 ± 1.86
	P	80 (76-88)	100 -	50 (45-56)	30 (26-36)	10 (6-15)	70 (65-74)	0	40 (35-47)	60 (54-66)	0	0
26	MA	210.4 ± 240	161 ± 161	8.6 ± 9.47	2.4 ± 5.64	0.2 ± 0.63	28.2 ± 28.30	0	2.1 ± 3.03	31.6 ± 39.86	0	0
	MI	263 ± 241	161 ± 161	17.2 ± 4.14	8 ± 8.71	2 ± 0	40.28 ± 25.17	0	5.25 ± 2.36	52.66 ± 39.09	0	0
	P	50 (45-56)	80 (76-88)	60 (54-66)	30 (26-36)	0 0	50 (45-56)	0	0	80 (76-88)	10 (6-15)	20 (16-26)
27	MA	47.2 ± 64.86	24 ± 21.41	3 ± 3.91	1.8 ± 3.32	0	7.6 ± 14.13	0	0	5.4 ± 6.94	0.1 ± 0.31	0.2 ± 0.42
	MI	94.4 ± 62.42	30 ± 19.59	5 ± 3.94	6 ± 3.46	0	15.2 ± 17.46	0	0	6.75 ± 7.18	1 ± 0	1 ± 0
	P	100 -	100 -	77 (71-83)	11 (7-16)	0	77 (71-83)	11 (7-16)	22 (19-26)	0	0	33 (28-37)
28	MA	62.88 ± 41.14	21.33 ± 16.30	4 ± 3	0.22 ± 0.66	0	7.77 ± 7.03	0.66 ± 2	0.22 ± 0.44	0	0	0.77 ± 1.39
	MI	60.88 ± 41.14	25.33 ± 16.30	5.14 ± 2.26	2 ± 0	0	10 ± 4.9	6 ± 0	1 ± 0	0	0	2.33 ± 1.52
	P	80 (76-88)	100 -	40 (35-47)	20 (16-26)	0	20 (16-26)	0	0	80 (76-88)	0	0
29	MA	2.2 ± 2.28	40.8 ± 24.35	0.6 ± 0.89	0.4 ± 0.89	0	10 ± 22.36	0	0	1.2 ± 1.09	0	0
	MI	2.75 ± 2.21	40.8 ± 24.35	1.5 ± 1	2 ± 0	0	50 ± 0	0	0	1.5 ± 1	0	0

Abbreviations: Cs, Cichlidogyrus sclerosus; Ct, Cichlidogyrus tilapiae; Cd, Cichlidogyrus dossoui; Cl, Cichlidogyrus longicornis; Cq,Cichlidogyrus quaestio; Ci, Cichlidogyrus sp.; Ch, Cichlidogyrus halli; Gy, Gyrodactylus cichlidarum; Tr, Trichodina sp.; Vo,Vorticella sp.; En, Enterogyrus malmbergi

Prevalence = P (CI-95 %), Mean abundance (± SD) = MA, Mean intensity (± SD) = MI, CI = Confidence intervals, SD = Standard deviation

*Expected apparent prevalence following [16]

Geographical Distribution of the Parasites Found in Nile tilapia Throughout Farms

The 29 operating Nile tilapia farms were roughly located along the northwestern coast and southwestern edge of Yucatán State. The map of the percentage of infected hosts, including the most prevalent parasite species, is shown in Fig. 1. For each bar chart (=farm), the larger the bar, the higher the percentage of infected hosts for each parasite species. One of the most important parasites was C. sclerosus (indicated by the bar number one in the chart in the map), with a prevalence between 80 and 100 % per farm (Table 2). The map showing the mean abundance of parasites in farmed Nile tilapia is presented in Fig. 2. In this map, the larger the radius of the pie chart, the higher the total number of parasites was at each farm and the larger the slice, the higher the mean abundance of parasites. The most abundant parasite species were the monogeneans C. sclerosus and C. tilapiae, which, in most farms, represented 50 % or more of the individual helminths collected (Table 2). Moreover, the greatest number of individual helminths was collected from farms along the coast and in northwestern Yucatán State. Gyrodactylus cichlidarum, a very important parasite due to its pathogenicity, was present in 26 of the 29 extant Nile tilapia farms in Yucatán, with a prevalence ranging between 5 % (in farms 5 and 21) and 70 % (in farm 13), and with a mean abundance ranging from 0.05 ± 0.22 (in farm 6) to 20.60 ± 44.96 (in farm 24) (Table 2). The names of the farms were not provided by request of the farm owners.

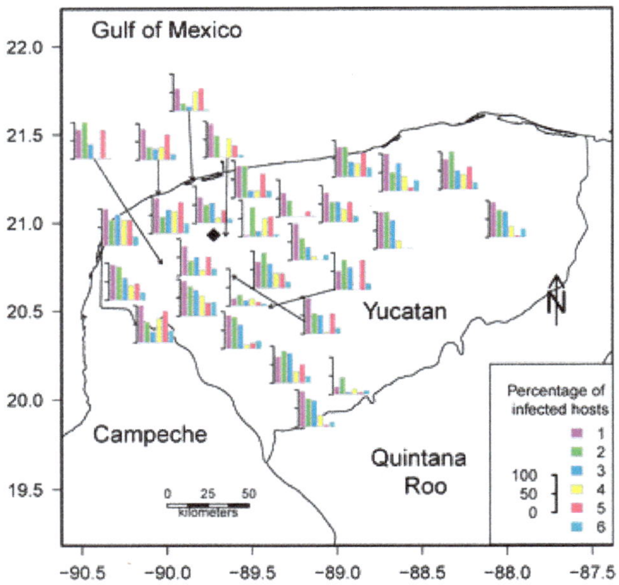

Figure 1. Map of the percentage of infected Nile tilapia Oreochromis niloticus (L.) from 29 farms in Yucatán, Southern México. The bar chart represents the percentage of fish infected with parasites in each tilapia farm. Each bar corresponds to a parasite species. The bigger the bar, the higher the percentage of infected hosts by each parasite species. The parasites are represented by a number as follows: 1, Cichlidogyrus sclerosus; 2, Cichlidogyrus tilapiae; 3, Cichlidogyrus dossoui; 4, Gyrodactylus cichlidarum; 5, Trichodina sp.; 6, other parasites (Cichlidogyrus longicornis, Cichlidogyrus quaestio, Cichlidogyrus sp., Cichlidogyrus halli, Vorticella sp., Enterogyrus malmbergi). ♦ = Mérida City

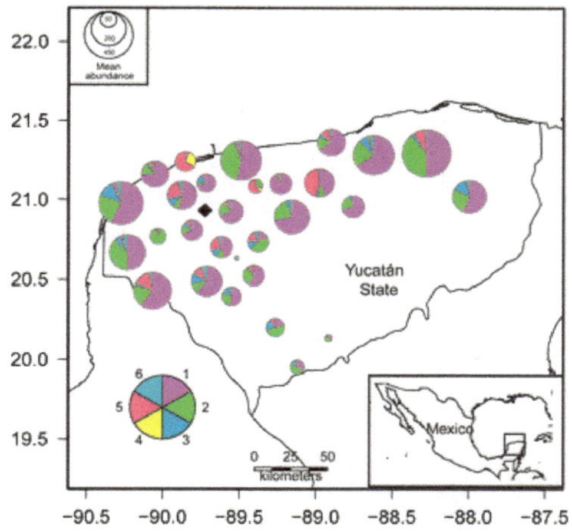

Figure 2. Map of mean abundance of the parasite species infecting the Nile tilapia Oreochromis niloticus (L.) from 29 farms in Yucatán, Southern México. The radius of each pie chart represents the total number of individual parasites of all species found in a Nile tilapia farm. Each slice corresponds to a parasite species. The bigger the slice, the higher the mean abundance of each parasite species. The parasites are represented by a number as follows: 1, Cichlidogyrus sclerosus; 2, Cichlidogyrus tilapiae; 3, Cichlidogyrus dossoui; 4, Gyrodactylus cichlidarum; 5, Trichodina sp.; 6, other parasites (Cichlidogyrus longicornis, Cichlidogyrus quaestio, Cichlidogyrus sp., Cichlidogyrus halli, Vorticella sp.,Enterogyrus malmbergi). ♦= Mérida City

Clustered patterns of the Parasitic helminths and Protozoans of Nile tilapia in Yucatán

The geographical distribution pattern for the mean density (= mean abundance) of the parasite species showed a clustered pattern (see Additional file 2). The red clusters in the maps of Fig. 3 present a high mean density for each one of the parasite species collected from Nile tilapia in this study, and the blue zones are related to low mean density of these parasites. Most of the clusters of parasites were found near the coast, mainly in northwestern Yucatán, where most farms are located. The highest concentration of C. sclerosus was located in the northwest area of Yucatán (Fig. 3a). The highest mean density for G. cichlidarum depicted a similar spatial distribution pattern to that of C. sclerosus but with lower values (Fig. 3f).

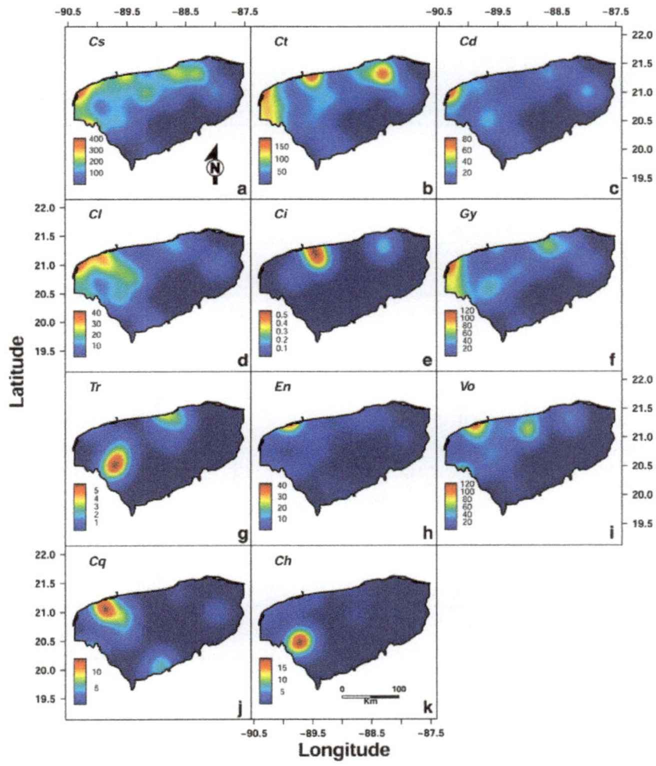

Figure 3. Spatial clustering of the density (mean abundance) of the ectoparasite species infecting the Nile tilapia Oreochromis niloticus (L.) in Yucatán, Southern México. Red coloured cluster zones indicate higher mean density of the parasite species. The lighter toned cluster zones indicate lower mean density of the parasite species. The parasites are represented by acronyms as follows: a Cs, Cichlidogyrus sclerosus; b Ct, Cichlidogyrus tilapiae; c Cd, Cichlidogyrus dossoui; d Cl, Cichlidogyrus longicornis; e Ci, Cichlidogyrus sp.; f Gy, Gyrodactylus cichlidarum; g Tr, Trichodina sp.; hEn, Enterogyrus malmbergi; i Vo, Vorticella sp.; j Cq, Cichlidogyrus quaestio; k Ch, Cichlidogyrus halli

The Questionnaire and the Redundancy Analysis

The results of the questionnaire applied to the farm owners (Table 3) indicated the presence of two groups of Nile tilapia farms in Yucatán. The first group had low technology with only a small number of concrete tanks (<4), low annual production (200–500 kg) with sub-standard sanitary measures, a lack of biosecurity knowledge, and represented 65 % of the total farmers intered. The second group included medium- to high-technology farms with better facilities including a higher numbers of tanks (>5), quarantine areas, guardhouses, better biosecurity measures in place and high annual production (1,000–25,000 kg). This group represented 35 % of the total farmers intered.

Table 3. List of environmental variables and management practices obtained from Nile tilapia production units between 2012 and 2013 in Yucatán State

Management practices (n=32)	**Low technology farms**	**Medium to high technology farms**
Environmental variables (7)		
Fish age (months)	2–4.5	2.5–6.0
Mean annual production (kg)	200–500	1,000–25,000
Water supply	<1 km	>1 km
Number of tanks on the farm	1–4	5–13
Tank capacity (m^3)	4–6.5	12–16
Fish density per tank	950–4,500	5,000–10,000
Number of workers	1–2	4–9
Nearby communities	<1 km	>1 km
Fingerlings source	Campeche	Campeche – Tabasco
Type of tanks	Concrete – geomembrane	Concrete – geomembrane
Epidemiological outbreaks	Parasites, bacteria, fungi	Parasites, bacteria
Prophylactic treatments	No	Salt, Crustabay
Quarantine area	No	Yes
Vector presence	pigs, cattle, poultry, dogs and cats	Sometimes dogs
Guardhouse	No	Yes
Footbaths	No use	Yes
Tank cleaning	No disinfection	Virkon, chlorine
Bathroom	Latrine / No use	Bathroom
Disinfection of equipment	No disinfection	Virkon, chlorine
Warehouse	Galley / Lacking	Cement
Drug use	No	Antibiotics
Incineration of organisms	No	Burying and incineration
Education level of workers	No education	Elementary school, sometimes preparatory
Drainage	No drainage	With drainage
Communication	Roadless	With road
Livestock areas	Poultry and pigs	Poultry
Agricultural areas	Yes	Yes

Accumulation of garbage	Yes	No
Accumulation of dead animals	Yes	Sometimes
Water exchange rate	Monthly and fortnightly	Weekly and fortnightly
Water discharge treatment	No	Sometimes
Other land uses	Husbandry and livestock	Sometimes livestock
Nitrates (mg/l)	4.21–7.57	0.10–5.0
Nitrites (mg/l)	0.002–2.97	1.83–3.10
Ammonium (mg/l)	1.85–3.38	0.01–1.68
Salinity (ppt)	0.30–0.70	0.83–2.11
Dissolved oxygen (mg/l)	1.98–7.68	7.78–13.61
Conductivity (μS/cm)	1044–1698	1831–4364
Temperature (T °C)	23.10–31	28.92–32

The results of the redundancy analysis (RDA) using the abundance of each parasite species in each individual host as a dependent variable concur with those obtained by the questionnaire, splitting the farms into two groups with different types of ectoparasites. The RDA accounted for 90 % of the total variance and was significant for the first axis (F=13.84; P=0.04; 4,999 permutations) and all canonical axes (F=2.65; P=0.03; 4,999 permutations) (Fig. 4). The first group of farms had medium to high technology and production (farms labelled with diamonds in Fig. 4), where the fish density per tank, number of workers, the presence of a quarantine area, bathrooms, and the concentration of nitrites (in black arrows) were positively associated with the abundance of G. cichlidarum,Trichodina sp. and Cichlidogyrus halli (in dashed arrows) (Fig. 4). Due to its pathogenicity, a further analysis of G. cichlidarum was undertaken using general additive models to determine the potential statistical relations between the independent management and environmental variables at the farms and the abundance of this parasite species. The second group (farms labeled with filled circles) included farms with low technology, which showed positive associations between the abundance of several species of Cichlidogyrus and Vorticella sp. and two management variables (ammonium and water exchange rate) (Fig. 4).

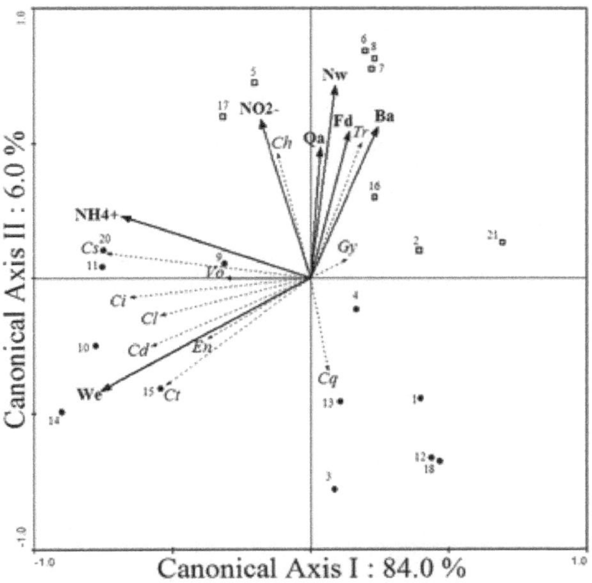

Figure 4. Redundancy analyses (RDA) showing the statistical association between the number of ectoparasite species of the Nile tilapia Oreochromis niloticus (L.) and the environmental and management variables of the farms. The acronyms for the parasite species were as follows: Cs, Cichlidogyrus sclerosus; Ct, Cichlidogyrus tilapiae; Cd, Cichlidogyrus dossoui; Cl,Cichlidogyrus longicornis; Cq, Cichlidogyrus quaestio; Ci, Cichlidogyrus sp.; Ch, Cichlidogyrus halli; Gy, Gyrodactylus cichlidarum; Tr, Trichodina sp.; Vo, Vorticella sp.; En, Enterogyrus malmbergi. Codes for management and environmental variables (in bold) are as follows: Fd, Fish density per tank; Nw, Number of workers; Qa, Quarantine; NO_2^-, Nitrites; Ba, Bathrooms; NH_4^+, Ammonium; We, Water exchange. The numbers associated with the circles and squares are the farm reference numbers in Table 3

Statistical Relations between the Management and Environmental Variables of Nile tilapia farms and the mean abundance of *Gyrodactylus cichlidarum*

A total of 39 environmental and management variables (Table 3) were originally included in the GAMLSS model. However, after the stepwise procedure, only four independent variables were selected for the final GAMLSS model: one environmental (dissolved oxygen) and three management variables (tank capacity, use of quarantine area and use of prophylactic treatments). The model including these four variables showed 63 % deviance and had the lowest values of global deviance (GD=48.05) and of the Akaike information criterion (AIC=74.05; Table 4). Additional file 3shows the nonlinear relations of each of the management and environmental independent variables with respect to the dependent variable (G. cichlidarum mean abundance). In regard to treatments, in Additional file 3, there were four treatments: formalin, salt, organophosphorus, and no treatment. However, only the last one (no treatment) was selected by the stepwise procedure.

Table 4. Generalized additive model of location, shape and scale (GAMLSS) fits for mean abundance of Gyrodactylus cichlidarum (MAG; dependent variable) in Nile tilapia farms in Yucatán state. The best model was selected using a stepwise procedure and the lowest values of Akaike information criterion (AIC) and global deviance. The final model was: MAG ~ cs(Do) + cs(Tc) + Qa + Pt where cs is a cubic spline smooth function, and those independent variables without smoother had a linear relationship with the dependent variable (MAG)

Model	Degrees of freedom (df)	Global deviance	Deviance explained (%)	AIC	*P* (0.05)
(Null model) MAG ~ 1	-	130.82	-	132.82	$5.4e^{-6}$*
MAG ~ cs(Do)	df=16	100.52	23.16	110.51	0.0002*
MAG ~ cs(Do) + cs(Tc)	df=16	92.65	29.17	110.65	0.0059*
MAG ~ cs(Do) + cs(Tc) + Qa	df=19	86.29	34.03	106.29	0.0208*
MAG ~ cs(Do) + cs(Tc) + Qa + Pt	df=17	48.05	63.27	74.05	0.0001*

Abbreviations: Do Dissolved oxygen (mg/l), Tc Tank capacity (m^3), Qa Use of quarantine area, Pt Use of prophylactic treatments

*P value < 0.05

DISCUSSION

The results showed a clustered pattern for the geographical distribution of the abundance for most of the parasite species of the Nile tilapia in Yucatán. Cichlidogyrus sclerosus, the most frequent and abundant monogenean species, showed a clustered pattern with the highest mean density (= mean abundance) in the northwestern corner of the state. A similar pattern was observed for the other species of Cichlidogyrus (C. tilapiae, Cichlidogyrus dossui, and Cichlidogyrus longicornis), and for G. cichlidarum (Fig. 3). This result, together with the RDA multivariate analysis, suggested that the high- and low-technology Nile tilapia farms in Yucatán have specific suites of ectoparasites, depending on their management and environmental variables. Thus, in medium- to high-technology Nile tilapia farms, G. cichlidarum, Trichodina sp., and C. halli were more prevalent and abundant, whereas the low-technology farms had more C. sclerosus, C. tilapiae, C. dossoui, C. longicornis, Cichlidogyrus quaestio,Cichlidogyrus sp., Vorticella sp. and Enterogyrus malmbergi.

Parasite Species Composition of Nile tilapia farms

The monogenean and protozoan parasites found in O. niloticus from the farms (Table 1) have previously been reported in Yucatán State [9]. In the specific case of the monogeneans of the genus Cichlidogyrus, they have also been recorded in farms and wild crops of tilapia throughout different regions of México and the world [6, 7, 9, 10,18]. The important point here was that these ectoparasites from the Yucatán aquaculture facilities were reported in relatively high numbers (Table 2). This is important because the few extant reports on the histopathology produced by Cichlidogyrus spp. in O. niloticus indicate that even at low numbers, these monogeneans can cause slight gill disease with hyperplasia, edema and epithelial sloughing [34]. Furthermore, the attachment structures (anchors) and cutaneous disorders (epithelial hyperplasia and proliferation of mucoid cells) can be an entry point for bacterial and fungal infections [34]. Although the present paper did not include a histopathological study, we consider that under the present circumstances of high prevalence and mean abundance of Cichlidogyrus spp. in most of the farms studied (Table 2), a histopathological study is necessary. Due to their direct life-cycles and attachment organs (hooks), these ectoparasites are likely to affect the intensive culture of Nile tilapia, particularly in combination with substandard management practices, high fish densities and favorable ecological conditions (e.g. high temperature) with regard to fish development [9]. Vidal-Martínez et al. [18, 19] have reported several species of the genus Cichlidogyrus in Oreochromis aureus (Steindachner) and O. niloticus from three southeastern Mexican states (Tabasco, Veracruz and Yucatán), suggesting that under deficient management conditions, these monogeneans can enhance parasite-induced host mortality [18]. Massive invasions of monogeneans of the genus Cichlidogyrus have also been reported in tilapia in Cuba, Colombia, the Philippines and Brazil [9].

Gyrodactylus cichlidarum was reported for the first time in Yucatán by [35] but only in one locality (Mérida). Consequently, all 26 farms in which this species has been reported in the present paper are new geographical records. Gyrodactylus cichlidarum infected up to 31 % of the Nile tilapia in these farms, whereas the protozoans Trichodina sp. and Vorticella had prevalences of 8 % and 41 %, respectively (Table 1). The high fish densities and substandard management practices in several Nile tilapia farms likely facilitate constant reinfection with these ectoparasites, which could be leading to host mortality. The presence of Gyrodactylus is important for Nile tilapia farming in Yucatán due to the pathogenicity of the species of this genus in other fish species [19, 35]. Only a single study has reported a fairly weak case of histopathology due to G. cichlidarum related to the mortality of fingerlings of farmed Nile tilapia in Peru [36]. However, Martins et al. [37] recently reported that tilapia vaccinated forStreptococcus iniae but parasitised with Trichodina heterodentata Duncan, 1977 and G. cichlidarum had significantly lower antibody levels and survival rates than tilapia that were not parasitised. Clearly, more histopathological and physiological studies on the response of tilapia to these species of ectoparasites are urgently needed.

Geographical Distribution of the Parasites of Nile tilapia among farms in Yucatán

Most of the sampled Nile tilapia farms were located on the northwestern border of the coast and in the southeastern Yucatán State. The farms located in the coastal zone and along the northwestern border had higher levels of infection than the farms located in southeastern Yucatán. This could be due to the higher fish densities managed by these farms and greater commercial interaction among the farms in comparison with those from other regions of the state. Nabeil-Salama & Murray [38] found that the rate of infection by pathogens in the smaller, more separated fish farms decreased, whereas in larger, nearby farms, the spread of infection throughout the individuals was almost instantaneous. However, within these medium- to high-technology farms, the parasites persist but are clearly not considered to be pathogenic by the owners. We believe that this assumption is a mistake because these parasites can become harmful under the challenging environmental circumstances typical of fish farms (e.g. high temperature and productivity, low water exchange rate, and high fish density) [39]. Mardones et al. [40] found that the geographical dispersion of the salmon anemia virus in Chile is linked to anthropogenic activities. Thus, the movement of live or harvested fish or their byproducts was suspected to have played a more important role than environment or passive transmission. We observed fluctuations in the levels of infection of C. sclerosus and C. tilapiae, which could be due to risk management practices that influence the life-cycles of these parasites. Most Nile tilapia farms in Yucatán have substandard health conditions such as poor filter cleaning, lack of disinfection of equipment and tanks, and poor water quality, which in turn could favor constant parasitic reinfection. Akoll et al. [41] found that frequent seining and pond flushing can interrupt C. sclerosus and C. tilapiae transmission by washing out the oncomiracidia. Several authors have suggested that the transport of live fish, high stocking density, low frequency of pond drainage and a lack of disinfection of canoes and angling equipment are some of the most relevant management practices that enhance the transmission of ectoparasites in farms [42, 43].

Clustered Patterns of the Parasitic Helminths and Protozoans of Nile tilapia in Yucatán

The estimated density of the clusters was consistent with the known abundance values of the parasite species in Nile tilapia farms (Table 2; Figs. 1 and 2). The most likely explanation for the high values of both estimated and actual abundance of these parasites in the farms were the high fish densities and deficient sanitation management, as well as increased anthropogenic and commercial interaction favouring parasite spread among farms throughout the coastal area of Yucatán. In fact, one of the main risk factors enhancing the dispersion and establishment of ectoparasites in the farms in Yucatán is likely to be the hatcheries producing infected Nile tilapia fingerlings and selling them without appropriate sanitary measures. Even when considering the difference between the discrete nature of the Nile tilapia farms based on the mainland Yucatán and those in open systems such as the Norway fjords, there are still some important similarities. The farms with high commercial interaction play a critical role in the spread of pathogenic infections in Norway [44]. In fact, fish farms located near (<5 km) to a harvest station have often been identified as a risk factor for disease transmission [44]. Harvest stations can be a source of infection to adjacent farms via other pathways such as fomites (e.g. fishing equipment and vehicles) [45, 46]. Clearly, there is a large need for the application of biosecurity measures such as those defined by the World Organization for Animal Health (OIE) [4] to diminish, as much as possible, the spread and establishment of the ectoparasites of Nile tilapia among farms in Yucatán.

The questionnaire and Statistical Associations between Management and Environmental Variables for Specific types of Parasites in the Nile Tilapia Farms

The results of the questionnaire applied at each Nile tilapia farm indicated that most workers and owners have very basic levels of education (elementary school = 6 years at most), and many of them are illiterate, often lacking technical training in aquaculture (Table 3). These people also have a poor understanding of fish health management, which is reflected in the deficient management conditions strongly associated with high nitrite concentration levels and specific types of ectoparasites for both types of farms in Yucatán. The multivariate redundancy analysis (RDA; Fig. 4) concurs with the results obtained by the questionnaire, suggesting that the mean abundance of G. cichlidarum, Trichodina sp. and C. halli in medium- to high-technology farms was positively statistically associated with high fish densities per tank, a large number of workers, the lack of a quarantine area and high concentrations of nitrites. The farms where production was highest may be raising fish at higher densities and could be employing proportionately more staff than the farms where production is lower [46]. However, even under these more favourable circumstances, these farms are still infected with ectoparasites due to the lack of technical training of the farmers.

The second group of farms (low-technology farms), where the Nile tilapia were infected with several species of Cichlidogyrus, also had deficient management practices and substandard sanitary conditions such as a lack of preventive treatments, insufficient disinfection equipment and tanks, poor fingerling sources, and deficient water quality (Table 3). The mean abundance of these ectoparasites was also positively associated with high ammonium concentrations and a deficient water exchange rate. Farms in this group were technologically disadvantaged compared to the first farm group. The owners' lack of awareness in this farm group concerning fish disease or the adverse effects of disease outbreaks is reflected in the poor preventive and treatment measures for eliminating ectoparasites in the fingerlings before they are introduced into the production areas. Several farmers used chemical products such as organophosphates for eliminating parasites in the fingerlings but had little understanding of their use, effectiveness and ecological consequences (Table 3). Better results might have been obtained by improved management practices such as the use of salt baths because of their low cost and effectiveness against bacteria, viruses and parasites [47, 48]. It is highly probable that most farms from both groups have a few main suppliers of infected fingerlings, which likely serve as the most serious threat of ectoparasite transmission. This situation, coupled with the deficient management conditions of each Nile tilapia farm, is reflected in the high parasite prevalence. Faruk et al. [49] found that small rural farms in Bangladesh had a high prevalence of diseases such as white spot, epizootic ulcerative syndrome and red spot, when compared to larger, highly technological farms. This situation can be explained because low-technology farms lack proper assistance from the government, medicine is unavailable, and there is a lack of training and facilities for the treatment of fish diseases. Nile tilapia fingerlings should be treated prior to being transported to farms; however, to apply such treatments, a sensitivity analysis needs to be undertaken to determine the appropriate concentrations of chemicals for killing the parasite without affecting the fish [47]. In addition, proper quarantine programs need to be established prior to introducing fingerlings into a farm. Wagner et al. [50] found that farms in the United States (Alabama, Arkansas, Louisiana, and Mississippi) that did not produce their own fingerlings were more likely to report enteric septicemia problems compared to those that produced their own fingerlings. A reason for this phenomenon could be that the sanitary conditions of the farms that produce their own fingerlings would be well known, and stocking decisions would then be based on that knowledge. In addition, handling or transporting can produce stress in fish, as well as stocking outside appropriate temperature ranges, causing immunosuppression conducive to disease outbreaks [51]. Because of the association between deficient management practices and the increase in ectoparasitic infections, technical training programs are needed to improve the management practices of both workers and owners of the Nile tilapia farms in Yucatán.

Effect of Management and Environmental Variables on the mean Abundance of *Gyrodactylus Cichlidarum*

The General Additive Model for Location Scale and Shape (GAMLSS; Table 4) suggested that the use of a quarantine area and prophylactic treatments, together with low values of dissolved oxygen (5–8 mg/l) in large tanks (12–14 m^3), increase the mean abundance of G. cichlidarum for high-tech farms. Thus, in those farms, the newcomers (fry or broodstock) are not being appropriately isolated in specific quarantine facilities, examined for parasites and, when necessary, treated before their release into the production areas. The absence of the adequate application of these management procedures clearly increases the probability of the transmission and reproduction of G. cichlidarum not only in the newcomers but also in the Nile tilapia already present in the farms. To the best our knowledge, there are no safe concentrations for sodium chloride or other substances for therapeutic control of G. cichlidarum. Clearly, the most effective concentrations for these substances for killing the parasite without killing the fish need to be determined [47]. In cultured Nile tilapia, outbreaks of gyrodactylosis have been suggested as being responsible for the high mortality of juvenile fish in a number of countries worldwide [52]. Studies in temperate latitudes have reported that risk factors such as the reuse of the equipment (especially hand nets without adequate disinfection) and the movement of farm equipment among tanks have caused the spread of Gyrodactylus salarisMalmberg, 1957 between salmon farms in Norway [45, 52]. Therefore, good sanitary practices together with proper therapeutic treatments are necessary to control the dispersion and establishment of G. cichlidarum within and between Nile tilapia farms in Yucatán.

CONCLUSION

The Nile tilapia farms in the state of Yucatán generally have substandard management conditions that play an important role in the high occurrence of parasitic infections. Both the geographical and multivariate analyses sought to identify high-risk locations and factors that might be associated with risky practices. Several of these practices were related to the distribution of infected fingerlings among farms, especially in the northwestern corner of Yucatán, the region where the most frequent and abundant parasites showed significant clusters. However, something that both high- and low-technology farms shared was the urgent need for proper technical training programs to improve the management and sanitary practices of both workers and owners of the Nile tilapia farms. We consider Yucatán a good model when trying to understand the sanitary circumstances of Nile tilapia farming in a developing country, with the results obtained here applying to rural areas of most countries in the Neotropics. We certainly acknowledge the consideration of FAO with respect to the fast growth of aquaculture worldwide [1], but if the benefit of this growth is to reach people living in the rural regions of these Neotropical countries, proper technical training for farmers is urgently needed to meet this expectation.

DECLARATIONS

Acknowledgements

This study is a part of the Ph. D. thesis of A.I.P.T who is grateful to CONACYT-México (213446) for granting a 4-year ship. Sincere and grateful thanks are extended to all at Laboratorio de Patología Acuática CINVESTAV-IPN Unidad Mérida including Clara Vivas, Gregory Arjona, Raúl Sima, Nadia Herrera, Francisco Puc, Geny Aíl, and Arturo Centeno for technical support. Thanks are also extended to the Yucatán Aquatic Animal Health Committee (CESAY), including Arturo González, Miroslava López and Mario Armenta who helped with field work and to the owners Nile tilapia farms from Yucatán state.

REFERENCES

1. FAO. El estado mundial de la pesca y la acuacultura 2010. ISBN 978-92-5-306675-9. Roma; 2010.
2. Engle R. Economics of tilapia aquaculture. In: Tilapia aquaculture in the Americas. J World Aquacult Soc. 1997;1:18–33.
3. Watanabe W, Losordo T, Fitzsimmons K, Hanley F. Tilapia production systems in the Americas: technological advances, trends, and challenges. Fisheries Sci. 2002;3:465–98.
4. Berthe F, Jie H, Vidal-Martinez V, Ernst I, Hjeltnes B, Gallardo-Lagno A. World Organization for Animal Health (OIE). ISBN: 978-92-9044-942-3277. Paris; 2014.
5. Conroy G, Conroy D. Importantes enfermedades detectadas en tilapias cultivadas en América Latina. Panorama Acuíc. 2004;6:20–5.
6. Florio D, Gustinelli A, Caffara M, Turci F, Quaglio F, Konecny R, et al. Veterinary and public health aspects in Nile tilapia (Oreochromis niloticus niloticus) aquaculture in Kenya, Uganda and Ethiopia. Ittiopatol. 2009;6:51–93.
7. Jiménez-García I, Rojas-García R, Castro-José C, Pavón-Suriano S, Lango-Reynoso F, Castañeda-Chávez R. Growth enhancement, survival and decrease of ectoparasitic infections in masculinized Nile tilapia fry in a recirculating aquaculture system. Trop Subtrop Agroecosyst. 2012;15:51–6.
8. Zamri-Saad M, Amal A, Siti A, Zulkafli A. Control and prevention of Streptococcosis in cultured tilapia in Malaysia: a re. Trop Agr Sci. 2014;37:389–410.
9. Flores-crespo J, Flores-Crespo R. Monogeneos, parásitos de peces en México: estudio recapitulativo. Rev Mex Cienc Pecu. 2003;41:175–92.
10. Jiménez-Garcia I, Vidal-Martínez V, López-Jiménez S. Monogeneans in introduced and native cichlids in México: evidence for transfer. J Parasitol. 2001;87:907–9.
11. Brooker S, Beasley M, Ndinaromtan M, Mobele E, Djenguinabe E, Simon I. Use of remote sensing and a geographical information system in a national helminth control programme in Chad World Health Organization. J Trop Medic Hyg. 2002;80:10–20.
12. Vanderstichel R, St-Hilaire S, Ibarra R, Lyngstad T, Rees E, Medina M. Space-time cluster analysis of the non-pathogenic infectious salmon anemia virus (HPR0 ISAV) in Chile, 2011–2012. Aquaculture. 2014;427:120–7.
13. Thrush A, Murray G, Brun E, Wallace S, Peeler E. The application of risk and disease modelling to emerging freshwater diseases in wild aquatic animals. Freshwater Biol. 2001;56:658–75.
14. Peeler E, Taylor N. The application of epidemiology in aquatic animal health-opportunities and challenges. Vet Res. 2011;42:94–8.
15. Bush AO, Lafferty KD, Lotz JM, Shostak AW. Parasitology meets ecology on its own terms. Margolis et al. Revisited. J Parasitol. 1997;83:575–83.
16. Des Clers S. Sampling to detect infections and estimate prevalence in aquaculture. Stirling, Pisces Press; 1994. p. 2-30.
17. Ojurongbe O, Adegbosin O, Taiwo S, Alli O, Olowe O, Ojurongbe T, et al. Assessment of Clinical Diagnosis, Microscopy, Rapid Diagnostic Tests, and Polymerase Chain Reaction in the Diagnosis of Plasmodium falciparum in Nigeria. Malaria Res Treat. 2013;2013:1–5.
18. Vidal-Martínez V, Aguirre-Macedo L, González-Solís D, Mendoza- Franco E. Atlas of the helminth parasites of cichlid fish of México. Prague: Academia; 2001.
19. Sanchez-Ramirez C, Vidal-Martínez V, Aguirre- Macedo L, Rodriguez-Canul R, Gold-Bouchot B. Cichlidogyrus sclerosus(Monogenea: Ancyrocephalinae) and its host, the Nile tilapia (Oreochromis niloticus), as bioindicators of chemical pollution. J Parasitol. 2007;93:1097–106.
20. Lane B. Operation manual YSI model 85. Yellow Spring, Ohio: YSI incorporated; 1998.
21. Douëllou L. Monogeneans of the genus Cichlidogyrus Paperna, 1960 (Dactylogyridae: Ancyrocephalinae) from cichlid fishes of Lake Kariba (Zimbabwe) with descriptions of five new species. Syst Parasitol. 1993;25:159–86.
22. Hassan M. Trichodiniasis in farmed freshwater tilapia in eastern Saudi Arabia. J Mar Sci. 1999;10:157–68.
23. Gerritsen H. Package mapplots. Reference Manual. 2015. https://cran.r-project.org/web/packages/mapplots/mapplots.pdf. Accessed 11 Mar 2015.
24. Baddeley A, Turner R. Implementing algorithms for spatial point pattern analysis in R. J Appl Stat. 2005;41:31–8.
25. Diggle P. Statistical Analysis of Spatial Point Patterns. London: Academic Press; 1983.
26. Baddeley A. Package spatstat. Reference Manual. 2015. https://cran.r-project.org/web/packages/spatstat/spatstat.pdf. Accessed 11 March 2015.
27. Brunette K, Nelson A. Aspectos básicos del análisis de datos: variables y distribución. Focus Field Epidemiol. 2003;3:5–10.
28. ter Braak C, Šmilauer P. CANOCO reference manual and user's guide to CANOCO for Windows: software for Canonical Community Ordination (version 4). New York: Microcomputer Power Ithaca; 1998. p. 254–350.
29. Sokal R, Rohlf F. Biometry. San Francisco: W.H. Freeman and Company; 1995. p. 130.
30. Jongman R, Ter Braak C, Van Tongeren O. Data analysis in community and landscape ecology. Cambridge: Cambridge University Press; 1995. p. 225.
31. Rigby R, Stasinopoulos D. Generalized additive models for location, scale and shape (GAMLSS) in R. J Stat Softw. 2005;23:507–54.
32. Akaike H. A new look at the statistical model identification. IEEE Trans Autom Control. 1974;19:716–23.

33. R Core Team R: A language and environment for statistical computing. R Foundation for Statistical Computing. 2014.http://www.R-project.org/. Accessed 2 january 2015.
34. Le Roux L: Aspects of the morphology, ecology and pathology of Cichlydogyrus philander collected from Pseudocrenilabrus philander philander in the Padda dam South Africa. University of Johannesburg; 2010. p. 103–105.
35. García-Vázquez A, Hansen H, Christison K, Rubio-Godoy M, Bron J, Shinn A. Gyrodactylids (Gyrodactylidae, Monogenea) infecting Oreochromis niloticus niloticus (L.) and O. mossambicus (Peters) (Cichlidae): A pan-global survey. Acta Parasitol. 2010;55:215–29.
36. Gonzales-Fernández J. Parasitofauna of tilapia cause mortalities in fingerlings in two fish farms, Lima, Peru. Neotrop Helminthol. 2012;6:219–29.
37. Martins M, Shoemaker C, Xu D, Klesius P. Effect of parasitism on vaccine efficacy against Streptococcus iniae in Nile tilapia. Aquaculture. 2011;314:18–23.
38. Nabeil-Salama K, Murray A. Farm size as a factor in hydrodynamic transmission of pathogens in aquaculture fish production. Aquat Environ Interac. 2011;2:61–74.
39. Ogut H, Reno P. Effects of fish density on spread of infectious Hematopoietic necrosis virus (IHNV) in rainbow trout,Oncorhynchus mykiss. Isr J Aquacult. 2005;218–255.
40. Mardones F, Perez A, Carpenter T. Epidemiologic investigation of the re-emergence of infectious salmon anemia virus in Chile. Dis Aquat Organ. 2009;84:105–14.
41. Akoll P, Fioravanti M, Konecny R, Schiemer F. Infection dynamics of Cichlidogyrus tilapiae and C. sclerosus (Monogenea,Ancyrocephalinae) in Nile tilapia (Oreochromis niloticus L.) from Uganda. J Helminthol. 2012;86:302–10.
42. Olstad K. Invasive Alien Species Fact Sheet Gyrodactylus salaris. 2013. NOBANIS http://www.nobanis.org. Accessed 2 january 2015.
43. Jansen P, Matthews L, Toft N. Geographic risk factors for inter-river dispersal of Gyrodactylus salaris in fjord systems in Norway. Dis Aquat Organ. 2007;74:139–49.
44. Munro P, Murray A, Fraser D, Peeler E. An evaluation of the risks of infectious salmon anemia transmission associated with different salmon harvesting methods in Scotland. Ocean Coast Manage. 2003;46:157–74.
45. Peeler E, Thrush A. Qualitative analysis of the risk of introducing Gyrodactylus salaris into the United Kingdom. Dis Aquat Organ. 2004;45:658–75.
46. Corsin F, Turnbull J. Risk factors associated with white spot syndrome virus infection in a Vietnamese rice-shrimp farming system. Dis Aquat Organ. 2001;47:1–12.
47. Sandoval G, Rodriguez R, Vidal-Martínez V. Determinación de la concentración terapéutica de la formalina y de respuesta al estrés en la tilapia Oreochromis niloticus infestada por Cichlidogyrus spp. Reunión Nacional de Innovación Acuícola y Pesquera. 2010.
48. Rasowo J, Okoth OE, Ngugi C. Effects of formaldehyde, sodium chloride, potassium permanganate and hydrogen peroxide on hatch rate of African catfish Clarias gariepinus eggs. Aquaculture. 2007;269:271–7.
49. Faruk M, Alan M, Sarker M, Kabir M. Status of fish disease and health management practices in rural fresh aquaculture of Bangladesh. Pak J Biol Sci. 2004;7:2092–8.
50. Wagner B. The epidemiology of bacterial diseases in food-size channel catfish. J Aquat Anim Health. 2002;14:263–72.
51. Portz D, Woodley C, Joseph J, Cech J. Stress-associated impacts of short-term holding on fishes. Rev Fish Biol Fisher. 2006;16:125–70.
52. Bakke T, Cable J, Harris P. The Biology of Gyrodactylid Monogeneans: the "Russian-Doll Killers". Adv Parasit. 2007;64:1–330.

SECTION V - ECOTOXICOLOGY ON FISH HABITATS AND ECOSYSTEMS.

CHAPTER 14

POPULATIONS OF A CYPRINID FISH ARE SELF-SUSTAINING DESPITE WIDESPREAD FEMINIZATION OF MALES

Patrick B Hamilton, Elizabeth Nicol, Eliane SR De-Bastos, Richard J Williams, John P Sumpter, Susan Jobling, Jamie R Stevens and Charles R Tyler

Keywords: Fishery, Genetic diversity, DNA microsatellites, Waste water treatment work, Ecotoxicology

BACKGROUND

Approximately two-thirds of the world's freshwater is used to dilute wastewater discharges. The demand for freshwater is expected to rise by 70% by 2050 [1] driving an urgent need to understand the impacts of treated waste effluent discharges on aquatic ecosystems. Wastewater treatment works (WWTW) effluents contain tens of thousands of chemicals, including pharmaceuticals and natural steroid estrogens that are biologically active at low (ng/L) exposure concentrations [2]. However, the long-term consequences of exposure to most of these chemicals on fish health and population sustainability are not known.

There is substantial evidence showing that experimental exposure of fish to WWTW effluents and the estrogens they contain can result in adverse health effects, including effects on reproductive development and breeding output. This has led to concerns that freshwater fish populations might also be affected with cascading consequences for freshwater ecosystems. Feminization of male fish is widespread in stretches of rivers downstream of WWTW outfalls as demonstrated in studies using wild [3, 4] and caged [5, 6] fish. Feminized phenotypes include the presence of vitellogenin, a female-specific protein in the blood of male fish [7] and the intersex condition: the presence of oocytes and/or female reproductive ducts in otherwise male gonads [3]. Feminization has been attributed to the presence of estrogens in effluents: estradiol (E_2) and estrone (E_1) from human excretion; 17 alpha-ethinylestradiol (EE_2), a component of the female contraceptive pill [8]; and a large number of other estrogenic chemicals from industrial and domestic effluents. WWTW effluents can also induce genotoxic effects [9], alterations in immune function [10], decreased reproductive output [11], altered stress response [12] and changes in reproductive behavior [13].

Concern about estrogens in rivers in the United Kingdom drove a £40 M programme to evaluate the efficacy of various tertiary treatment processes in the removal of estrogens [14, 15]. Implementation of such processes will, however, incur considerable costs and a greater carbon footprint for WWTW [14, 16], emphasising the need to understand better the population-level consequences for exposure to estrogenic and other so-called endocrine disrupting chemicals (EDCs).

Citation: B Hamilton, Elizabeth Nicol, Eliane SR De-Bastos, Richard J Williams, John P Sumpter, Susan Jobling, Jamie R Stevens and Charles R Tyler, Populations of a cyprinid fish are self-sustaining despite widespread feminization of males, DOI: 10.1186/1741-7007-12-1.

A critical question is whether chronic exposure to estrogenic effluents negatively impacts the viability of wild fish populations, but this has been difficult to address experimentally as it requires controlled experiments extending over periods of several years. Limited studies suggest that high concentrations of EE_2 (between 3 to 6 ng/L) in the aquatic environment could be a threat to the sustainability of fish populations. For example, a controlled exposure of an entire lake to EE_2 in Canada resulted in the collapse of the fathead minnow (*Pimephales promelas*) population within three years [17]. Likewise, long-term (>204 days) laboratory exposures of a range of fish species have resulted in the absence of breeding males [18, 19, 20] and a three-year exposure of roach (*Rutilus rutilus* L.) to an undiluted WWTW effluent in large tanks resulted in an all-female population [21]. It is not known, however, if this occurs in rivers contaminated by effluents. Female fecundity can also be reduced through estrogen exposure, which can potentially reduce population growth rates [22]. Although the exposure concentrations in these studies were high compared to those typically experienced by wild fish populations [23], exposures to EE_2 at concentrations below 1 ng/L during the period of sexual development, have been shown to result in feminized gonads in roach [19] and decreased egg fertilization and female-skewed sex ratios in fathead minnows [24]. Evidence from wild roach living in UK rivers has similarly shown that feminized fish (generally less than 10% of males) with large numbers of eggs in their gonads have impaired semen quality [25] and severely (up to 76%) reduced reproductive success [26].

While these studies suggest that exposure to high concentrations of effluent could threaten the viability of fish populations, aggregates of cyprinid fish, including roach, are often found in effluent contaminated rivers. However, numbers alone may provide a misleading assessment of population sustainability as these could be sink populations maintained by substantial immigration from less contaminated locations where successful reproduction still occurs. Likewise, effective population sizes (N_e) – related to the breeding population of fish – may be decreased without necessarily impacting on population sizes [27], as effluent exposure can affect the number of reproducing individuals and can skew reproductive success [21, 26]. Density-dependent growth and survival can also play an important role [28], so a few reproducing individuals can potentially maintain large adult population sizes. Indeed, studies in several species of marine fish with high fecundity have shown that N_e can be several orders of magnitude smaller than census population sizes. One study found two populations of the exploited New Zealand snapper (*Pagrus auratus*) to have values of N_e less than 1,000 despite adult census population sizes in the millions [29]. Similarly, a study of striped bass (*Morone saxatilis*), a freshwater fish species, found cohorts to consist of a few, full sib families, despite an adult census size of over 300,000 [30]. Critically, N_e influences long-term sustainability as it determines the rate at which genetic diversity is lost from a population through genetic drift [31]. High genetic diversity increases the long-term potential for populations to adapt to changes in the environment and also acts to reduce the risk of inbreeding [32]. Small N_e, however, may act to increase the chances of losing some lethal or sub-lethal mutations through genetic purging.

Understanding the impact of estrogenic effluents on the sustainability of fish populations is, therefore, paramount, but has been limited to date by the logistical challenges involved in undertaking long-term exposures to realistic effluent concentrations, and understanding the demographic history of wild fish populations at highly contaminated sites. In this study, we examine evidence for population impacts on wild roach (*R. rutilus*), a fish species in which feminization is widespread, in southern England. Southern England has some of the highest proportions of WWTW effluent in rivers known globally, and numerous weirs and locks which potentially confine fishes to heavily polluted stretches of river. We have used this system to evaluate whether stretches of river highly contaminated with estrogenic effluents have impaired breeding populations of roach. To do this we undertook analysis of population genetic structures of roach in the region using DNA microsatellite analysis. Microsatellite data were also used to calculate N_e and estimate levels of gene flow to determine the extent to which these populations are maintained through immigration of fish from less contaminated stretches of river.

RESULTS

Genetic Diversity and Genetic Bottlenecks

A total of 1,769 roach, constituting 39 samples (roach sampled from one location within a river – typically a stretch of approximately 100 m – at one time point) from 32 different geographic locations in England were genotyped (Figure 1). Data for 14 microsatellite loci [see Additional file 1 for details] revealed high genetic diversity in all 39 samples (Table 1). Allelic richness (*AR*) ranged from 6.8 to 8.9, and expected heterozygosity (H_e) ranged from 0.69 to 0.75 (Table 1, see Additional file 2 for diversity statistics for each locus). Nevertheless, significant differences in*AR* (analysis of variance (ANOVA), $F_{(38,532)} = 2.1398$, $P=0.00014$) and observed heterozygosity (H_o) (ANOVA, $F_{(38,532)}=1.8677$, $P=0.0017$) among samples were detected. Roach sampled at two relatively unpolluted sites (LamSha and LeeHUS'95) exhibited comparatively low *AR* and LeeHyd'95 exhibited comparatively relatively low H_o[see Additional file 3]. No significant differences in H_e were found. Additionally, there was evidence for genetic bottlenecks at two relatively unpolluted sites sampled within the rivers Arun (AruHUS) and Lee (LeeHUS), and at a polluted site in the Lee (LeeWhe) (Table 1).

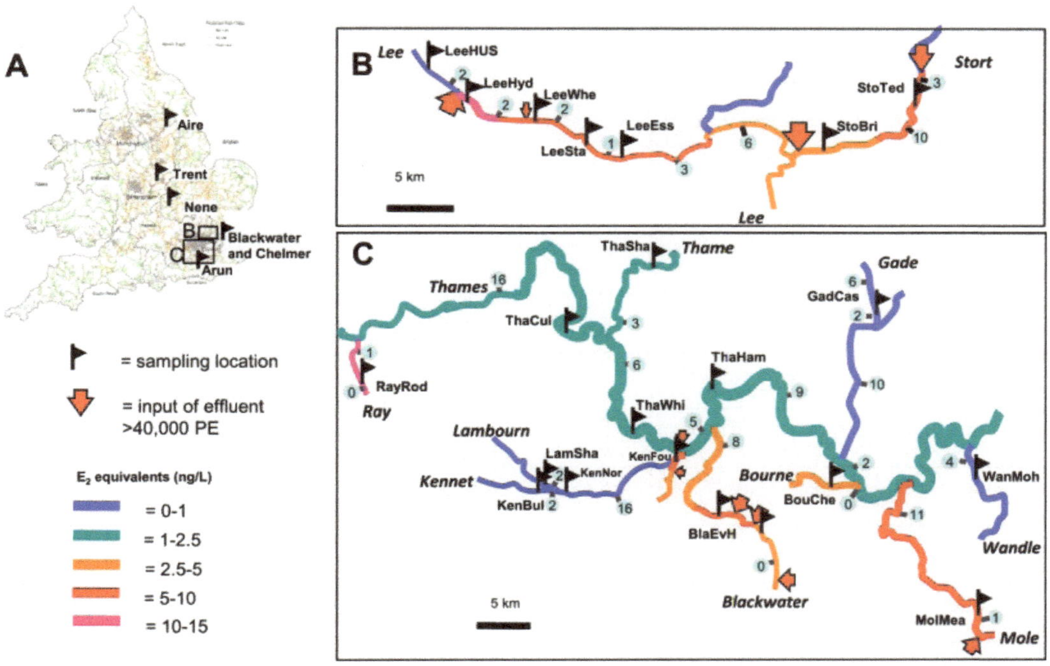

Figure 1. Locations of sample sites in England. (A) Modified from Williams *et al.* [33]. For **(B)** and **(C)**, numbers in circles represent the number of obstructions to fish movement (either weirs or locks). Locks in the Kennet and Thames below ThaWhi have fish passes for salmon movements, but are likely to represent a barrier to movement of roach. In the upper river Lee, only weirs over 1 m are shown. PE=population equivalents, which relates to the size of the population served by the waste water treatment works. The different colours used to depict the rivers represent predicted mean estrogenicity (E2 equivalents in ng/L) [34].

Table 1: Sampling locations, genetic diversity statistics (allelic richness (*AR*) and expected heterozygosity (H_e)) for each population sample

Sample code[a]	River	Year	Lat.	Long.	E_2Eq (ng/L)[b]	No. fish genotyped	Bottleneck (TPM)[c]	*AR*	H_e	Ref.[d]
Blackwater										
BlaBlM	Blackwater	2010	51.78529	0.651013	4.1	55	0.40	7.9	0.71	
BlaSti	Blackwater	2010	51.88792	0.606184	7.1	55	0.60	8.2	0.72	
CheAbB	Chelmer	2010	51.84596	0.423716	1.2	50	0.06	7.7	0.75	
Nene										
NenBro’95	Nene	1995	52.24627	-0.77987	1.2	47	0.69	8.3	0.72	[3]
NenBro’99	Nene	1999	”	”	1.2	47	0.80	8.3	0.73	[25]
NenEct	Nene	2007	52.24234	-0.80281	4.2	51	0.79	8.6	0.73	
Aire										
AirDar	Aire	2011	53.79336	-1.54911	2.7	43	0.73	8.6	0.75	
Arun										
AruHor’95	Arun	1995	51.05516	-0.36197	4.1	54	0.90	7.8	0.73	
AruHor’00	Arun	2000	51.0556	-0.36124	4.1	34	0.29	7.6	0.73	[25]
AruHor’08	Arun	2008	“	“	4.1	69	0.73	7.9	0.74	[26]
AruHUS	Arun	1995	51.05953	-0.35326	0.2	48	**0.01**	7.4	0.73	[3]

Thames catchment										
BlaEvH'10	Blackwater	2010	[e]		4.2	41	0.98	8.5	0.72	
BlaEvH'00	Blackwater	2000	51.327864	-0.769635	8.8	47	0.98	8.3	0.71	[25]
BouChe'11	Bourne	2011	51.38086	-0.47711	4.8	56	0.62	8.3	0.74	
BouChe'02	Bourne	2002	51.40286	-0.54223	5.8	31	0.55	8.8	0.75	
BouChe'06	Bourne	2006	51.40330	-0.54150	5.8	48	0.25	8.3	0.75	[26]
GadCas	Gade	2010	51.65893	-0.42559	NM [f]	56	0.38	8.1	0.74	
KenBul	Kennet	2010	51.39707	-1.28485	0.6	51	0.82	8.4	0.74	
KenFou	Kennet	2010	51.43564	-0.97664	8.1	32	0.64	8.7	0.75	
KenNor	Kennet	2010	51.40165	-1.33725	0.2	52	0.62	8.5	0.73	
LamSha	Lambourn	2011	51.40816	-1.30843	0.03	41	0.10	6.8	0.72	
LeeEss	Lee	2010	51.77279	-0.18818	6.6	56	0.60	8.3	0.73	
LeeHyd	Lee	2010	51.83958	-0.35825	10.3	28	0.36	8.2	0.73	
LeeHyd	Lee	1995	51.84751	-0.37111	11.6	44	0.62	7.7	0.70	[3]
LeeHUS	Lee	1995	51.84935	-0.37395	NM	37	**0.001**	7.0	0.73	[3]
LeeSta	Lee	2010	51.7894	-0.22496	6.6	31	0.85	8.4	0.71	
Lee'00	Lee	2000	[g]		NM	41	0.90	8.1	0.70	[25]
LeeWhe	Lee	2010	51.81424	-0.28903	6.6	55	**0.05**	8.5	0.75	
MolMea	Mole	2010	51.19028	-0.18581	5.8	42	0.84	8.4	0.73	
RayRod	Ray	2003	51.57093	1.81815	10.9	30	0.40	7.7	0.72	
StoBri	Stort	2010	51.77989	0.05024	4.1	52	0.92	8.1	0.71	
StoTed	Stort	2010	51.83115	0.16892	6.0	30	0.97	8.2	0.69	
ThaCul	Thames	2010	51.65046	-1.26739	1.6	44	0.45	8.8	0.74	
ThaHam	Thames	2010	51.55989	-0.87347	1.8	44	0.45	8.2	0.74	
ThaWhi	Thames	2010	51.48662	1.08974	1.5	60	0.88	8.2	0.74	
ThaSha	Thame	2010	51.75281	-1.03319	1.9	50	0.38	7.9	0.74	
WanMoh	Wandle	2011	51.40329	0.18821	3.3	48	0.33	8.2	0.75	
Trent										
TreWol	Trent	1995	52.781178	-1.971789	3.7	45	0.55	8.0	0.72	
TreNot	Trent	2007	[g]			24	0.50	8.9	0.75	

Full diversity statistics for each locus are given in Additional file 8 and statistical differences in genetic diversity are given in Additional file 3. [a]Headings in bold indicate different catchments; [b]Estradiol equivalents, the predicted average estrogenicity at the sample site, calculated from the predicted concentrations of E_1, E_2 and EE_2; [c]TPM=two phase model for microsatellite evolution used for this test; [d]Source of samples; [e]A composite sample of fish caught at BlaEve (51.354119, -0.8584853) and BlaHaw (51.324119, -0.7665606). Effluent concentrations are an average between the two sites for statistical analysis;

[f]NM=not modelled. For this site there are no major upstream discharges. For statistical analysis, and average E_2Eq value for sites with no major discharges was used. [g]The exact locations for these sample sites are unknown. Numbers in bold indicate significance ($p \leq 0.05$).

Population Genetic Structure of roach in ENGLISH RIVERS

We undertook an analysis of population genetic structure in order to examine the genetic similarity of roach between and within catchments. Analysis of molecular variance (AMOVA) indicated the majority of variation was partitioned among individuals within river locations, with river location accounting for a small 2.27%, but highly significant proportion of the genetic variation (Table 2). Average pairwise F_{ST} between roach samples from different catchments was 0.028 and comparisons were consistently highly significant (Table 3, Additional file 4). The population tree (Figure 2) shows distinct clusters of samples in different catchments: the Arun, the Nene, the Anglian Blackwater and the Trent, supported with moderate–high bootstrap values (>64%). Samples from the Arun and the Nene also group in the principal component (PCA) and the STRUCTURE analyses [see Additional file 5 and Additional file 6] demonstrating a distinct genetic identity of fish at these sites. Utilising the method of Evanno *et al.*[35], the inferred most likely number of genetically distinct clusters in the STRUCTURE analysis was three, comprising: the Arun, the two most upstream sample sites in the river Lee, and all remaining sites genotyped. However, from visual examination of STRUCTURE plots run with higher levels of K [see Additional file 6] other possible groups are apparent. We found no evidence that roach in the Thames catchment constitute a distinct genetic group, as samples failed to group together in any analysis (Figure 2, Additional file 5 and Additional file 6). This may reflect a true lack of genetic distinctiveness of roach in this catchment, but may also result from the limited ability of the microsatellite markers used to resolve population genetic structure at this level.

Table 2. Analysis of molecular variance (AMOVA) testing for partitioning of genetic variation among roach samples, grouped according to geography

Source of variation	d.f.	Sum of squares	Variance	% Total	*P*-value
Geographical partition, location [a]					
Among locations	32	584.063	0.11631	2.27	<0.00001
Among samples within locations	6	35.197	0.00953	0.19	0.03226
Within samples	3,499	17,489.296	4.99837	97.54	<0.00001
Total	3,537	18,108.556	5.12421		
Geographical partition, catchment [b]					
Among groups	5	189.529	0.06309	1.22	<0.00001
Among samples within groups	33	429.731	0.08909	1.73	<0.00001
Within samples	3,499	17,489.296	4.99837	97.05	<0.00001
Total	3,537	18,108.556	5.15055		

[a]Samples from same location caught in different years are grouped; [b]Samples grouped by catchment (Aire, Arun, 'Blackwater and Chelmer', Nene, Thames and Trent).

Table 3. ummary of pairwise F_{ST} **and** D_{est} among roach samples (see Additional file 4 for full table of values)

	F_{ST}	D_{est}	*P*-value[a]
Between catchments			
Thames/Arun	0.033, (0.002-0.074)[b]	0.0617, (0.0124-0.1612)[b]	
Thames/Nene	0.026, (0.007-0.076)	0.0453, (0.0126-0.1458)	
Thames/Trent	0.026, (0.008-0.069)	0.0447, (0.0059-0.1394)	
Thames/Chelmer, Blackwater	0.022, (0.005-0.057)	0.0444, (0.0035-0.1191)	
Within catchments			
Thames	0.022, (-0.004-0.090)	0.0376, (-0.0058-0.1914)	
Arun	0.007, (0.002-0.014)	0.0064, (0.0015-0.0104)	
Nene	0.002, (0.000-0.003)	0.0010, (-0.0001-0.0020)	
Blackwater/Chelmer	0.022, (0.005-0.057)	0.0454, (0.0058-0.0652)	

Neighbouring samples in the river Lee/Stort			
LeeHUS/LeeHyd'95	-0.002	0.0017	0.78
LeeHyd'10/LeeWhe	0.020	0.0352	>0.00001
LeeWhe/LeeSta	0.009	0.0189	>0.00001
LeeSta/LeeEss	0.009	0.0127	>0.00001
LeeEss/StoBri	0.015	0.0349	>0.00001
Other neighbouring stretches			
BlaBlM/BlaSti	0.007	0.0058	>0.00001
AruHUS/AruHor'95	0.002	0.0043	0.22

[a]For F_{ST} estimate; [b]average, (range). Pairwise values for neighbouring stretches in the Lee/Stort, the Anglian Blackwater and the Arun are also given.

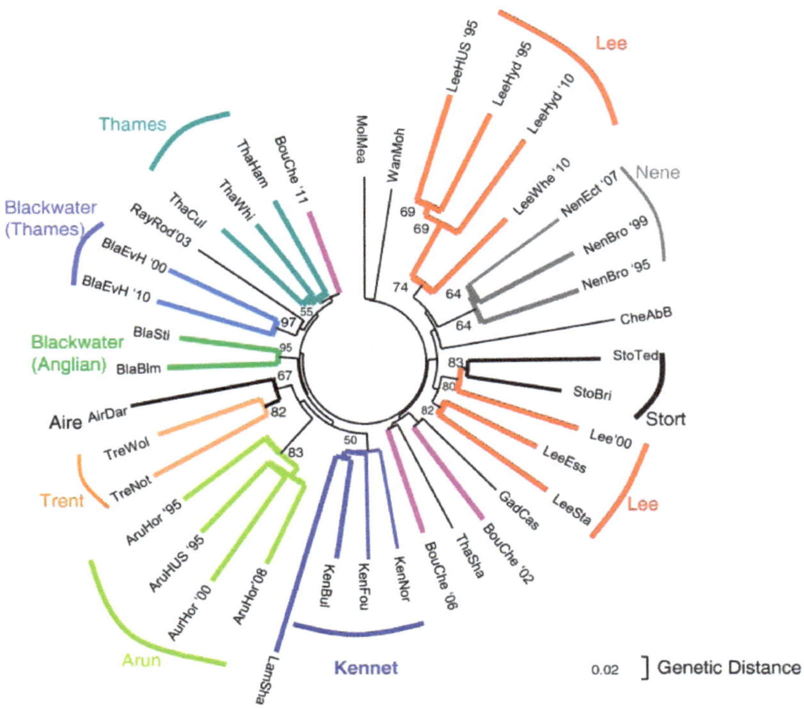

Figure 2. Neighbour-joining phylogenetic tree for the 39 roach population samples. The tree is based on the data from 14 microsatellite loci using chord distance from Cavalli-Sforza and Edwards [36]. Only bootstrap values above 50% are shown. Numbers at the end of sample codes indicate years that populations were sampled.

Population Genetic Structure of roach within Rivers and Catchments

Despite Thames catchment roach appearing not to constitute a single genetic unit, distinct from roach in other regions, the study did find evidence for significant genetic structuring in roach populations within the Thames catchment. This suggests the existence of local subpopulations exchanging a limited number of effective migrants (breeding individuals) rather than panmixia (where all individuals are potential partners). For example, average F_{ST} between samples in the Thames catchment was 0.022, only slightly lower than the average for between-catchment comparisons (0.028) for the study as a whole, while 262 of the 325 pairwise F_{ST} comparisons in the catchment were highly significant. There was a weak, but significant, relationship between genetic and geographic distance ($r^2=0.1089$, $P=0.010$) within the catchment, indicating a tendency for individuals to produce offspring with fish from nearby populations rather than distant populations [see Additional file 7]. Additionally, the population tree (Figure 2) and PCA analyses [see Additional file 5] showed groups comprising samples from neighbouring Thames sites: three from the main Thames; four from the Kennet and its tributary (Lambourn); samples from the Stort and the Lee; and samples from the Wandle and Mole. Samples from the Thames Blackwater collected in the years 2000 and 2010 (approximately two to three generations) clustered with very high bootstrap support (98%) in the population tree (Figure 2). This indicates that this roach population is largely restricted to this stretch of river, which includes both moderately and highly polluted sites, and has no substantial uncontaminated upstream stretch (Figure 1).

Despite the proximity of some populations in the PCA and tree, we also found significant genetic differentiation between samples from some neighbouring stretches of the same river, sometimes occurring over small distances of separation (<10 km), for example, within the upper Lee (see below), between the Lee and the Stort, between the Blackwater and main Thames, between the Lambourn and the Kennet and within the Anglian Blackwater. In other cases, despite the separation of sampling locations by in-river impoundments such as weirs, we found no significant genetic differentiation between sites, for example, within the Stort, the main Thames, the Kennet, the Arun, the Nene and the Trent. Thus, patterns of within-river genetic structure differed between river stretches. For some other fish species analysis of genome-wide SNP data has provided greater resolution in population structure than that achievable using microsatellite data [37], and it is possible that some fine-scale genetic structure in the roach populations has not been detected with the microsatellites used in the current study.

Relationship between Exposure to Estrogenic Effluents and Effective Population Size

Estimates of N_e calculated from the microsatellite data using the approximate Bayesian computation method ($N_{e(ABC)}$), ranged from 54 to 301 for each sample, with higher precision for small N_e estimates (Figure 3A). We found no evidence for a correlation between $N_{e(ABC)}$ and predicted E_2 equivalents (E_2Eq), a measure of total estrogenicity of the river water due to contamination by sewage effluent generalized linear models (GLM), $F_{(1,20)}=0.7468$, $P=0.40$) or for an interaction between sample site and estrogen exposure (GLM, $F_{(6,19)}=1.9954$, $P=0.14$) across the 28 sample sites where no recent restocking had occurred and had sufficient sample sizes for robust $N_e$calculation. However, the 95% confidence intervals (CI) for the model coefficient indicated $N_{e(ABC)}$ could decrease by a maximum of 5.6% for each incremental increase in exposure of 1 ng/L E_2Eq, or 65% at 11.6 ng/L E_2Eq, equivalent to the most polluted river stretch included in this study. The inclusion of roach density as an additional covariate within the model also produced a non-significant result (GLM, $F_{(1,16)}=1.3966$, $P=0.26$), albeit for a reduced number of sites (19). Similarly, there was no significant correlation between the other variables included in the statistical analyses (average flow rate, geographic/phylogenetic group and roach density) with $N_{e(ABC)}$. This analysis makes the assumption that immigration of fish from remote sites is limited and this is discussed below.

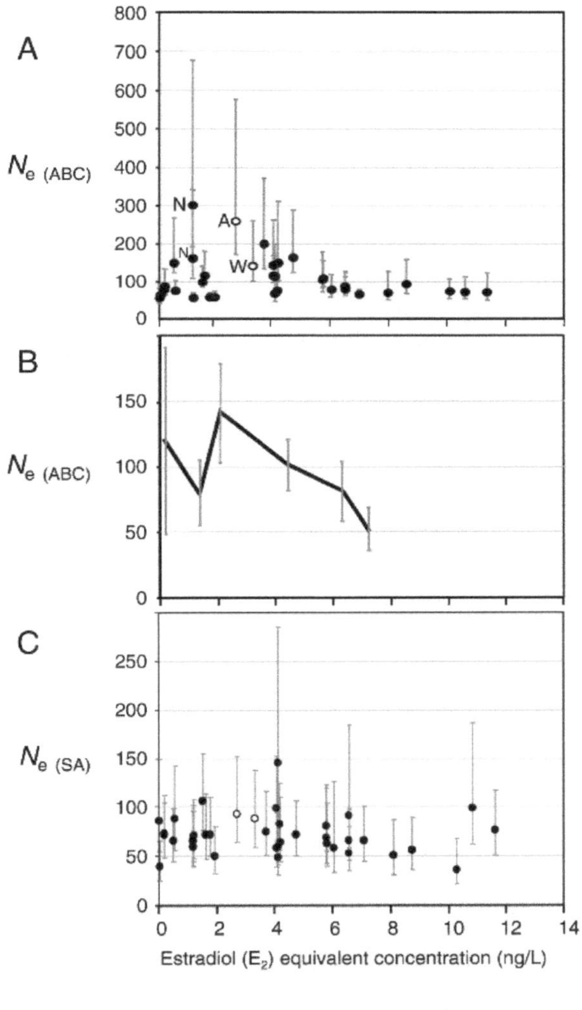

Figure 3. Effective population size (N_e) plotted against predicted estrogen exposure for 37 population samples of*Rutilus rutilus* . (A) N_e calculated using the Approximate Bayesian Computation (ABC) method in the program OneSAMP [38]. Tests for homogeneity of variances: Bartlett's, $P=0.0036$, Levene's, $P=0.17$. **(B)** Results of binning analysis for data shown in A; each bin which encompasses all data points starting at the lower point represented by the mean and standard error up to, but not including, the next bin. **(C)** N_e calculated using the sibling assignment (SA) method in Colony [39]. In A and C, error bars are 95% confidence intervals. In cases in which more than one population had similar values, data points overlie each other; thus, individual data points are not always visible. These plots include estimates from sample sites sampled in different years, for example, in the River Nene (N) (which were averaged for statistical analysis) and sites where recent restocking had occurred (open circles: River Aire (A), River Wandle (W)), which were excluded from the statistical analyses.

There was limited evidence for reduced variation in N_e in roach sampled from more contaminated stretches of river compared to those sampled from less contaminated sites; all estimates above 6 ng/L E_2Eq were below 100, whereas there was greater variation (54 to 301) where E2Eq was below 6 ng/L (Figure 3B). Estimates of N_e using other methods were of the same order of magnitude but had wider confidence intervals for each estimate; $N_{e(SA)}$ ranged from 36 to 145 and also showed no relationship with E_2Eq (Figure 3C). Temporal estimates of N_e, calculated from allele frequency changes over several generations using the Jorde and Ryman method [40], varied from 14 to 265, but were available for too few locations to make meaningful comparisons (Table 4). Overall, the relatively small variations in N_e observed in this study could not be explained by any of the environmental or other variables measured in this study.

Table 4. Temporal estimates of effective population size (N_e) among roach samples

Site	Time interval	G[a]	N_e(95% CI)[b]	N_e(95% CI)[c]
Nene	1995 to 1999	1	265 (66-∞)	619 (82-∞)
AruHor	1995 to 2000	1	14 (8–79)	48 (26–137)
	2000 to 2008	2	60 (31–1733)	321 (106-∞)
	1995 to 2008	3	73 (43–247)	232 (123–669)
BouChe	2002 to 2006	1	32 (17–219)	51 (25–151)
	2006 to 2011	1	45 (23–9572)	63 (35–162)
	2002 to 2011	2	87 (44–3897)	495 (295–1405)
LeeHyd	1995 to 2010	3	137 (81-∞)	346 (145–35518)
BlaEvH	2000 to 2010	2	141 (68-∞)	206 (83-∞)

[a]G = assumed number of generations between sampling points; [b]Calculated using the Jorde and Ryman method [40]; [c]Calculated using the classical moment-based method of Waples [41]. CI, confidence interval.

Population Genetic Structure within the River Lee, a High Effluent River

The average proportion of effluent in the upper Lee downstream from Harpenden and East Hyde WWTWs ranges from 28% to 70% in different stretches. Significant genetic differentiation was detected between fish sampled from four of the five locations in this stretch of river (F_{ST} values ≥0.009), and between these samples and two from its tributary, the Stort (F_{ST} ≥0.015) shown in Table 3. The presence of numerous large weirs (Figure 1) likely confines fish to particular areas of this river; nonetheless, samples from the Lee and the Stort did cluster together in some analyses (Figure 2, Additional file 5). A sample from the upstream, unpolluted sample site (LeeHUS) grouped with two samples (collected in 1995 and 2010) from the most polluted river stretch immediately downstream (LeeHyd). The next sample site downstream, LeeWhe, was distinct from these (Figure 2, Additional file 5 and Additional file 6), indicating restricted movement of fish between LeeHyd and LeeWhe over at least three to five generations (Figure 2). Analysis using the program IM_A2 [42] suggested that there was less than one effective (breeding) migrant per generation between LeeHyd and LeeWhe in either direction, and about one migrant per generation from LeeWhe downstream to LeeSta (Figure 4). Collectively, these data suggest that roach populations at LeeWhe and those downstream do not rely on migration from the uncontaminated stretch of this river. Despite this, $N_{e(ABC)}$ estimates for these polluted sites in the upper Lee ranged from 70 to 84 (95% CI: 50 to 127) compared to only 54 (95% CI: 42 to 82) for the upstream uncontaminated location, suggesting no substantial impact of the effluent on the effective population size of these roach.

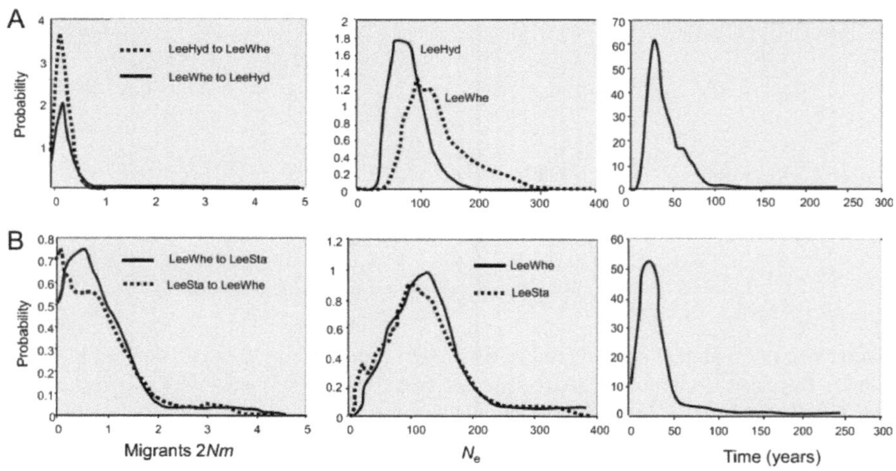

Figure 4. Posterior probabilities of migration rates, effective population sizes (N_e) and time since divergence estimated using IM_A2 [[42]] for two pairs of roach populations in the River Lee. (A) Estimates for parameters calculated for the LeeHyd'10 and LeeWhe and **(B)** for LeeSta and LeeWhe. These N_e values are influenced by average N_e since the initial split of the populations.

DISCUSSION

In this study, analyses of population genetic structure *via* the analysis of DNA microsatellite loci identified distinct subpopulations of roach in two tributaries of the Thames, the rivers Lee and the Blackwater, that were largely restricted to high-effluent stretches of the rivers over multiple generations. This is despite evidence for widespread feminization of male fish in the studied rivers and previous evidence that feminization alters breeding capabilities [21, 26]. Both of these tributaries contain feminized fish [25]), with predicted average exposure of between 4 and 9 ng/L E_2Eq. We also found no statistically robust evidence for a substantial impact of estrogenic sewage effluents on N_e of roach. The possibility of a reduction in N_e of up to 65% for roach living in the most polluted river stretches (E_2Eq of 11.6 ng/L) could not be ruled out, due to the wide 95% confidence intervals associated with the statistical model. Moreover, our analysis included relatively few samples from rivers in the highest risk category, largely because these sites are rare.

Caveats

As with any modelling exercise, this analysis makes assumptions that may affect the interpretation of the results. One of these assumptions is that migration between sites with different pollution profiles is limited over two to three generations, the time frame likely to have the greatest influence on $N_{e(ABC)}$[43]. This was ensured by selecting sites with physical obstructions between them. However, quantifying migration rates over this timescale was not always possible because all potential source populations could not be sampled and, in some cases, we found no significant genetic differentiation between roach at sites distant from one another. Genetic differentiation can take many generations to manifest with low levels of migration [44]. Histological data from the Arun and the Lee show that feminized gonads in roach were approximately 6-fold (Lee) and approximately 2.5-fold (Arun) more prevalent in populations living in the stretches downstream of major WWTW inputs compared with those living upstream [3]. This demonstrates that migration in these rivers was indeed restricted to stretches delimited by physical barriers, despite no significant genetic differentiation observed between river stretches (F_{ST} <0.002, Table 3).

A second assumption is that no restocking of the rivers sampled had occurred or that the effect of restocking activities on $N_{e(ABC)}$ was relatively minor. Approximately 500,000 hatchery-reared roach just over one-year-old (so called '1+' fish) have been introduced into the Thames catchment since 2000; broodstock for hatchery fish originate from the river Trent. The influence of restocking activities on $N_{e(ABC)}$, however, is likely to be relatively minor, as the sites sampled in this study were separated by major physical barriers from sites where these introductions had occurred. Moreover, introductions prior to 2000 are unlikely to have had a large influence on $N_{e(ABC)}$ as this is primarily affected by the size and variance in reproductive success of the parental generation, which would have spawned between 2004 and 2007 for most of the samples in this study. However, we cannot exclude some influence of introductions prior to 2000 as some of the summary statistics used to calculate $N_{e(ABC)}$ are known to be affected by demographic processes over a longer time period [38]. The effects of introductions on genetic diversity, the detection of bottlenecks and population structure are likely to be greater, as these factors are affected by demography over many generations. However, neither the success of the reintroduced fish nor the size of the roach population in the Thames is currently known. In salmonids, restocking success is highly variable and has been attributed to local adaptation [45]. Using our microsatellite dataset, 73% of 48 individual roach from a stretch of the Wandle (restocked in 2007, 2009, 2010) assigned to the Thames reporting regions. Only 5% (two fish) assigned to the Trent (the source of the parents of introduced fish), which may be mis-assignments, as 5% also assigned to the Arun, 10% to the Anglian Blackwater and 10% to the Nene, from where no restocking had taken place. Thus, the success of the re-introduced fish may be low, but this requires further investigation.

Evidence for Self-sustaining Populations in Effluent Contaminated Rivers

While this study does not exclude the possibility that estrogenic effluents reduce N_e of fish populations, it suggests that roach populations can be self-sustaining despite exposure to estrogens over several generations. These findings are consistent with the fact that the prevalence of male fish with moderate to severely feminized gonads (that have been shown to have substantially reduced reproductive competitiveness in controlled breeding studies) is generally less than 10% in English rivers [3, 25, 26, 46]. The reproductive competitiveness of fish with the more common mild-intersex condition is similar to those of fish without gonadal feminization [26]. In roach, the gonads of male fish exposed to estrogens become progressively feminized with age [46], so gonadal feminization could theoretically increase N_e by reducing the reproductive dominance of large older males in estrogen-contaminated rivers. While the effects on females are less well studied, female roach exposed to an undiluted effluent for three years in large tanks were able to breed, despite the fact that this exposure caused complete gonadal feminization of males [21]; similarly, the majority of females collected from two effluent-polluted rivers examined in this study were also able to breed [26].

Population risks of long-term Exposures to Estrogenic Effluents

The results of this study on wild roach populations seem to contrast with studies that have assessed population risk through long-term exposures to estrogens, where exposure to concentrations between 3 and 6 ng EE_2/L [17, 18, 19,20] or to a full-strength effluent [21] resulted in all-female populations and/or reproductive failure. The apparent difference between the wild populations and those experiments designed to simulate 'real world' exposure, however, may be because the fish living in the effluent-contaminated rivers examined in this study have been exposed to a lower level of estrogen or because all of the estrogen is not bioavailable; organic pollutants can bind to particulates and dissolved organic matter [47]. The most contaminated river in this study has a mean proportion of effluent of approximately 70%, although the majority of contaminated English rivers average approximately 10% to 30% [34]. While EE_2 has been measured up to approximately 4 to 8 ng/L in English WWTW effluents [33, 48], for the most part, they are lower [23, 49], and estrogen concentrations vary greatly over short periods. For instance, EE_2 was detected in only 21 of 135 water samples from the Lee, although occasionally reaching 4 ng/L [33]. Considering the totality of estrogen content, the predicted average estrogenicity of the most contaminated site in this study is 12 ng/L E_2Eq and would be below 21 ng/L for 90% of the time. Only 1% to 3% of 10,313 individual river reaches in the UK receiving WWTW effluent were predicted to have average E_2Eq >10 ng/L, and, of these, many are ditches composed almost entirely of sewage effluent [34]. As E_2 is approximately 10 times less potent than EE_2 in inducing gonadal feminization in fish [34, 50], it is probable that average life-time exposure to estrogens in the wild does not currently reach the concentrations shown to cause sex-reversal and population collapse in controlled experimental exposures. Green *et al.* [51] recently predicted a doubling of estrogen exposure concentrations in some rivers with population growth and climate change by 2050 suggesting an increased likelihood of population level effects of estrogenic effluents in the future, unless mitigated by substantial improvements in sewage treatment processes.

Influences on the Population Genetic Structure of Roach

The population genetic structure of roach in southern England observed in this study may have been influenced by historical biogeography, migratory behaviour, human translocations and in-river barriers. Roach can be highly mobile and can migrate over 10 km, particularly in the spawning period April to June, if migration is not obstructed [52]; additionally, there is some evidence that roach show fidelity in migration and return to spawning sites they have used previously [53]. Within the Thames catchment, the observed population genetic structure likely results, at least in part, from the large number of physical barriers, such as weirs and locks (Figure 1); these have been recognized as major factors restricting movement (including downstream) of roach [54]. Similarly, the importance of barriers in driving intra-catchment genetic variation is well documented in other fish species, for example, brown trout [55]. Only obstructions in the main River Thames and the Kennet are equipped with fish passes and, although some passes can be used by roach [56], the effectiveness of these passes in allowing fish movement has not been studied. As we identified significant genetic differentiation between roach from the Kennet and the Thames, despite being connected by fish passes, these passes may represent major physical separation barriers to this fish species.

CONCLUSIONS

Despite the widespread feminization of male roach in effluent-contaminated rivers of southern England, using nuclear DNA microsatellites we were able to identify some populations that have been confined to stretches of river with moderate to high exposure to estrogenic effluents over multiple generations. We also found no evidence of a correlation between the N_e of roach populations and predicted exposure to estrogens, although because of the wide confidence intervals, a reduction in N_e of up to 65% is still possible at the most contaminated sites.

METHODS

Study Location

Southern England, particularly the region within the Thames catchment, was chosen for this study for four reasons. Firstly, it is a densely populated region with relatively low rainfall and, therefore, includes some river stretches with some of the highest concentrations of WWTW effluents in the United Kingdom [34]. Secondly, feminization of roach has been widely reported in the region [3, 46]. Thirdly, many rivers in the region have locks, dams or weirs which are likely to limit movement of fish species between stretches of river with different pollution profiles. Fourthly, the effluent concentrations and risk of estrogenic endocrine disruption have been modelled [34]. Sample sites are shown in Figure 1 and Table 1 and were selected to span the full range of predicted estrogen concentrations in English rivers and where obstructions are likely to restrict fish movements [3, 46].

Roach Study Species

Roach was selected as the study species because it is native and widely distributed in the United Kingdom, including in rivers polluted with WWTW effluents. Additionally, widespread feminization has been reported in wild populations and with a proven association with exposure to estrogenic effluents [14, 19, 21, 57]. Roach generally reach sexual maturity between two and three years and spawn annually in the spring. Adult roach can migrate considerable distances, but where weirs obstruct upstream and downstream movement they are able to complete their lifecycles in a single stretch of river [54].

Population-genetic Analyses

To understand the extent to which roach populations are restricted to various stretches of river, several approaches were used to investigate population genetic structure. We analysed microsatellite loci variation in 1,769 fish sampled between 1995 and 2011. Each fish was genotyped at between 14 to 19 microsatellite loci. Microsatellite genotypes are provided in Additional file 8. Protocols for DNA extraction and details of amplification of the microsatellite loci are illustrated in Additional file 1. Data for 14 microsatellite loci were used to calculate three measures of genetic diversity: observed heterozygosity (H_O) and expected heterozygosity (H_e) using GenAlEx 6 [58]; allelic richness (*AR*) was calculated using Fstat v2.9.3 [59] – see Additional file 1 for full details. The programme BOTTLENECK [60, 61] was used to test for recent genetic bottlenecks. This programme tests for a relative excess in heterozygosity that is apparent for a few generations after a bottleneck and develops because allelic diversity declines faster than heterozygosity, due to loss of rare alleles. Pairwise genetic differentiation between the sampled sites was estimated using F_{ST}, calculated using Arlequin 3.5 [62] and Jost's D, D_{est}[63], calculated using SMOGD [64]. The significance of the F_{ST} estimates was assessed based on 10,000 permutations. AMOVA was performed using Arlequin. In order to test whether fish are more likely to produce offspring with local mates, compared to mates in geographically distant locations within the Thames catchment, isolation by distance analysis was performed using the Mantel test [65] in GenAlEx

6 [58]. Genetic similarity between populations was investigated using population based trees, calculated in POPULATIONS, v1.2.30beta [66], PCA in GenAlEx 6 [58] and a Bayesian clustering approach in STRUCTURE [67]. Finally, the program IM_A2 [42, 68] was used to estimate migration rates between adjacent populations within high effluent stretches of the Lee, giving relatively high pairwise F_{ST} values (LeeHyd, LeeWhe, and LeeSta). See Additional file 1 for further details. To investigate the influence of restocking, genetic assignment of fish from the Wandle was undertaken using the 'leave one out test' in the computer program, ONCOR [69], based on their microsatellite genotypes. The reporting regions comprised: (1) the Wandle, (2) Lee/Stort, (3) rest of the Thames, (4) Trent, (5) Nene, (6) Arun, (7) Chelmer and (8) Anglian Blackwater. All animals used in this research were treated humanely and with regard for the alleviation of suffering; all procedures were subject to approval by the local ethical re process as required under the U.K. Animals (Scientific Procedures) Act (1986).

Effective Population Size

To test whether WWTW effluents substantially reduce the size of breeding populations, effective population sizes (N_e), which relate to the number of breeding fish and skews in breeding success, were estimated using the microsatellite genotypes. We compared N_e from sites ranging from little/no upstream WWTW effluent inputs to those where the majority of the flow can comprise WWTW effluent. Two single sample (generation) methods, that use different aspects of the microsatellite data, were used to estimate N_e for each population; the Approximate Bayesian Computation (ABC) method using ONeSAMP 1.2 [38], hereafter referred to as $N_{e(ABC)}$; and the sibling assignment method (SA), $N_{e(SA)}$[39]. Temporal estimates for N_e, which are calculated from the change in allele frequencies between generations, were also estimated for sites where fish had been sampled more than once using TempoFs [40] and NeEstimator [70]. For further details see Additional file 1.

Statistical Analysis

GLM were used to examine the relationship between predicted exposure to estrogenic effluents and $N_{e(ABC)}$. For further details see Additional file 1. Differences in genetic diversity among sampled populations were tested using ANOVA. All statistical analysis was performed using the software R 2.13.0 [71].

ACKNOWLEDGEMENTS

This work was funded by Natural Environmental Research Council (NERC) grant numbers NE/G019355/1 and NE/K004263/1 and the Environment Agency (England and Wales). Contains Environment Agency information © Environment Agency and database right. We thank the Environment Agency for support from their Fisheries Teams and databases, Jinliang Wang and David Tallmon for their advice on using and interpreting the output of Colony and ONeSAMP, and Tiago Antao for advice on calculating and interpreting N_e. We are grateful to Anke Lange for providing EST sequences, Andrew Griffiths and Andy King for advice on population-genetic analysis, and Veronica Vinciotti, Marta Söffker, Venura Perera and James Creswell for support on statistical aspects of this work.

REFERENCES

1. Gilbert N: Water under pressure. Nat Biotechnol. 2012, 483: 256-257.
2. Heberer T: Occurrence, fate, and removal of pharmaceutical residues in the aquatic environment: a re of recent research data. Toxicol Lett. 2002, 131: 5-17. 10.1016/S0378-4274(02)00041-3.
3. Jobling S, Nolan M, Tyler CR, Brighty G, Sumpter JP: Widespread sexual disruption in wild fish. Environ Sci Technol. 1998, 32: 2498-2506. 10.1021/es9710870.
4. Hinck JE, Blazer VS, Schmitt CJ, Papoulias DM, Tillitt DE: Widespread occurrence of intersex in black basses (*Micropterus*spp.) from US rivers, 1995–2004. Aquat Toxicol. 2009, 95: 60-70. 10.1016/j.aquatox.2009.08.001.
5. Harries JE, Sheahan DA, Jobling S, Matthiessen P, Neall M, Sumpter JP, Taylor T, Zaman N: Estrogenic activity in five United Kingdom rivers detected by measurement of vitellogenesis in caged male trout. Environ Toxicol Chem. 1997, 16: 534-542. 10.1002/etc.5620160320.
6. Lu GH, Song WT, Wang C, Yan ZH: Assessment of *in vivo* estrogenic response and the identification of environmental estrogens in the Yangtze River (Nanjing section). Chemosphere. 2010, 80: 982-990. 10.1016/j.chemosphere.2010.05.038.
7. Purdom CE, Hardiman PA, Bye VVJ, Eno NC, Tyler CR, Sumpter JP: Estrogenic effects of effluents from sewage treatment works. Chem Ecol. 1994, 8: 275-285. 10.1080/02757549408038554.
8. Desbrow C, Routledge EJ, Brighty GC, Sumpter JP, Waldock M: Identification of estrogenic chemicals in STW effluent. 1. Chemical fractionation and *in vitro* biological screening. Environ Sci Technol. 1998, 32: 1549-1558. 10.1021/es9707973.
9. Gravato C, Santos MA: *Dicentrarchus labrax* biotransformation and genotoxicity responses after exposure to a secondary treated industrial/urban effluent. Ecotoxicol Environ Saf. 2003, 55: 300-306. 10.1016/S0147-6513(03)00024-1.
10. Filby AL, Neuparth T, Thorpe KL, Owen R, Galloway TS, Tyler CR: Health impacts of estrogens in the environment, considering complex mixture effects. Environ Health Perspect. 2007, 115: 1704-1710. 10.1289/ehp.10443.
11. Thorpe KL, Maack G, Benstead R, Tyler CR: Estrogenic wastewater treatment works effluents reduce egg production in fish. Environ Sci Technol. 2009, 43: 2976-2982. 10.1021/es803103c.
12. Pottinger TG, Henrys PA, Williams RJ, Matthiessen P: The stress response of three-spined sticklebacks is modified in proportion to effluent exposure downstream of wastewater treatment works. Aquat Toxicol. 2013, 126: 382-392.
13. Sebire M, Katsiadaki I, Taylor NGH, Maack G, Tyler CR: Short-term exposure to a treated sewage effluent alters reproductive behaviour in the three-spined stickleback (*Gasterosteus aculeatus*). Aquat Toxicol. 2011, 105: 78-88.

14. Baynes A, Green C, Nicol E, Beresford N, Kanda R, Henshaw A, Churchley J, Jobling S: Additional treatment of wastewater reduces endocrine disruption in wild fish-a comparative study of tertiary and advanced treatments. Environ Sci Technol. 2012, 46: 5565-5573. 10.1021/es204590d.
15. Filby AL, Shears JA, Drage BE, Churchley JH, Tyler CR: Effects of advanced treatments of wastewater effluents on estrogenic and reproductive health impacts in fish. Environ Sci Technol. 2010, 44: 4348-4354. 10.1021/es100602e.
16. Owen R, Jobling S: Environmental science: The hidden costs of flexible fertility. Nature. 2012, 485: 441-10.1038/485441a.
17. Kidd KA, Blanchfield PJ, Mills KH, Palace VP, Evans RE, Lazorchak JM, Flick RW: Collapse of a fish population after exposure to a synthetic estrogen. Proc Natl Acad Sci U S A. 2007, 104: 8897-8901. 10.1073/pnas.0609568104.
18. Lange R, Hutchinson TH, Croudace CP, Siegmund F, Schweinfurth H, Hampe P, Panter GH, Sumpter JP: Effects of the synthetic estrogen 17 alpha-ethinylestradiol on the life-cycle of the fathead minnow (*Pimephales promelas*). Environ Toxicol Chem. 2001, 20: 1216-1227. 10.1002/etc.5620200610.
19. Lange A, Paull GC, Coe TS, Katsu Y, Urushitani H, Iguchi T, Tyler CR: Sexual reprogramming and estrogenic sensitization in wild fish exposed to ethinylestradiol. Environ Sci Technol. 2009, 43: 1219-1225. 10.1021/es802661p.
20. Nash JP, Kime DE, Van der Ven LT, Wester PW, Brion F, Maack G, Stahlschmidt-Allner P, Tyler CR: Long-term exposure to environmental concentrations of the pharmaceutical ethynylestradiol causes reproductive failure in fish. Environ Health Perspect. 2004, 112: 1725-1733. 10.1289/ehp.7209.
21. Lange A, Paull GC, Hamilton PB, Iguchi T, Tyler CR: Implications of persistent exposure to treated wastewater effluent for breeding in wild roach (*Rutilus rutilus*) populations. Environ Sci Technol. 2011, 45: 1673-1679. 10.1021/es103232q.
22. Grist EP, Wells NC, Whitehouse P, Brighty G, Crane M: Estimating the effects of 17 alpha-ethinylestradiol on populations of the fathead minnow *Pimephales promelas*: Are conventional toxicological endpoints adequate?. Environ Sci Technol. 2003, 37: 1609-1616. 10.1021/es020086r.
23. Hannah R, D'Aco VJ, Anderson PD, Buzby ME, Caldwell DJ, Cunningham VL, Ericson JF, Johnson AC, Parke NJ, Samuelian JH, Sumpter JP: Exposure assessment of 17 alpha-ethinylestradiol in surface waters of the United States and Europe. Environ Toxicol Chem. 2009, 28: 2725-2732. 10.1897/08-622.1.
24. Parrott JL, Blunt BR: Life-cycle exposure of fathead minnows (*Pimephales promelas*) to an ethinylestradiol concentration below 1 ng/L reduces egg fertilization success and demasculinizes males. Environ Toxicol. 2005, 20: 131-141. 10.1002/tox.20087.
25. Jobling S, Coey S, Whitmore JG, Kime DE, Van Look KJ, McAllister BG, Beresford N, Henshaw AC, Brighty G, Tyler CR, Sumpter JP: Wild intersex roach (*Rutilus rutilus*) have reduced fertility. Biol Reprod. 2002, 67: 515-524. 10.1095/biolreprod67.2.515.
26. Harris CA, Hamilton PB, Runnalls TJ, Vinciotti V, Henshaw A, Hodgson D, Coe TS, Jobling S, Tyler CR, Sumpter JP: The consequences of feminization in breeding groups of wild fish. Environ Health Perspect. 2011, 119: 306-311.
27. Cotton S, Wedekind C: Population consequences of environmental sex reversal. Conserv Biol. 2009, 23: 196-206. 10.1111/j.1523-1739.2008.01053.x.
28. Lorenzen K, Enberg K: Density-dependent growth as a key mechanism in the regulation of fish populations: evidence from among-population comparisons. Proc Biol Sci. 2002, 269: 49-54. 10.1098/rspb.2001.1853.
29. Hauser L, Adcock GJ, Smith PJ, Ramirez JH, Carvalho GR: Loss of microsatellite diversity and low effective population size in an overexploited population of New Zealand snapper (*Pagrus auratus*). Proc Natl Acad Sci U S A. 2002, 99: 11742-11747. 10.1073/pnas.172242899.
30. Liu JX, Ely B: Sibship reconstruction demonstrates the extremely low effective population size of striped bass *Morone saxatilis* in the Santee-Cooper system, South Carolina, USA. Mol Ecol. 2009, 18: 4112-4120. 10.1111/j.1365-294X.2009.04343.x.
31. Charlesworth B: Effective population size and patterns of molecular evolution and variation. Nat Rev Genet. 2009, 10: 195-205. 10.1038/nrg2526.
32. Reed DH, Frankham R: Correlation between fitness and genetic diversity. Conserv Biol. 2003, 17: 230-237. 10.1046/j.1523-1739.2003.01236.x.
33. Williams RJ, Johnson AC, Smith JJ, Kanda R: Steroid estrogens profiles along river stretches arising from sewage treatment works discharges. Environ Sci Technol. 2003, 37: 1744-1750. 10.1021/es0202107.
34. Williams RJ, Keller VD, Johnson AC, Young AR, Holmes MG, Wells C, Gross-Sorokin M, Benstead R: A national risk assessment for intersex in fish arising from steroid estrogens. Environ Toxicol Chem. 2009, 28: 220-230. 10.1897/08-047.1.
35. Evanno G, Regnaut S, Goudet J: Detecting the number of clusters of individuals using the software STRUCTURE: a simulation study. Mol Ecol. 2005, 14: 2611-2620. 10.1111/j.1365-294X.2005.02553.x.
36. Cavalli-Sforza L, Edwards A: Phylogenetic analysis: models and estimation procedures. Am J Hum Genet. 1967, 19: 233-257.
37. Corander J, Majander KK, Cheng L, Merila J: High degree of cryptic population differentiation in the Baltic Sea herring*Clupea harengus*. Mol Ecol. 2013, 22: 2931-2940. 10.1111/mec.12174.
38. Tallmon DA, Koyuk A, Luikart G, Beaumont MA: ONeSAMP: a program to estimate effective population size using approximate Bayesian computation. Mol Ecol Resour. 2008, 8: 299-301. 10.1111/j.1471-8286.2007.01997.x.

39. Jones OR, Wang J: COLONY: a program for parentage and sibship inference from multilocus genotype data. Mol Ecol Resour. 2009, 10: 551-555.

40. Jorde PE, Ryman N: Unbiased estimator for genetic drift and effective population size. Genetics. 2007, 177: 927-935. 10.1534/genetics.107.075481.

41. Waples RS: A generalized approach for estimating effective population-size from temporal changes in allele frequency. Genetics. 1989, 121: 379-391.

42. Hey J: Isolation with migration models for more than two populations. Mol Biol Evol. 2010, 27: 905-920. 10.1093/molbev/msp296.

43. Neel MC, McKelvey K, Ryman N, Lloyd MW, Bull RS, Allendorf FW, Schwartz MK, Waples RS: Estimation of effective population size in continuously distributed populations: there goes the neighborhood. Heredity. 2013, 111: 189-199. 10.1038/hdy.2013.37.

44. Mills LS, Allendorf FW: The one-migrant-per-generation rule in conservation and management. Conserv Biol. 1996, 10: 1509-1518. 10.1046/j.1523-1739.1996.10061509.x.

45. Griffiths AM, Ellis JS, Clifton-Dey D, Machado-Schiaffino G, Bright D, Garcia-Vazquez E, Stevens JR: Restoration versus recolonisation: the origin of Atlantic salmon (*Salmo solar* L.) currently in the River Thames. Biol Conserv. 2011, 144: 2733-2738. 10.1016/j.biocon.2011.07.017.

46. Jobling S, Williams R, Johnson A, Taylor A, Gross-Sorokin M, Nolan M, Tyler CR, van Aerle R, Santos E, Brighty G: Predicted exposures to steroid estrogens in UK rivers correlate with widespread sexual disruption in wild fish populations. Environ Health Perspect. 2006, 114: 32-39.

47. Gong J, Ran Y, Chen DY, Yang Y, Zeng EY: Association of endocrine-disrupting chemicals with total organic carbon in riverine water and suspended particulate matter from the Pearl River, China. Environ Toxicol Chem. 2012, 31: 2456-2464. 10.1002/etc.1961.

48. Rodgers-Gray TP, Jobling S, Morris S, Kelly C, Kirby S, Janbakhsh A, Harries JE, Waldock MJ, Sumpter JP, Tyler CR: Long-term temporal changes in the estrogenic composition of treated sewage effluent and its biological effects on fish. Environ Sci Technol. 2000, 34: 1521-1528. 10.1021/es991059c.

49. Williams RJ, Churchley JH, Kanda R, Johnson AC: Comparing predicted against measured steroid estrogen concentrations and the associated risk in two United Kingdom river catchments. Environ Toxicol Chem. 2012, 31: 892-898. 10.1002/etc.1756.

50. Caldwell DJ, Mastrocco F, Anderson PD, Länge R, Sumpter JP: Predicted-no-effect concentrations for the steroid estrogens estrone, 17β-estradiol, estriol, and 17α-ethinylestradiol. Environ Toxicol Chem. 2012, 31: 1396-1406. 10.1002/etc.1825.

51. Green C, Williams R, Kanda R, Churchley J, He Y, Thomas S, Goonan P, Kumar A, Jobling S: Modeling of steroid estrogen contamination in UK and South Australian Rivers predicts modest increases in concentrations in the future. Environ Sci Technol. 2013, 47: 7224-7232.

52. Baade U, Fredrich F: Movement and pattern of activity of the roach in the River Spree, Germany. J Fish Biol. 1998, 52: 1165-1174. 10.1111/j.1095-8649.1998.tb00963.x.

53. Vollestad LA, Labeelund JH: Reproductive biology of stream-spawning roach, *Rutilus rutilus*. Environ Biol Fishes. 1987, 18: 219-227. 10.1007/BF00000361.

54. Geeraerts C, Ovidio M, Verbiest H, Buysse D, Coeck J, Belpaire C, Philippart JC: Mobility of individual roach *Rutilus rutilus*(L.) in three weir-fragmented Belgian rivers. Hydrobiologia. 2007, 582: 143-153. 10.1007/s10750-006-0561-x.

55. Griffiths AM, Koizumi I, Bright D, Stevens JR: A case of isolation by distance and short-term temporal stability of population structure in brown trout (*Salmo trutta*) within the River Dart, southwest England. Evol Appl. 2009, 2: 537-554. 10.1111/j.1752-4571.2009.00092.x.

56. Knaepkens G, Baekelandt K, Eens M: Fish pass effectiveness for bullhead (*Cottus gobio*), perch (*Perca fluviatilis*) and roach (*Rutilus rutilus*) in a regulated lowland river. Ecol Freshw Fish. 2006, 15: 20-29. 10.1111/j.1600-0633.2005.00117.x.

57. Lange A, Katsu Y, Ichikawa R, Paull GC, Chidgey LL, Coe TS, Iguchi T, Tyler CR: Altered sexual development in roach (*Rutilus rutilus*) exposed to environmental concentrations of the pharmaceutical 17 alpha-ethinylestradiol and associated expression dynamics of aromatases and estrogen receptors. Toxicol Sci. 2008, 106: 113-123. 10.1093/toxsci/kfn151.

58. Peakall R, Smouse PE: GenAlEx 6: genetic analysis in Excel. Population genetic software for teaching and research. Mol Ecol Notes. 2006, 6: 288-295. 10.1111/j.1471-8286.2005.01155.x.

59. Goudet J: FSTAT, a program to estimate and test gene diversities and fixation indices. 2.9.3 edition. 2002, Available at:http://www2.unil.ch/popgen/softwares/fstat.htm

60. Cornuet JM, Luikart G: Description and power analysis of two tests for detecting recent population bottlenecks from allele frequency data. Genetics. 1996, 144: 2001-2014.

61. Piry S, Luikart G, Cornuet JM: BOTTLENECK: a computer program for detecting recent reductions in the effective population size using allele frequency data. J Hered. 1999, 90: 502-503. 10.1093/jhered/90.4.502.

62. Excoffier L, Laval G, Schneider S: Arlequin (version 3.0): an integrated software package for population genetics data analysis. Evol Bioinform Online. 2005, 1: 47-50.

63. Jost L: G_{ST} and its relatives do not measure differentiation. Mol Ecol. 2008, 17: 4015-4026. 10.1111/j.1365-294X.2008.03887.x.

64. Crawford NG: SMOGD: software for the measurement of genetic diversity. Mol Ecol Resour. 2010, 10: 556-557. 10.1111/j.1755-0998.2009.02801.x.

65. Mantel NA: The detection of disease clustering and a generalized regression approach. Cancer Res. 1967, 27: 209-220.

66. Langella O: Populations 1.2.30; Population genetic software: individuals or populations distances based on allelic frequencies, phylogenetic trees, file conversions. 1999, Available at: http://www.bioinformatics.org/project/?group_id=84

67. Pritchard JK, Stephens M, Donnelly P: Inference of population structure using multilocus genotype data. Genetics. 2000, 155: 945-959.

68. Hey J, Nielsen R: Integration within the Felsenstein equation for improved Markov chain Monte Carlo methods in population genetics. Proc Natl Acad Sci USA. 2007, 104: 2785-2790. 10.1073/pnas.0611164104.

69. Kalinowski ST, Manlove KR, Taper ML: ONCOR: a Computer Program for Genetic Stock Identification, Volume 2. 2008, Bozeman, MT: Department of Ecology, Montana State University

70. Ovenden JR, Peel D, Street R, Courtney AJ, Hoyle SD, Peel SL, Podlich H: The genetic effective and adult census size of an Australian population of tiger prawns (*Penaeus esculentus*). Mol Ecol. 2007, 16: 127-138.

71. R Core Team: R: A Language and Environment for Statistical Computing. 2012, Vienna, Austria: R Foundation for Statistical Computing

72. Nei M, Tajima F, Tateno Y: Accuracy of estimated phylogenetic trees from molecular data. II. Gene frequency data. J Mol Evol. 1983, 19: 153-170. 10.1007/BF02300753.

73. Earl DA: Structure Harvester v0.3. 2009, Available at http://taylor0.biology.ucla.edu/structureHarvester/

CHAPTER 15 INCREASED IMMUNOGLOBULIN PRODUCTION IN SILVER CATFISH (RHAMDIA QUELEN) EXPOSED TO AGRICHEMICALS

L.C. Kreutz, T.R. Pavan, A.G. Alves, A.G. Correia, B. Barriquel, E.D. dos Santos, L.J.G. Barcellos

Keywords: Fish, Humoral response, Antibody production, Herbicides, Adjuvant

INTRODUCTION

Research supported by CNPq (#473195/2010-0). T.R. Pavan is a Master Student with a CAPES fellowship (#2013/1236). A.G. Alves was supported by a Probic/FAPERGS fellowship (#11/0461-6). L.J.G. Barcellos was the recipient of a CNPq research fellowship (#302073/2011-6).

Fish comprise one of the most diverse animal species and inhabit highly different environments in which a wide range of pathogens might be found. As such, fish responses and defense mechanisms to the challenge of microorganisms are also highly diverse. Hence, the ability of fish to cope with infecting pathogens relies on a delicate balance between the environment and the components of the innate and acquired immune systems.

In Teleostean fish, tissue macrophages are the main cells of the innate immune system and represent one of the first and most important barriers to bacterial colonization. Macrophages also provide important signaling to lymphocytes, which play a central role in acquired immunity ([1]), which is mediated in part by soluble tetrameric, IgM-like immunoglobulins. Other classes of immunoglobulin have also been detected in fish, but their roles in specific defense have not yet been clearly demonstrated ([2]). Because of their major role in protection, the ideal condition for immunoglobulin production has been investigated in several fish species and, in some instances, it has been shown to be affected by environmental and management practices or the presence of xenobiotics such as agrichemicals ([3]).

Water bodies and soil samples located in the vicinity of agricultural areas might be contaminated with several types of herbicides and pesticides at concentrations that can be harmful to a variety of non-target organisms including fish. Atrazine (6-chloro-*N*2-ethyl-*N*4-isopropyl-1,3,5-triazine-2,4-diamine) and glyphosate (*N*-phosphonomethyl glycine) are among the most widely used agrichemicals in soybean and corn fields, and their toxicity to silver catfish (*Rhamdia quelen*), an indigenous fish species endemic in South American rivers and lakes, has already been demonstrated. Exposure of silver catfish to sublethal concentrations of herbicides alters hormonal profiles ([4]), causes oxidative stress ([5]), liver and kidney damage ([6]), and affects blood cells and innate immunological functions ([7,8]). Current data indicate that these agrichemicals have the potential to alter silver catfish general metabolism and the

Citation: L.C. Kreutz, T.R. Pavan, A.G. Alves, A.G. Correia, B. Barriquel, E.D. dos Santos and L.J.G. Barcellos, Increased immunoglobulin production in silver catfish (Rhamdia quelen) exposed to agrichemicals, doi.org/10.1590/1414-431X20143890.

ability of the innate immune system to protect against bacterial infections.

Immunoglobulin production in fish might be altered by environmental factors, and previous studies indicated a deleterious effect of selected agrichemicals on catfish defense mechanisms ([7]). Here, we aimed to investigate the kinetics of immunoglobulin secretion and whether different water temperatures, boosting vaccination, and the presence of atrazine and glyphosate-based herbicides would have any impact on the production of antibodies to*Aeromonas hydrophila*, a Gram-negative bacterium frequently isolated from fish presenting hemorrhagic and ulcerative syndromes. We found that the kinetics of immunoglobulin production was similar in fish kept at the two different water temperatures used in the experiment and that no recall response was obtained following booster vaccination. However, an intriguing finding was that both atrazine and glyphosate had a stimulating effect on immunoglobulin production, in that higher titers of specific antibodies were observed in the serum of vaccinated, agrichemical-exposed fish.

MATERIAL AND METHODS

The first experiment was designed to investigate the kinetics of antibody production and the effect of water temperature and booster immunization. During the acclimatization period of 7 days, and up to the end of the experiment, fingerlings (32.2±6.7 g/14.3±1.1 cm) were kept in self-cleaning tanks containing 2000 liters of continuously running water at the indicated temperatures. One group of fingerlings (n=70) was inoculated intraperitoneally with 0.1 mL inactivated*A. hydrophila* (2×10^8 CFU/mL) and placed in water at non-controlled, environmental temperature (*Aeromonas*-inoculated, 17.4±0.4°C). A second group of fingerlings (n=70) was inoculated with the same amount of antigen and placed in heated water (*Aeromonas*-inoculated, 21.3±0.3°C). Saline-inoculated control fish were tagged and placed in water at environmental temperature (n=18; saline-inoculated, 17.4±0.4°C) or heated water (n=17; saline-inoculated, 21.3±0.3°C). At 21 days after primary immunization, 21 fingerlings from each *Aeromonas*-inoculated group were boosted with the same amount of bacteria and transferred to separate tanks with water at the respective temperatures.

Blood samples were collected from at least 7 fish of each group prior to inoculation on day 0, and at 7-day intervals up to 35 days, when the experiment was finished. At each time, fish captured for blood sampling were removed from the experiment. Boosted fingerlings were also sampled at 7-day intervals (days 28 and 35).

The second experiment consisted of a static bioassay that aimed to investigate the effect of atrazine or glyphosate on the production of specific antibodies. Prior to immunization, fingerlings were acclimatized for 7 days in tanks containing chlorine-free, continuously aerated tap water (22.1±0.7°C). During the entire experiment, fish density was kept lower than 1 g fish per liter of water. The concentration of atrazine and glyphosate added to the water was equivalent to 10% of the previously reported lethal concentration ($LC_{50\text{-}96h}$) ([8]) for silver catfish (1.02 mg/L for atrazine and 0.730 mg/L for glyphosate).

After the acclimation period, 30 fingerlings were inoculated with sterile saline solution and placed in tanks containing atrazine (saline-inoculated atrazine, n=15) or glyphosate (saline-inoculated glyphosate, n=15) and evaluated as the saline-inoculated herbicide-exposed (S-HE) control group. The remaining fingerlings (n=90) were inoculated intraperitoneally with 0.1 mL inactivated *A. hydrophila* (2×10^8 CFU/mL) and distributed equally in nine tanks (10 fish/tank) as three treatments in triplicate: tanks without herbicides (*Aeromonas*-inoculated herbicide-free,*Ah*-HF), tanks containing atrazine (*Aeromonas*-inoculated atrazine, *Ah*-Atr), or tanks with glyphosate (*Aeromonas*-inoculated glyphosate,*Ah*-Gly). All tanks were cleaned every 2 days by bottom suction, and the amount of water removed by cleaning (approximately 10% of the volume) was replaced using water from tanks containing the same concentration of herbicides, or herbicide-free water. Water conditions in all tanks were monitored throughout the experiment and were maintained within normal parameters as reported previously ([7]).

After 21 days, all fish were captured for blood sampling. Culturing and inactivation of *A. hydrophila*, blood sampling, and the serum agglutination assay were performed as reported previously ([7]). Total serum protein and immunoglobulin concentrations were determined by standard methods as already published ([9]). The results obtained were analyzed by the Shapiro-Wilk test (P>0.05) using SPSS statistics version 19 (IBM Corp., USA) and were found to have a normal distribution. Differences among treatments were analyzed by two-way ANOVA followed by the Bonferroni multiple comparisons test, and plotted using the GraphPad Prism Statistics Software v. 5 (GraphPad Software, Inc., USA). P values of 0.05 or less were considered to be significant. Results are reported as means±SE.

The experiments were carried out according to the guidelines of Colégio Brasileiro de Experimentação Animal and were approved by the Ethics Committee for the Care and Use of Experimental Animals of Universidade de Passo Fundo, Brazil.

RESULTS AND DISCUSSION

Here we provide the first report related to the kinetics of antibody production in silver catfish (*R. quelen*) in response to inoculated antigen. Fish immunized with *A. hydrophila* bacterin had a sharp and significant (P<0.05) increase of specific serum antibodies as soon as 7 days post-inoculation (Figure 1). Antibody titers continued to increase steadily for up to 35 days, when the experiment was concluded. Anti-aeromonid agglutinating antibodies in saline-inoculated fish remained at basal levels throughout the experiment (Figure 1) and were within the range reported in previous studies ([7]) of non-immunized fish.

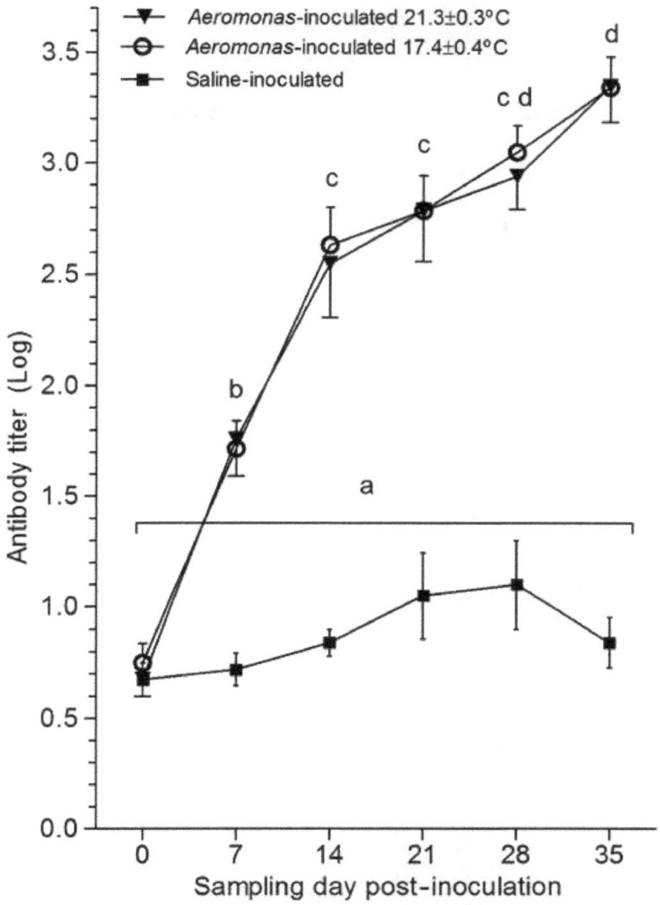

Figure 1 Kinetics of specific antibody production in silver catfish following intraperitoneal inoculation with inactivated *Aeromonas hydrophila* or sterile saline and kept at different water temperatures. Serum samples were collected prior to immunization (day 0) and at 7-day intervals. The results obtained from saline-inoculated fish kept at different temperatures were similar and were combined in a single group. Data are reported as the logarithm (± SE) of the last serum dilution that caused bacterium agglutination and were compared using two-way ANOVA and the Bonferroni post-test. Significant differences ($P<0.05$) are indicated by different letters at each time point.

The kinetics and timing of maximum antibody response to inoculated antigen is central to vaccine development and vaccination strategies. In general, fish antibody response to bacteria-derived antigens occurs within a few days and reaches maximum levels 4 to 5 weeks following priming, as reported here. A similar response has been observed in Indian major carp (*Labeo rohita*) and turbot (*Scophthalmus maximus*) inoculated with different combinations of adjuvants and bacterial antigens, including *A. hydrophila* ([10],[11]). Higher antibody titers, however, are usually obtained following inoculation with antigens emulsified in Freund's complete adjuvant, indicating that antibody secretion depends partly on the nature of antigen and proper presentation to B lymphocytes. Fish immunized with T-cell-dependent antigens, such as some viral or parasite-derived proteins might have a lower and delayed humoral response ([12]).

Because antibody titers in fish are believed to be short-lived, there is major concern related to the requirements for antigen reinoculation. A recall response to vaccination might be demonstrated by rising antibody titers following reinoculation of antigens. Immunological memory has been demonstrated in rainbow trout (*Oncorhynchus mykiss*) inoculated with a T-cell-dependent antigen ([12]). Here, however, reinoculation of*A. hydrophila* at 21 days after priming had no effect on subsequent antibody titers at both temperatures tested (data not shown). Because at 21 days after priming, anti-aeromonid antibodies were already relatively high and steadily increasing, as shown in the kinetics curve (Figure 1), they could have inhibited further stimulation of memory B lymphocytes. In addition, whole bacteria such as inactivated *A. hydrophila*, comprise mostly T-independent antigens, and, therefore, a booster inoculation might not be required to raise antibody levels. Further experiments, using non-related bacterial antigens free of endotoxins and different reinoculation schedules and adjuvant combinations, should provide more information that might improve our understanding on this subject.

Because antibody production might be influenced by several different factors, we investigated whether water temperature would impact the humoral response. As depicted in Figure 1, we were unable to find differences in anti-aeromonid antibody titers at any sampling time in immunized fish kept at different temperatures. Even though the difference of water temperature in the tanks (17.4±0.4° *vs* 21.3±0.3°C) was relatively small (±4°C), it represented realistic temperature variations found during the end of fall, throughout winter, and in early spring. Indeed, seasonal differences in water temperature at a local hydroelectric plant (latitude 28°15′S; longitude 52°24′W, 687 m above sea level) ranged from 26.2° to 28.9°C in summer, 24.2° to 26.2°C in fall, 16.5° to 17.7°C in winter, and 22.6° to 25.1°C in spring ([13]). The effect of water temperature on immune response is central to vaccine development. In general, lower water temperatures reduce antibody production in most fish species studied. In non-vaccinated Nile tilapia (*Oreochromis niloticus*) kept at 28°C for 2 weeks, the immunoglobulin levels were higher than in fish kept at 23°C ([14]). In the channel catfish (*Ictalurus punctatus*) no antibodies to *Ichthyophthirius multifilis* could be detected in immunized fish kept for 21 days at 15°C or at temperatures cycling from 15° to 25°C ([15]). However, at higher temperatures (25° and 30°C) a robust and protective immune response was observed. In Japanese flounder (*Paralichthys olivaceus*) intraperitoneally vaccinated with inactivated virus, maximum immunoglobulin production was found in fish kept at 21°C; lower (9° to 15°C) or higher temperatures (26°C) had a deleterious effect on the production of circulating immunoglobulin ([16]). In fact, water temperature below 15°C decreases immunoglobulin production in several other fish species such as sea bass (*Dicentrarchus labrax*), Atlantic halibut (*Hippoglossus hippoglossus*), summer flounder (*Paralichthys dentatus*), Japanese flounder, Nile tilapia and carp (*Cyprinus carpio*) ([17]) when compared with higher temperatures (18° to 23°C). It should be noted, however, that, for fish species adapted to colder temperatures, increasing the temperature does not improve the acquired immune response. Thus, current data indicate that immunoglobulin production is indeed affected by water temperature but that each fish species has a temperature range at which the acquired immune response might be optimal. The silver catfish is perfectly adapted to seasonal temperature variation but is more active in the warmer months of the year. Thus, we cannot rule out the possibility that, at higher temperatures, differences in antibody production might be detected in immunized fish. However, the temperatures used in our experiments were within the range found throughout the year. In addition, we used bacterin antigens that *per se* are excellent particulate antigens in fish and might not require boosting immunization to maintain high levels of serum antibodies. Hence, the effect of temperature on the kinetics of immunoglobulin production should be further investigated using purified soluble antigens, adjuvant combinations, and fish kept at a wider range of temperatures, such as those found during late summer. However, regardless of season and water temperature, fish should always be immunized prior to any possible encounter with a pathogenic microorganism, which is usually dictated by husbandry management at each farm.

In previous research, we demonstrated that the presence of glyphosate and atrazine reduced fish survival after challenging with *A. hydrophila*, and had a deleterious effect on innate immune functions ([7,8]). Here, we wanted to know whether antibody production would be affected under similar experimental conditions. Unexpectedly, fish inoculated with *A. hydrophila* and exposed to atrazine (*Ah*-Atr) or glyphosate (*Ah*-Gly) had significantly (P<0.05) higher antibody responses compared to inoculated but non-exposed fish (*Ah*-HF; Figure 2). Although total serum proteins were similar among all groups (Figure 3), immunoglobulin concentrations were significantly higher (P<0.05) in the *Ah*-Atr and*Ah*-Gly groups.

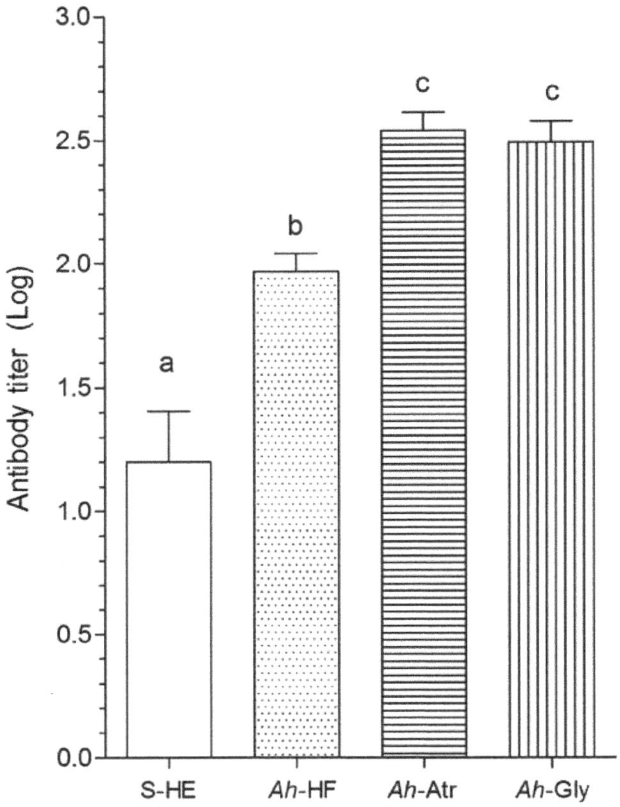

Figure 2 Exposure to atrazine or glyphosate increases the production of antibodies in silver catfish. Fingerlings were inoculated with inactivated*Aeromonas hydrophila* and kept on herbicide-free water (*Ah*-HF), exposed to atrazine or glyphosate (*Ah*-Atr and *Ah*-Gly, respectively) or inoculated with saline and exposed to herbicides (S-HE). Data from the S-HE represents all data from saline-inoculated fish exposed to either atrazine or glyphosate. Data are reported as means±SE. The results are shown as the logarithm of the last serum dilution that caused bacterium agglutination. One-way ANOVA and the Bonferroni post-test were used to compare the data and significant differences (P<0.05) are indicated by different letters.

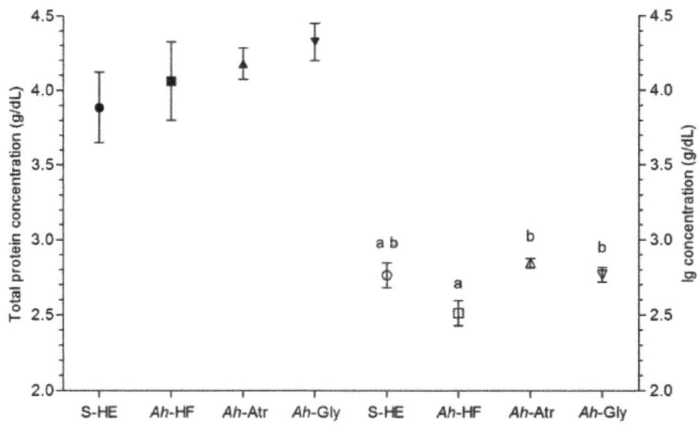

Figure 3 Total serum proteins (closed symbols) and immunoglobulins (open symbols) from silver catfish fingerlings (for details, seeFigure 2). Data are reported as means±SE. One-way ANOVA and the Bonferroni post-test were used to compare the data and significant differences (P<0.05) are indicated by different letters.

The intriguing effect of atrazine and glyphosate on acquired humoral immune responses is in sharp contrast with the deleterious effect they have on the innate immune system [7,8]. An insight into the mechanisms underlying the immune stimulating effect of herbicides was recently reported in Nile tilapia (*O. niloticus*). Following inoculation with *A. hydrophila*, Nile tilapia exposed to endosulfan had higher anti-aeromonid antibody titers than the non-exposed inoculated group [18]. In addition, the endosulfan-exposed group had higher levels of interleukin-2 (IL-2)-like factor that could have stimulated lymphocytes in a nonspecific manner and contributed to the increased immunoglobulin production. In fact, IL-2 might be important for immunoglobulin production in fish, in that Nile tilapia and rainbow trout challenged with *A. hydrophila* had increased levels of IL-2 compared to non-challenged groups. The effects of atrazine and glyphosate on immune cells and cytokine expression have not been widely investigated. However, in a mammalian model, atrazine-exposed BALB/c mice had higher levels of IgM-secreting cells in the spleen and increased cytotoxic lymphocyte function and proliferative response of splenocytes after *in vivo* stimulation [19], indicating a potential for nonspecific immune stimulation. Increased levels of IL-1β and IL-1 receptor mRNA, mainly in spleen and head kidney, were detected in common carp exposed to atrazine [20]. Thus, because IL-1β is a cytokine with pleiotropic effects and central to early stimulation of the immune response, exposure of fish to atrazine could induce a nonspecific stimulation of lymphocytes and contribute to an increase of antibody titers. In fact, we noticed that serum agglutination activity against *A. hydrophila* was relatively higher in saline-inoculated herbicide-exposed fish (S-HE; Figure 2) compared to saline-inoculated fish in experiment 1 (Figure 1) and similar to that found in previous experiments [8]. Nonspecific antibodies that could have been induced in the presence of the agrichemical *per se*, as previously discussed, might account for this agglutination activity. Consistent with this, S-HE fish had higher immunoglobulin concentrations than*Ah*-HF fish (Figure 3). Because total serum proteins in S-HE fish were similar to those in the other groups, the possibility that atrazine or glyphosate induced a general increase in protein synthesis might be ruled out.

Therefore, even though the stimulatory effect of atrazine and glyphosate should be of concern because their presence in the water might lead to continuous production of pro-inflammatory cytokines, the molecular mechanism behind this stimulation might be of general interest to immunologists. Indeed, the concept that even toxic substances at low doses might have beneficial effects on exposed individuals is becoming widely recognized. Nonetheless, it should be kept in mind that nonspecific stimulation of lymphocytes and the generation of nonspecific antibodies might cause cell damage, mainly in the spleen and head kidney, and trigger or lead to autoimmune diseases.

In summary, our results demonstrate that silver catfish inoculated with *A. hydrophila* bacterin produce a quick and robust humoral response that is not affected by small differences in water temperature or recall vaccination at 21 days after priming. An understanding of the requirements for optimal antibody secretion is central to the development of effective vaccines and vaccination strategies in Neotropical fish species such as catfish. However, the most intriguing finding was that the presence of atrazine or glyphosate in the water increased the production of antibodies to *A. hydrophil a* in vaccinated fish, extending our previous observation that atrazine or glyphosate-exposed silver catfish have higher serum immunoglobulins compared with fish kept in herbicide-free water. The effect of agrichemicals on the expression of cytokines and immune-related genes should be a subject of further investigation aiming to better understand the effects of agrichemicals on the fish immune system.

ACKNOWLEDGMENTS

Research supported by CNPq (#473195/2010-0). T.R. Pavan is a Master Student with a CAPES fellowship (#2013/1236). A.G. Alves was supported by a Probic/FAPERGS fellowship (#11/0461-6). L.J.G. Barcellos was the recipient of a CNPq research fellowship (#302073/2011-6).

REFERENCES

1. Aoki T, Takano T, Santos MD, Kondo H. Molecular innate immunity in teleost fish? Review and future perspectives. Fish Glob Welf Environ 2008; 263-276.http://www.terrapub.co.jp/onlineproceedings/fs/wfc2008/pdf/wfcbk_263.pdf. Accessed June 19, 2013.
2. Fillatreau S, Six A, Magadan S, Castro R, Sunyer JO, Boudinot P. The astonishing diversity of Ig classes and B cell repertoires in teleost fish. Front Immunol 2013; 4: 28, doi: 10.3389/fimmu.2013.00028.
3. Bowden TJ. Modulation of the immune system of fish by their environment. Fish Shellfish Immunol 2008; 25: 373-383, doi: 10.1016/j.fsi.2008.03.017.
4. Cericato L, Neto JG, Kreutz LC, Quevedo RM, da Rosa JG, Koakoski G, et al. Responsiveness of the interrenal tissue of Jundia (*Rhamdia quelen*) to an *in vivo* ACTH test following acute exposure to sublethal concentrations of agrichemicals. Comp Biochem Physiol C Toxicol Pharmacol 2009; 149: 363-367, doi: 10.1016/j.cbpc.2008.09.002.
5. Ferreira D, da Motta AC, Kreutz LC, Toni C, Loro VL, Barcellos LJ. Assessment of oxidative stress in *Rhamdia quelen* exposed to agrichemicals. Chemosphere 2010; 79: 914-921, doi: 10.1016/j.chemosphere.2010.03.024.
6. Mela M, Guiloski IC, Doria HB, Randi MA, de Oliveira Ribeiro CA, Pereira L, et al. Effects of the herbicide atrazine in neotropical catfish (*Rhamdia quelen*). Ecotoxicol Environ Saf 2013; 93: 13-21, doi: 10.1016/j.ecoenv.2013.03.026.
7. Kreutz LC, Gil Barcellos LJ, de Faria V, de Oliveira Silva T, Anziliero D, Davi dos Santos E., et al. Altered hematological and immunological parameters in silver catfish (*Rhamdia quelen*) following short term exposure to sublethal concentration of glyphosate. Fish Shellfish Immunol 2011; 30: 51-57, doi: 10.1016/j.fsi.2010.09.012.
8. Kreutz LC, Barcellos LJ, Marteninghe A, Dos Santos ED, Zanatta R. Exposure to sublethal concentration of glyphosate or atrazine-based herbicides alters the phagocytic function and increases the susceptibility of silver catfish fingerlings (*Rhamdia quelen*) to *Aeromonas*

hydrophila challenge. Fish Shellfish Immunol 2010; 29: 694-697, doi: 10.1016/j.fsi.2010.06.003.

9. Siwicki AK, Anderson DP, Rumsey GL. Dietary intake of immunostimulants by rainbow trout affects non-specific immunity and protection against furunculosis. Vet Immunol Immunopathol 1994; 41: 125-139, doi: 10.1016/0165-2427(94)90062-0.

10. Swain P, Behura A, Dash S, Nayak SK. Serum antibody response of Indian major carp, *Labeo rohita* to three species of pathogenic bacteria; *Aeromonas hydrophila, Edwardsiella tarda* and *Pseudomonas fluorescens*. Vet Immunol Immunopathol 2007; 117: 137-141, doi: 10.1016/j.vetimm.2007.02.010.

11. Zheng Z, Yingeng W, Qingyin W, Nannan D, Meijie L, Jiangbo Q, et al. Study on the immune enhancement of different immunoadjuvants used in the pentavalent vaccine for turbots. Fish Shellfish Immunol 2012; 32: 391-395, doi: 10.1016/j.fsi.2011.11.014.

12. Arkoosh MR, Kaattari SL. Development of immunological memory in rainbow trout (*Oncorhynchus mykiss*). I. An immunochemical and cellular analysis of the B cell response. Dev Comp Immunol 1991; 15: 279-293, doi: 10.1016/0145-305X(91)90021-P.

13. Hahn L, Camara LF da. Qualidade da água do reservatório da usina hidrelétrica de Passo Fundo, RS. Tractebel Energia-Suez 2003; 3: 3. http://www.tractebelenergia.com.br/wps/wcm/connect/5dfad7eb-fbcf-4f45-a209-2f317da8927b/8.pdf?MOD=AJPERES&CACHEID=5dfad7eb-fbcf-4f45-a209-2f317da8927b.pdf. Accessed August 3, 2013.

14. Dominguez M, Takemura A, Tsuchiya M, Nakamura S. Impact of different environmental factors on the circulating immunoglobulin levels in the Nile tilapia, *Oreochromis niloticus*. Aquaculture 2014; 241: 491-500, doi: 10.1016/j.aquaculture.2004.06.027.

15. Martins ML, Xu DH, Shoemaker CA, Klesius PH. Temperature effects on immune response and hematological parameters of channel catfish*Ictalurus punctatus* vaccinated with live theronts of*Ichthyophthirius multifiliis*. Fish Shellfish Immunol 2011; 31: 774-780, doi: 10.1016/j.fsi.2011.07.015.

16. Xu G, Sheng X, Xing J, Zhan W. Effect of temperature on immune response of Japanese flounder (*Paralichthys olivaceus*) to inactivated lymphocystis disease virus (LCDV). Fish Shellfish Immunol 2011; 30: 525-531, doi: 10.1016/j.fsi.2010.11.026.

17. Bowden TJ, Thompson KD, Morgan AL, Gratacap RM, Nikoskelainen S. Seasonal variation and the immune response: a fish perspective. Fish Shellfish Immunol 2007; 22: 695-706, doi: 10.1016/j.fsi.2006.08.016.

18. Tellez-Banuelos MC, Santerre A, Casas-Solis J, Zaitseva G. Endosulfan increases seric interleukin-2 like (IL-2L) factor and immunoglobulin M (IgM) of Nile tilapia (*Oreochromis niloticus*) challenged with*Aeromona hydrophila*. Fish Shellfish Immunol 2010; 28: 401-405, doi: 10.1016/j.fsi.2009.11.017.

19. Rowe AM, Brundage KM, Schafer R, Barnett JB. Immunomodulatory effects of maternal atrazine exposure on male Balb/c mice. Toxicol Appl Pharmacol 2006; 214: 69-77, doi: 10.1016/j.taap.2005.12.006.

20. Wang X, Xing H, Li X, Xu S, Wang X. Effects of atrazine and chlorpyrifos on the mRNA levels of IL-1 and IFN-gamma2b in immune organs of common carp. Fish Shellfish Immunol 2011; 31: 126-133, doi: 10.1016/j.fsi.2011.04.015.

INDEX

R

S

T